S0-DUV-422

Productivity of World Ecosystems

Proceedings of a Symposium

Presented August 31–September 1, 1972,
at the V General Assembly of the
Special Committee for the International Biological Program

Seattle, Washington

Symposium sponsored by

U.S. National Committee for the International Biological Program
Division of Biological Sciences
Assembly of Life Sciences
National Research Council

NATIONAL ACADEMY OF SCIENCES
WASHINGTON, D.C. 1975

NOTICE: The symposium reported herein was undertaken under the aegis of the National Research Council with the express approval of its Governing Board. Such approval indicated that the Board considered the problem to be of national significance, that its elucidation required scientific or technical competence, and that the resources of NRC were particularly suitable to the conduct of the project. The institutional responsibilities of the NRC were then discharged in the following manner:

The participants in the symposium were selected for their individual scholarly competence and judgment with due consideration for the balance and breadth of disciplines. Responsibility for all aspects of this report rests with them and with the organizing committee, to whom sincere appreciation is expressed.

Although proceedings of symposia are not submitted for approval to the Academy membership or to the Council, each is reviewed according to procedures established and monitored by the Academy's Report Review Committee. Such reviews are intended to determine, *inter alia*, whether the major questions and relevant points of view have been addressed and whether the reported findings and conclusions arose from the available data and information. Distribution is approved, by the President, only after satisfactory completion of this review process.

This symposium was supported by the National Science Foundation.

Library of Congress Cataloging in Publication Data

Main entry under title:

Productivity of world ecosystems.

Includes bibliographies.

1. Biological productivity–Congresses. I. International Council of Scientific Unions. Special Committee for the International Biological Programme. II. U.S. National Committee for the International Biological Program. [DNLM: 1. Ecology–Congresses. 2. Environment–Congresses. QH541 P9635 1972]

ISBN 0-309-02317-3

Available from
Printing and Publishing Office, National Academy of Sciences
2101 Constitution Avenue, N.W., Washington, D.C. 20418

Printed in the United States of America

PREFACE

This symposium reflects one of the central objectives of the International Biological Program: the worldwide study of organic production on the land, in fresh waters, and in the seas, and the potentialities and uses of new as well as of existing natural resources. It examines the productivity of oceans, fresh water, grassland, desert, temperate forests, and tundra. It seeks to respond to these questions: What are the ranges of productivity in each ecosystem? What are the factors that provide the main controls in each? What are the potentials for utilizing and increasing productivity in man's interest?

Since the symposium, many of the contributors have had occasion to update their material and to provide formal papers that enlarge upon the initial presentations.

DAVID E. REICHLE
JERRY F. FRANKLIN
DAVID W. GOODALL

EXECUTIVE COMMITTEE OF THE U.S. NATIONAL COMMITTEE FOR THE IBP

John F. Reed, *Chairman*
Stanley I. Auerbach
Paul T. Baker
Stanley A. Cain
Everett S. Lee
Terence A. Rogers
George F. Sprague
Kenneth V. Thimann
Frederic H. Wagner

ORGANIZING COMMITTEE FOR THE V GENERAL ASSEMBLY, IBP

W. Frank Blair, *Chairman*
James S. Bethel
Everett S. Lee
Jerry S. Olson

SUBCOMMITTEE ON PRODUCTIVITY SYMPOSIUM

Jerry F. Franklin, *Chairman*
David W. Goodall
David E. Reichle

CONTENTS

PRODUCTIVITY OF TUNDRA ECOSYSTEMS

F. E. WIELGOLASKI

ABSTRACT

A review of productivity studies in tundra ecosystems shows that primary production is the best known ecosystem parameter. Results from both repeated harvestings and physiological measurements are given. Yearly production ranges from less than 10 $g/m^2/yr$ to 400 to 500 $g/m^2/yr$ aboveground vascular plants. On a daily basis primary productivity in tundra may be considerable; values of 5 to 6 $g/m^2/day$ are found when above- and belowground parts and vascular plants and cryptograms are combined. Some estimates of the tundra microflora are also given.

The ratio of aboveground : belowground parts of vascular plants decreases with temperature, oligotrophy, and water content of soil, i.e., with decreasing decomposition rate. Lowest ratios, down to about 1 : 20, are found on sedge-dominated wet areas with limited amounts of shrubs in arctic and alpine tundra.

The influence of the small mammal population cycles on primary production is mentioned as well as the interaction between grazing by large mammals, especially reindeer, and plant growth. The direct influence of invertebrates on plant biomass is small, but invertebrates in the vegetation layer are important for birds, and invertebrates in the soil influence the decomposition rate. Invertebrate biomass is difficult to measure, but some crude estimates are given. Nematoda, Enchytraeidae, Acari and Collembola are found to be the most important invertebrate groups in the tundra. Vertebrate biomass is summarized for various tundra areas. Vertebrates generally constitute a smaller energy flow component of the tundra ecosystem than the plant–soil organic-decomposer cycle, although they are still important. Some preliminary values for decomposition rates in tundra are given. For some aboveground plant materials weight losses may be 25 to 35 percent during the first year, usually it is less. Although energy and carbon flow through tundra ecosystems is the primary aim of the tundra studies, work on nutrient cycles is also in progress, particularly on problems of nitrogen supply.

INTRODUCTION

Tundra is often defined as areas with permafrost in the soil. This situation may occur in both polar and mountain regions. Within IBP, studies are carried out in both Arctic and subarctic tundra ecosystems in the U.S.S.R., Canada, U.S.A., Finland and Greenland, in Antarctic ecosystems at South Georgia and Macquarie Island and in mountain areas with or without permafrost in Sweden, Norway and Austria, as well as in tundra-like blanket bogs in Great Britain and Ireland. The studies are coordinated by the International Tundra Biome Steering Committee.

Tundra areas may be divided into different zones. The biology of Arctic, Antarctic and alpine tundra will vary, and the moorland areas included in the IBP tundra studies are in many ways different from the polar areas, e.g., lacking the permafrost. There also may be a zonation of the tundra for the continents of Eurasia and North America, from north (Arctic areas) to south (subarctic areas) in relation to solar radiation as well as from west to east because of changes in continentality of climate (mainly changes in temperature and precipitation). In the U.S.S.R. several attempts have been made to differentiate regions within the tundra (see the review by Alexandrova, 1970). Climate combined with vegetation studies has been the main criterion for tundra classification in the U.S.S.R. Similar methods have been used in Canada (e.g., see Beschel, 1970). The nutrient content of the soil could also be used for categorizing various eutrophic and oligotrophic tundra types. A rough classification of the tundra sites within IBP is based on vegetation studies (Wielgolaski, 1972a). More advanced techniques could also be used, e.g., principle component analysis, which has been used for IBP sites (e.g., Moore, personal communication, 1969) and for some areas in Canada (Beschel, 1970).

Most tundra is biologically fragile. Tundra areas usually share a severe environmental situation with low temperatures and a short growing season. Therefore, annual biological productivity is usually low as is the number of species both of plants and animals. As Bliss (1970) states it is generally accepted that a small biomass and low annual production favor ecological and taxonomic simplicity and hence system instability. Daily productivity may be as high in tundra areas as in other parts of the world, however. There may be great differences in productivity from year to year caused both by variations in environment, small mammal populations, and the carnivores preying upon them. The consumer component of tundra ecosystems is especially important for the nutrient cycling and energy flow through the system in years with high small mammal populations. Even then, however, the decomposer pathway is the major route for breakdown of plant materials as in most other terrestrial ecosystems. Still, the systems will not function without consumers—both in the soil and aboveground.

Before IBP, studies of tundra ecosystems were carried out mainly in the U.S.S.R. and U.S.A. A survey of Russian tundra literature is found in Alexandrova (1970); Firsova *et al.*, (1969) and Tikhomirov (1971) are additional exemplary papers on tundra productivity in U.S.S.R. Literature concerning extensive studies of lemming cycles at Point Barrow, Alaska, is cited by Bliss (1970). Rodin and Bazilevich (1967) and Bliss (1962, 1966) have summarized Arctic research on primary production. A review of IBP tundra studies and some preliminary results are given in Heal (1971) and in Wielgolaski and Rosswall (1972), and results on soil organisms and decomposition from IBP tundra studies are given in Holding *et al.* (1974). It can be expected that within a few years much more information will be available on productivity of tundra ecosystems [e.g., Moore, in preparation; Wielgolaski, in press (a)].

PRIMARY PRODUCTION

Most work on tundra ecosystems concerns biomass and production of plants, especially the aboveground biomass of higher plants (Bliss, 1962, 1966, 1970; and Andreev, 1966; Khodachek, 1969; Alexandrova, 1970; Wielgolaski, 1972; Bliss and Wielgolaski, 1973).

Biomass

The biomass of higher plants may be very small in zones classified as polar desert by Alexandrova (1970), e.g., 6 g/m^2 in moss–lichen polygons at Franz Josef Land (Table 1). Low biomass values are also found in other areas with only patchy vegetation, while higher values, 15–50 g/m^2, may be found in *Salix herbacea* snow beds in Norway at 1,300 m above sea level and 60°N latitude [Wielgolaski, 1972, in press (a)].

Even in the polar deserts the amount of cryptogams may be rather high. Alexandrova (1970) reports 123 g/m^2, while Andreev (1966) reports only 9 g/m^2 in polar semideserts, a northern variant of arctic tundra with more than 50 percent bare soil. In the Norwegian snow bed mentioned above the cryptogam biomass was ~40 g/m^2. The highest amounts of living aboveground biomass of cryptogams reaches 1,345 g/m^2 (Alexandrova, 1970) in shrub tundra, 800 g/m^2 for a polygonal bog (Khodachek, 1969; Shamurin *et al.*, 1972), and about 400 g/m^2 for a flat palsa bog (Pospelova, 1972).

When both above- and belowground biomass of phanerogams are summed with cryptogams (Alexandrova, 1970) a total biomass of ~150 g/m^2 occurs in the polar desert and ~200 g/m^2 in alpine snow beds in Norway. This suggests that even in closed vegetation in alpine areas, at a low latitude the total amount of biomass may be nearly as low as in polar deserts at a high latitude.

Vassiljevskaja and Grishina (1972) found that organic matter reserves in both the total plant biomass (living and dead) and soil organic matter increased from eluvial to accumulative landscapes in Western Taimyr, e.g., from spotted *Dryas*-moss tundra to a marshy brush-sedge-moss tundra (Table 2). The biomass ratio between the vegetation types was 2.77 and in soil organic matter, 2.12. At the Norwegian IBP tundra sites at Hardangervidda the total carbon content of soil organic matter in a dry meadow with *Dryas* is comparable to that found in the Russian *Dryas* tundra. Similar carbon values to the marshy tundra in Taimyr are found in soil organic matter in a Norwegian wet peaty sedge meadow (Veum, personal communication, 1972). The ratio of total biomass between the two types in Norway (3.33) was somewhat higher than in U.S.S.R., while the ratio in soil organic matter was slightly lower (2.06). The ratios between carbon in soil organic matter and biomass were highest at the driest sites in both U.S.S.R. and Norway. The ratios were higher in Norway than at similar vegetation sites in U.S.S.R., e.g., 19.2 at the dry meadow in Norway and 12.2 at the *Dryas* tundra in the U.S.S.R. which may reflect influences of the oceanic climate.

The maximum aboveground living biomass of vascular plants differs very much from site to site but could be grouped in 5 to 6 groups (Table 1). There is, of course, a gradient from the most Arctic sites with low temperatures to sites with more favorable conditions for growth. For example, at Hardangervidda, Norway, the average yearly temperature is slightly below 0°C, while at Macquarie Island it is 4.5°C (Jenkin and Ashton, 1970); both of these sites have much higher temperatures than Taimyr at about –15°C (Matveyeva, 1972). At Moor House, a moorland site in United Kingdom at about 54°30′N, the maximum aboveground vascular living biomass was about 850 g/m^2 (Forrest,

TABLE 1 Total Live Aboveground Vascular Plant Biomass (g/m² dry weight) at the Time of Maximum Aboveground Biomass

Country	Area	Site Types	g/m²	References
U.S.S.R.	Franz Josef Land	Polar desert	6	Alexandrova, 1970
Norway	Hardangervidda	Alpine snow bed	15–50	Wielgolaski, 1972, in press (a)
Canada	Devon Island	Plateau	20	Bliss, 1972
Norway	Hardangervidda	Lichen heath	30–60	Wielgolaski, 1972, in press (a)
U.S.S.R.	Northeastern Europe	Polar semidesert	40	Andreev, 1966
Canada	Devon Island	Sedge meadow	60–90	Muc, 1973
U.S.S.R.	New Siberian Island	Arctic tundra	71	Alexandrova, 1970
Canada	Devon Island	Raised beach ridge	90–130	Svoboda, 1973
Norway	Hardangervidda	Dry and wet meadow	90–140	Wielgolaski, 1972, in press (b)
U.S.S.R.	Taimyr	Moss-*Dryas*-sedge tundra	100	Shamurin *et al.*, 1972
U.S.A.	Point Barrow	Wet sedge meadow	100	Tieszen, 1972
Sweden	Abisko	Bog	110	Rosswall, 1972
U.S.S.R.	Kola	Spotted eutrophic alpine tundra	110	Chepurko, 1972
Finland	Kevo	Subalpine heath	182	Kallio and Kärenlampi, 1971
U.S.S.R.	Western Taimyr	Dwarf shrub–sedge moss–lichen tundra	188	Pospelova, 1972
U.S.S.R.	Taimyr	Polygonal bog	190	Shamurin *et al.*, 1972
U.S.S.R.	Salekhard	Moss–shrub–hummock	204	Gorchakovsky and Andreyashkina, 1972
Ireland	Glenamoy	Open bog	214	Moore, personal communication, 1970
U.S.S.R.	Western Taimyr	Flat palsa bog	513	Pospelova, 1972
U.S.S.R.	Kola	Alpine meadow	528	Chepurko, 1972
Australia	Macquarie Island	Herbfield	615	Jenkin and Ashton, 1970
Norway	Hardangervidda	Willow thicket	800	Kelvik and Kärenlampi, in press
U.S.S.R.	Eastern Europe	Shrub tundra	817	Alexandrova, 1970
United Kingdom	Moor House	Blanket bog	846	Forrest, 1971
Australia	Macquarie Island	Grassland	1,138	Jenkin and Ashton, 1970
Austria	Patseherkofel	Loiseleurietum	1,150	Larcher *et al.*, 1973

TABLE 2 Total Carbon in Biomass (Living and Dead Above- and Belowground Components) and in Soil Organic Matter for the U.S.S.R. and Norway Tundra Sites

Sites		Biomass (gC/m²)	Soil Organic Matter (gC/m²)	Soil Organic Matter: Biomass Ratio
U.S.S.R. Western Taimyr	Spotted *Dryas*–moss tundra	1,049	12,784	12.2
	Marshy brush–sedge–moss tundra	2,907	27,036	9.3
Norway Hardangervidda	Dry meadow (with *Dryas*)	825	15,795	19.2
	Wet peaty sedge meadow	2,750	32,506	11.8

1971). At herbfield and grassland sites on Macquarie Island at a similar latitude in the southern hemisphere (Jenkins and Ashton, 1970) even higher values were found, which ranged up to above 1,100 g/m². Naturally, the biomass is also considerable where the amount of woody plants is high, running up to 817 g/m² in a shrub tundra in the East European forest–tundra region (Alexandrova, 1970), and still higher in alpine regions in Austria [850 to 1,200 g/m² in IBP studies (Larcher *et al.*, 1973)].

The biomass of belowground parts is sometimes high even in Arctic and high alpine areas. This is especially true of some of the sites investigated in U.S.S.R. (Khodachek, 1969; Alexandrova, 1970; Pospelova, 1972; Shamurin *et al.*, 1972) and for some sedge meadows in Canada (Muc, 1973) and Norway (Wielgolaski, 1972b). Generally it seems that the amount of belowground relative to aboveground parts is especially high in wet areas with relatively low amounts of shrubs compared to herbs.

Ratios of Aboveground to Belowground Biomass (A/B)

The ratios at peak aboveground biomass are of the order 1:20 and 1:10 for live, and live and dead material at the wet sedge dominated meadow in Norway (Wielgolaski and Kjelvik, in press). For some of the sedge meadows at Devon Island, Canada, A/B ratios of live material about 1:12 are calculated (Muc, 1973). In a polygonal bog at Taimyr, U.S.S.R. (Shamurin *et al*., 1972), the ratio is 1:17 for total organic matter (live and dead), and in spotted *Dryas*-sedge-mossy tundras in the same area the ratio is 1:12 (Pospelova, 1972; Shamurin *et al*., 1972). These low ratios are of the same order as the lowest ones found by Dennis (1968) and by Alexandrova (1958, 1970). Somewhat higher ratios (about 1 : 10) are usually found if only living parts of above- and belowground vegetation are considered (Wielgolaski, 1972). Higher ratios are found in tundra areas with higher percentages of shrubs. In a spotted *Dryas*-sedge-mossy tundra in the Ary-Mas forest in Taimyr, Ignatenko *et al*. (1972) reported an aboveground:belowground ratio for total vegetation of about 1 : 5.3 (ranging from about 1 : 3.9 at the ridges to 1 : 6.2 in the depressions). Still this was a lower ratio than for forest vegetation in the same area (ratios 1 : 2 to 1 : 3). Alexandrova (1970) reports a ratio of 1 : 6.9 for total biomass or 1 : 4.5 for living biomass in a shrub tundra in an East European forest tundra region. Pospelova (1972) has found values for above- and belowground total biomass in shrubby vegetation types in Western Taimyr giving ratios from 1 : 6 to 1 : 4, while Chepurko (1972) at Kola found ratios of about 1 : 3. The same ratio was calculated from preliminary results from a bog in Northern Sweden. Dennis (1958) has reported the highest ratios aboveground : belowground to be about 1 : 5 and Alexandrova (1958) about 1 : 4 which is of the same order as the other ratios mentioned. From values reported by Tieszen (1972) for the wet sedge meadow at Point Barrow, Alaska, a ratio of 1 : 6 could be calculated for living parts.

Still higher ratios of living biomass are found at IBP tundra sites in Northern Finland dominated by *Empetrum* and *Vaccinium* species as well as at the alpine sites in Austria with similar vegetation (*Vaccinium, Calluna* and *Loiseleura*). The ratios at these sites range from 1 : 2 to 1 : 0.5. Within the same range are the ratios at the moorland sites in the United Kingdom and Ireland. These ratios are, however, lower than ratios reported from grassland vegetation in temperate regions. For instance, studies of grassland areas in New Zealand at even 1,000 m above sea level show an aboveground : belowground ratio for living biomass of about 1 : 0.05. Relatively high ratios for herbaceous vegetation are also reported by Bray (1963). These data confirm the hypothesis that a typical tundra environment is relatively more severe for aboveground than for belowground parts of the plants (Bliss, 1970). Even if the yearly growth of belowground compartments is normally slow, the mortality and decomposition of roots are also low. Consequently, the belowground biomass is great. The decomposition rate is usually slowest at the lowest temperatures; however, oxygen availability in the soil is also important. The belowground biomass will, therefore, normally be highest in relation to aboveground in wet peaty soil with poor aeration. This is supported by data from Alexandrova's (1970) studies, for example. Decomposition rates will also be somewhat higher in nutrient-rich soils than in oligotrophic soils. That may be the reason for the relatively high aboveground : belowground ratio (1 : 4.0) found in an alpine meadow on the Kola peninsula (Chepurko, 1972).

These hypotheses are confirmed in Norwegian IBP tundra studies (Wielgolaski, 1972b). The highest A/B ratios of living biomass (understory 1:3 to 1:4, total 1:0.6) are found in a birch wood at a lower altitude and, therefore, with relatively higher temperatures. Above the tree line ratios of the same order (1:4 to 1:7) are found in a poor, but relatively warm, lichen health with sandy, well-aerated, and often dry soil deficient in humus as in a somewhat more humid and richer dry meadow at about the same altitude where decomposition is relatively fast because of good nutrient conditions. In the *Salix herbacea* snow bed decomposition is relatively slow because of both the short period without snow and the nutrient deficient soil. Here, therefore, the ratio aboveground:belowground is relatively low in spite of the shrub dominated vegetation (1:5 to 1:9), but still noticeably higher than in the frequently saturated wet sedge meadows cited earlier.

Production

Primary production can be determined by repeated harvestings, preferably of both above- and belowground biomass. Estimates of the production can also be developed from phytosynthesis-respiration values when translocation within the plant is taken into account. Chlorophyll measurements and carbohydrate analyses are a third method used in primary production studies. Harvesting has been the major method used in tundra production studies.

Primary production is often described as the difference between biomass at the time of peak living aboveground vascular plant biomass and the biomass of the same parts before the growth season begins. This gives only a very rough estimate of the plant production, however. While the green parts of vascular plants have the greatest increment in biomass in the early summer, the root biomass most often decreases in spring because of translocation of food reserves for new green growth. To a certain extent the same pattern exists in tundra areas for nongreen, living, aboveground parts during periods of high respiration in spring, before photosynthesis of green parts is high enough to compensate for respiration. Usually, tundra root mass increases most in the autumn when the green parts decrease. Lichens and, to some extent, bryophytes also continue growth until relatively late autumn in tundra areas.

Decreases in biomass between two summer harvestings may be found for all plant compartments. This can result from harvesting errors, but mortality of plant parts, decomposition, and animal consumption may also be responsible. In the IBP tundra studies, for example, decreases in living biomass are sometimes found for belowground material between the time of maximum aboveground living material and the start of the growing season at some wet sedge meadow sites. This indicates simply that most belowground growth takes place in relatively late autumn after aboveground biomass peaks. "Production" of green components may be taken as the increase in green biomass, plus any increase in mass of standing dead aboveground and of litter. This assumes that the increased weight of these dead parts comes mostly from the green material; if this mortality had not occurred during the growth period, the green parts would have increased accordingly. When dead parts of the plants have lower weight at later harvestings, decomposition can be considered responsible.

Annual primary productivity is normally low in tundra which is frequently a consequence of a very short growing season—less than two months in extreme cases. In Taimyr the growing season lasts about 80 days, at Hardangervidda in Norway about 100 days, and at Moor House in the United Kingdom about 180 days. The highest yearly dry matter production is found at tundra sites in the lowest latitudes, i.e., the Austrian IBP sites in the northern hemisphere and at Macquarie Island in the southern hemisphere, as well as on moorlands in Ireland and the United Kingdom. At those sites the green vascular plant production ranged from 100 to 400 g/m^2. When belowground production was added, the total biomass of vascular plants at Moor House increased annually by the order of 600 to 700 g/m^2, with about half of the productivity belowground (Forrest, 1971). Bryophyte production at Moor House was also considerable, i.e., up to 300 g/m^2 in *Sphagnum* (Clymo, 1970). Relatively high primary productivity was also found at some sites in the U.S.S.R., such as in relatively dry alpine meadows on the Kola Peninsula (Chepurko, 1972), as well as in Norway [Wielgolaski, in press (b)]. The total yearly production (vascular plants) was about 500 g/m^2, 225 g/m^2 aboveground and 275 g/m^2 belowground. Even higher primary production have been calculated for wet eutrophic alpine meadows in Norway (vascular plants about 650 g/m^2). Relatively high values were also found at a marshy brush-sedge-moss site in Western Taimyr (Pospelova, 1972); total yearly production was about 400 g/m^2, but only 60 g/m^2 was aboveground.

Many tundra sites have an aboveground vascular plant production of 40 to 100 g/m^2 and a total vascular plant accretion of 100 to 200 g/m^2. For example, such values were attained at many Russian tundra sites (Andreev, 1966; Chepurko, 1972), on sedge meadows at Devon Island in Canada (Muc, 1973) and at some Finnish sites in the understory of sub-alpine woodlands (Kallio and Kärenlampi, 1971).

Low aboveground vascular plant production (less than 30 g/m^2) was found in snow beds in Norway (Wielgolaski, unpubl.), the northern arctic tundra of the U.S.S.R. (Andreev, 1966), some spotted tundras in the U.S.S.R. (Chepurko, 1972; Pospelova, 1972), and in beach ridges and a plateau at Devon Island (Svoboda, 1973). Including lichens and bryophytes, a lichen heath in Norway showed a production of more than 150 g/m^2 aboveground, however (Kjelvik and Kärenlampi, in press). Cryptogams also contributed significantly to primary production at other sites, 30 to above 200 g/m^2 by mosses in meadows in Norway and at Devon Island, Canada, for example [Pakarinen and Vitt, 1973; Wielgolaski, in press (b)].

Primary production may vary considerably from year to year for several reasons including lemming cycles and climatic variations. Dennis (1968) found variations in aboveground dry matter production from 60 to 97 g/m^2 in 1964 and from 3 to 48 g/m^2 in 1965, when lemming populations were high. Several years are therefore necessary for productivity estimates. The values cited earlier are mostly from only a few years of IBP-tundra studies; they are, however, mostly within the 50 to 200 g/m^2 productivity range found in other tundra investigations (Bliss, 1970). The extremely low yearly production at a *Salix artica*-dominated barren site (3 g/m^2) on Cornwallis (Warren Wilson, 1957) lies below the values reported in this paper.

The daily aboveground primary productivity may be rather high in tundra areas, Bliss (1970) having recorded up to 3 $g/m^2/day$. Incoming radiation may be high and the energy balance is often positive during the whole 24-hour period in polar areas during parts of the growing season, e.g., until August 6th in Taimyr, U.S.S.R. (Zalenskij *et al.*, 1972). Tieszen (1972) found positive photosynthesis over 24-hour periods at Point Barrow, Alaska, on most days up to August 2nd. Bliss (1972) reported that *Dryas* is photosynthetically active within a few days of snow melt. Photosynthetic values for *Dryas* were quite comparable to temperate zone grasses and tree seedlings. On clear days most *Dryas* production took place at night because of high temperatures during the day.

Still, effective utilization of solar energy by plants may be rather low in tundra areas, e.g., 0.7 percent on a spotted *Dryas*-moss tundra and 1.8 percent on a marshy tundra in Western Taimyr (Vassiljevskaya and Grishina, 1972). At the latter site daily total primary production was 5 to 6 g/m^2 (Pospelova, 1972), but only about 1 g/m^2 was aboveground productivity. Daily production ranged to as low as 2 g/m^2 at other sites in the same area, i.e., in a spotted *Dryas*-moss tundra (0.25 $g/m^2/day$ in aboveground production). Based on maximum values for aboveground living biomass and biomass of similar components at the beginning of the vegetative period, an average total daily production of about 2 g/m^2 is found at the tundra dry meadow in Norway. Considering the different growth periods for tops and roots and for vascular plants and cryptogams, the daily total primary production of the same site (without compensation for consumption, but for decomposition) was about 5 g/m^2; about half was aboveground parts [Wielgolaski, in press (b)], ranging from about 2.5 to 6 g/m^2 in different years. Even somewhat higher values were calculated for wet, eutrophic alpine meadows in Norway. At Moor House daily aboveground production was about 1.6 g/m^2 (Forrest, 1971).

Measurements of photosynthesis and respiration by tundra plants relevant to productivity estimates have been performed by Hadley and Bliss (1964), Scott and Billings (1964) and Johnson and Kelley (1970), among others. Within the IBP-tundra group the same processes are being studied in several countries. Tieszen (1972) has provided preliminary data on wet sedge meadow at Point Barrow, Alaska. Net CO_2 incorporation by photosynthesis is estimated to be 9 to 12 g/m^2/day. This converts to 6 to 8 g/m^2/day of dry matter which is comparable to daily production calculated by harvesting at the wet sedge meadow in Norway which usually has higher temperatures but shorter days [Wielgolaski, in press (b)]. Zalenskij *et al.* (1972) studied photosynthesis of tundra plants in Taimyr, U.S.S.R., and found that a deficiency of CO_2 in the atmosphere may restrict plant assimilation, especially at high light intensities. In their analyses maximum apparent photosynthesis was 6 mg CO_2 per gram dry weight per hour, which they say supports the concept of low levels of apparent photosynthesis in Arctic plants. Data on photosynthesis and respiration in Norweigian alpine tundra (Skre, in press) indicate higher maximum apparent photosynthesis in some vascular plants in moist, eutrophic communities (partly above 15 mg CO_2/g/h at 15° C and 20,000 lux) early in the growing season. There is relatively good correlation between production estimated from harvesting data. Temperatures on the day before the photosynthetic measurement seem to influence the apparent photosynthetic values, however (Nygaard, in press).

Chlorophyll content of tundra plants might be used to estimate dry matter production after calibration with dry weight data (Bliss, 1970). Chlorophyll contents for some vascular plant species and tundra vegetation types are given in Table 3; obviously, the values expressed as chlorophyll content in mg/m^2 and in mg/g dry weight are not necessarily strongly correlated. Kärenlampi (1972) has investigated distribution of chlorophyll within the lichen *Cladonia alpestris.*

At Point Barrow maximum amounts of chlorophyll per g dry weight occurs in mid-July (Tieszen, 1972), just as it does at the various tundra sites in Norway (Berg, in press). Chlorophyll per m^2 reaches its maximum somewhat later—about July 25th at Point Barrow and about August 1st in the dry and wet meadow in Norway. Chlorophyll data for mosses and lichens are also available in Norway. Understandably maximum aboveground biomass occurs some days after the peak of chlorophyll per m^2. At Point Barrow maximum aboveground biomass was found on August 4th in 1970 and a few days later in Norway.

TABLE 3 Total Chlorophyll *a* and *b* Content of Different Vegetation Types (Vascular Plants Only) for Tundra Ecosystems

Sites	mg/m^2	Dry Weight (mg/g)
Northern Alaska[a]		
Dry sedge	130	3.8
Willow	330	5.4
Wet sedge	760	8.8
Wet meadow community (maximum value)	450	5.8
Mt. Washington[b]		
Heath	540	1.9
Diapensia	180	2.7
Wet sedge	820	4.7
Norway[c]		
Lichen heath	90	2.1
Dry meadow	580	5.0
Wet sedge meadow	390	5.5

[a]Tieszen, 1972.
[b]Bliss, 1966.
[c]Berg, in press.

The maximum leaf area index (LAI) of 1.0 at Point Barrow occurred concurrently with maximum chlorophyll content per m^2 (Tieszen, 1972). Bliss (1970) found LAI's ranged from 0.94 (heath-rush community) to 3.30 (snowbank community) on Mt. Washington, although LAI of most communities was between one and two. In Norway LAI of green leaves was 0.3 on the lichen heath site at Hardangervidda and 1.1 at the wet sedge meadow in early August (Berg *et al.*, in press).

Microflora

The microflora may be substantial in some tundra soils. Direct counts revealed over 10^{10} bacteria per g dry soil in the upper few cm of frost-boil tundra spot crusts and steep river banks in Taimyr, U.S.S.R. (Aristovskaya and Parinkina, 1972). There were also about 0.5×10^5 fungi and similar amounts of actinomycetes. Low values were found for all organisms in hummocky tundra, less than 10^8 bacteria by direct counts. Usually the microbial activity is restricted to surface layers of soil, but in some cases an increase in number of bacteria was registered at the permafrost level. In Alaska a maximal bacteria count of about 10^9 per g (direct counting) was found shortly after thaw (Brown, 1972). Similar values were obtained by the same methods at Hardangervidda, Norway, with highest values (somewhat above 10^{10} per g dry soil) at the eutrophic dry and wet meadows (Charholm *et al.*, in press).

At Taimyr crude estimates of bacterial productivity were from 0.05 g to 0.25 g/g soil (Matveyeva, 1972). The average number of generations per month was 15.5 in the frost-boil tundra and steep river banks, but only 1 to 4 in polygon bogs (Aristovskaya and Grishina, 1972). At Hardangervidda, Norway, the bacterial dry-weight biomass ranged from estimated 550 g/m^2 (lichen heath) to 940 g/m^2 (wet meadow) in the upper 35 cm of the soil layer (Clarholm *et al.*, in press).

Fungi are an important part of reindeer diets in Northern Finland (Kallio, personal communication, 1971) and recent subjects of biomass and productivity studies. In the U.S.S.R. (Stepanova and Tomlin, 1972) dry-weight biomass of mushrooms varied between 0.2 g/ha (spotted tundra) and 1.3 g/ha (dwarf-sedge-mossy tundra). At Hardangervidda, Norway, estimates of dry weight biomass of fungal hyphae in the upper 10 cm of the soil ranged from 50 g/m^2 (lichen heath) to 110 and 180 g/m^2 (dry and wet meadows) and about 200 g/m^2 in subalpine birch forest (Hanssen and Goksøyr, in press).

PLANT-ANIMAL INTERACTIONS

The most striking influence of animals on tundra primary productivity is the result of small mammal consumption. Dennis (1968) has shown how productivity varies with lemming cycles, i.e., low primary production in a lemming high year and high primary production the year before. He also noted changes in plant species during the cycling period. Selective consumption by the small mammals was one factor but differences between reproductive potential of plant species in dense stands (with high amounts of litter) and more open vegetation (with less litter) was also important; such differences in stand density were largely caused by animal grazing. Schultz (1964) reported a 50 percent reduction in vigor and yield of plants in one half to two thirds of the tundra and a near 90 percent reduction in the remainder during the 1960 lemming high in Alaska; thus resulting in a reduced plant productivity the following year. In the U.S.S.R. experiments on the influence of small mammal consumption on vegetation have been carried out. When about 10 percent of aboveground vegetation was consumed, increased shoot growth resulted the following summer (Smirnov and Tokmakova, 1972). These results may differ from those cited for the U.S.A. because of the lower grazing pressure in the Russian experiments. At Hardangervidda, Norway, plant consumption by small mammals was about 10 g/m^2/year in 1969–70, a small mammal high year. This means about 10 percent of the aboveground biomass was grazed by small mammals in their preferred vegetation types. The following year consumption was only 3 to 4 percent of the previous summers values (Østbye, personal communication, 1972).

Large mammals may be important consumers of plant biomass in some areas. In Canada the summer removal by muskox is estimated to be 1.5 percent of aboveground biomass. Each native reindeer at Hardangervidda, Norway, consumes 1,200 to 1,900 g/day on an average during the winter months and above 100 g more per day during summer (Garre, personal communication, 1972). Most of the winter consumption is in the windblown plant community Loiseleurio–Arctostaphylion with some in the Phyllodoco–Myrtillion. The lichens *Cladonia mitis* and *Cetraria nivalis* (Gaare and Skogland, in press) preferred by the animals during the winter. The lichen health at Hardangervidda (belonging to the alliance Loiseleurio–Arctostaphylion) had an aboveground yearly primary production of vascular plants and cryptogams of above 150 g/m^2 (the lichens accounting for about 50 to 90 g/m^2). Obviously a minimum of 5,000 m^2 of lichen heath is required for one reindeer during the seven winter months (November to May) if the lichen heath is to be kept in steady state. This does not take trampling damage into account; if this is done, the minimum area might be 10 times higher. Makhaeva (1959) found each reindeer (domestic) in the Murmansk area grazed 65 m^2 each day in lichen tundra. If this estimate is applied to Hardangervidda the daily consumption per reindeer would be about 20 to 25 g/m^2 of the lichen community during winter.

The influence by invertebrates on plant biomass is difficult to estimate. In the U.S.S.R. some estimates of insect consumption of willows are available (Danilov, 1972; Bogatschova, 1972); values range from 1.3 to 3.3 percent of the green biomass in tundra and from 5 to 9 percent on river banks. These values are, of course, higher than estimated consumption by psyllids at Moor House which is 0.1 to 1 percent of the annual shoot production by *Calluna* (Hodkinson, 1971).

ANIMAL PRODUCTION

Invertebrates

The number, biomass and production of invertebrates vary greatly yearly, seasonally and between vegetation types. Furthermore, comparisons of values provided by various authors are difficult because of variations in measuring techniques. In the U.S.S.R. differences within one tundra zone were sometimes 10 times greater than between zones. Collembola are very important in the Taimyr watersheds (Matveyeva, 1972) accounting for about 2/7 of the total maximal invertebrate zoomass. At the same site Enchytraeidae account for about 1/2 and Nematoda 1/7 of the zoomass. These two groups are also important in other tundra communities at Taimyr. In a xeromorphic dwarf shrub community (mainly *Dryas*) the Oligochaeta are also important accounting for a zoomass of 25 g/m^2 (possibly fresh weight) out of a total of 40 g/m^2. Tipulidae are an important group in spots of bare ground and in wet bogs (the last site having a total maximum zoomass of only 6 g/m^2). Collembola are abundant in the dwarf shrub communities and grass meadows. The highest invertebrate zoomass (90 g/m^2) is found in the grass meadow. Danilov (1972) reports the total biomass (possibly dry weight) of invertebrates (excluding Collembola and flying insects) in soil and vegetation (Table 4) near Salekhard in the U.S.S.R. About 3.5 g/m^2 is the maximum for arthropods and Oligochaeta in the water banks, while the same groups total less than 2 g/m^2 in the tundras of the area. The biomass, particularly of sawfly and leaf-bettle larvae, is high in the bushes on the water banks—22 times higher than in the tundra areas. The biomass of flying insects (not included in the above mentioned values) is also considerable; e.g., mosquitoes along the banks total about 4 g/m^2 early in summer (most of which is later consumed by birds). These values indicate that flying insects as well as invertebrates developed in water have to be included to get the total invertebrate biomass influencing the rest of the tundra ecosystem.

Invertebrate predators have the greatest biomass (56 percent) in the tundra studies cited above, but this group was relatively less important on the water banks (29 percent of the biomass) even though predatory forms dominated numerically (57 percent) in this vegetation type. At the tundra sites spiders accounted for about 27 percent of the arthropod biomass and 10 percent on the water banks. The respective percentage biomass in the two vegetation types was 20 and 4 percent for Carabidae, 11 and 22 percent for Tipulidae larvae and 5 and 22 percent for bush-dwelling sawfly and leaf-beetle larvae.

In the Norwegian tundra project soil invertebrates have been studied except for nematodes and protozoans (Kauri *et al.*, personal communication, 1972) but quantitative data on flying insects are not available. The highest dry weight of Enchytraeidae occurred in the middle of August (about 600 mg/m^2) at the dry meadow site and in winter (above 2,000 mg/m^2) at the wet sedge meadow site. The maximum dry weight of arthropods (200 mg/m^2 by quick-trap sampling) was in the middle of July. The dominating group at that time was Diptera larvae (28 percent), while Acari (18.4 percent), Carabidae (12.5 percent), Lepidoptera larvae (8 percent) and Collembola (6 percent) were other important associates. Later in the season (mid-September) Lepidoptera larvae (29 percent) and Collembola (26 percent) had the highest dry weight totals, while Acari (17 percent) and Carabidae (12.5 percent) contributed about the same percentage to the total dry weight biomass as in the summer. Collembola had a minimum biomass in August, and maximum in early spring and late autumn. The biomass of Lepidoptera larvae was minimum in August between maxima in July and September. Acari and Collembola dominate at both the dry and wet meadow sites (Acari maximum about 150,000 per m^2 in August and Collembola maximum about 100,000 per m^2 in July and in early September at the dry meadow). Maximum numbers of Enchytraeidae occur in late summer (about 50,000 per m^2 in wet meadow and 30,000 per m^2 in dry meadow sites).

The same invertebrate groups were important at tundra sites on Devon Island, Canada, as at the alpine tundra sites in Norway (Ryan, 1972), while 80 percent of the soil fauna in the moorlands in United Kingdom consisted of Tipulidae larvae and Enchytraeidae (Cragg, 1961). Recognizing the great differences in invertebrate fauna the most important groups in tundra soil and low vegetation layers generally are Nematoda, Enchytraeidae, Acari and Collembola; at certain periods Diptera, Lepidoptera, Carabidae and Tipulidae and, in some places, Aranidae are also important.

Vertebrates

Birds have been included in IBP tundra studies in the U.S.S.R. (Danilov, 1972; Matveyeva, 1972; Vinokurov *et al.*, 1972), Canada (Pattie, 1972), U.S.A. (Brown, 1972) and Norway (Wielgolaski, 1972b). Vinokurov *et al.* (1972) report from 1.7 to 3.7 birds/ha in tundra during the breeding period. Matveyeva (1972) reports a density of 2.8 adult birds/ha in an optimum year. The biomass of herbivorous birds was estimated to be 246 g/ha, of insectivorous birds 628 g/ha and of predators 13 g/ha. Loss of eggs and young birds amounted to 25 percent of reproduction. Danilov (1972) discusses the eating habits of the insectivorous birds which change throughout the summer; e.g., the most abundant food for passerines during the breeding season was spiders, mosquitoes and flies, while in August 73 per-

TABLE 4 Invertebrate Biomass (mg/m^2) in Tundra Soil and Vegetation from Southern Jamal (N.W. Salekhard, U.S.S.R.)[a]

	Total	Oligochaeta Worms	Arthropods (excl. Collembola and Flying Insects)	Sawfly and Leaf-Beetle Larvae in Bushes	Spiders	Carabidae	Tipulidae
Tundra	1,780	1,190	590	30	160	115	65
Water banks	3,465	1,865	1,600	650	150	65	350

[a]Excluding Collembola, flying insects, and all insects developing in water (Danilov, 1972).

cent of the food was sawfly larvae. Arthropod biomass consumed by birds was about 3 kg/ha in tundra and between 6 and 12 kg/ha in wet areas close to rivers and lakes. Even when the young birds were fairly large, not more than about 2 percent of available food was utilized by birds, and their influence on the invertebrate population was insignificant. Great annual variation is found in the avian fauna. The variation partly follows the 3 to 4 year cycle of small mammals, but the bird cycles are not as regular and vary with the species (Vinokurov *et al.*, 1972). The cycles of avian predators in tundra areas, such as snow owls and jaegers in Alaska, mostly follow the small mammals cycles.

The yearly fluctuations in small mammals is high. Thompson (1955) says that lemming densities may exceed 100/ha in peak years at Point Barrow, Alaska, and Vinokurov *et al.* (1972) reports fluctuations from zero to 725 lemmings/ha in favorable habitats at Taimyr. At Hardangervidda, Norway, 93 small mammals/ha (mostly *Microtus oeconomus*) were caught in September 1970, while only 2 were found in September 1971 (Østbye, personal communication, 1972). The biomass of small mammals in September 1970 was estimated to be 3.2 kg/ha at Hardangervidda while that of 119 lemmings/ha at Taimyr, U.S.S.R. in 1967 was 8.3 kg/ha (Matveyeva, 1972). In Canada the 1970 standing crop of small mammals was low, i.e., only 4.6 g/ha for lemmings in August (Speller, 1972). At Point Barrow, Alaska, an increase in brown lemmings occurred over the winter of 1970–1971, and in June 1971 the estimated density was 75 individuals/ha (Brown, 1972).

The role of lemmings in tundra ecosystems is the subject of much additional study. As an example, a population dynamics and feeding model is being constructed which combines data on physiology and behavior (such as food and habitat selection). In Canada small mammals prefer the plants *Dryas integrifolia* and *Salix arctica.* In Norway *Dryas octopetala* and *Salix herbacea* are favored as food. The small mammal production from September 1969 to September 1970 was 2,650 kcal/ha with a plant consumption of 366,000 kcal/ha/year. Arctic fox populations are dependent on the number of lemmings (Bannikov, 1970; Vinokurov *et al.*, 1972). The latter authors reported the maximum density of arctic foxes to be 0.032 per ha; Bannikov (1970), using data from various authors, gives values of 0.011 to 0.035 per ha with increases in population from west to east due to higher quality habitat. The Devon Island arctic fox population was estimated to have an average biomass of 1.5 g/ha (Riewe, 1972). In the same area the standing crop of muskox averaged about 250 g/ha for the three summer months (Hubert, 1972).

The reindeer population at Taimyr is estimated to be 1.3 individuals/ha with a biomass up to 1,000 kg/ha (Matveyeva, 1972). Reindeer populations at Hardangervidda, Norway, fluctuate widely. In 1970 the population was about 3 individuals/km^2; only about 1/10 of the area was useful as winter grazing and, consequently, the lichen heaths were highly overgrazed during winters. In the winter of 1970 to 1971 a heavy ice crust covered many of the lichen heaths drastically reducing the area available for grazing. A reduction of the reindeer population to about 1/4 of its earlier size followed.

Decomposition

Energy flow diagrams have been constructed for tundra ecosystems (e.g., see Dahl and Gore, 1968; Bliss, 1972; Brown, 1972). According to Bliss (1972) the plant–soil organic–decomposer cycle seems to be the main pathway, with herbivores and carnivores being only minor pathways. Heal (1972) also stresses this for the moorland sites in the United Kingdom. Some of the IBP tundra data have shown, however, that many soil invertebrates are important for decomposition of organic material.

The decomposition rate of litter in tundra has been studied by the litter bag technique (Heal, 1972). In the moorlands of United Kingdom about 50 percent weight loss of *Calluna* shoots and *Eriophorum* leaves occurred over a five-year period. In Norway a weight loss of above 74 percent of *Carex nigra* leaf litter was found over a period of one year, with lower values for most other materials; while in Sweden the decomposition rate for *Rubus* was about 25 percent in the first year (Rosswall *et al.*, in press). A study of cellulose decomposition at different soil depths indicates that the rate of decay in the surface 2 cm may be 5 to 6 times higher than the rate of decay of the same material at 20 cm depth when water is not limiting the process (Heal *et al.*, 1974; Berg *et al.*, in press).

NUTRIENT RELATIONSHIPS

Even though knowledge of energy and carbon flow has been the main aim of most IBP-tundra projects, emphasis has also been placed on nutrient cycling. Studies of nitrogen fixation have been foremost because of the hypothesis that N limits productivity. In Norway both asymbiotic and symbiotic N-fixation (mostly lichens and blue-green algae) was found at most tundra sites. A rough estimate sets N-fixation at up to 1 to 2 g/m^2/yr (Granhall and Lid-Torsvik, in press). In Sweden values up to 5 to 6 g/m^2/yr were found for blue-green algae (Granhall and Selander, 1972, 1973; Rosswall, 1972). In Finland the lichens *Nephroma arcticum, Solorina crocea, Peltigera* sp. and *Stereocaulon* sp. are important in N-fixation (Kallio and Suhonen, 1972; Kallio and Kallio, in press). *Solorina crocea* seems to be the lichen species in Norway with highest N-fixation. From Point Barrow, Alaska, Brown (1972) reports N-fixation of about 15 μg/m^2/hour (about 20 mg/m^2/year) in areas with *Peltigera aphthosa*. On Devon Island, Canada, considerable N-fixation was found in mesic meadow soils with decreasing amounts in hydric meadows and on beach ridges (possibly too cold and too dry respectively). Soil microorganisms appear to be the principal nitrogen fixers, although *Nostoc commune* is common in the mesic meadows (Stulz, 1972). In the U.S.S.R. N-fixing blue-green algae, such as *Nostoc microscopica*, are also found in tundra areas along with the N-fixing Cyanophyceae (Novichkova-Ivanova, 1972). The algal flora of tilled areas was found to be twice as rich as the flora of virgin tundra.

Besides these intensive studies of nitrogen, chemical analyses of precipitation, soil ground water (input and output), soil, various components of plants and animals have been performed for several elements at many sites. Fertilizer experiments combined with lysimeter studies are carried out at some sites to obtain better values for rates of nutrient cycling in tundra ecosystems.

ACKNOWLEDGEMENTS

I would like to thank the individual workers in all IBP tundra projects who contributed ideas for organization, as well as information and data, to this paper, and also to thank the members of the IBP Tundra Biome Steering Committee, which reviewed the paper and made constructive suggestions.

REFERENCES

Alexandrova, V. D. 1958. An attempt to determine the aboveground productivity of plant communities in the Arctic tundra. Bot. Zh. 43(12):1748–1761. (in Russian).

Alexandrova, V. D. 1970. The vegetation of the tundra zones in the USSR and data about its productivity, p. 93–114. *In* W. A. Fuller and P. G. Kevan (ed.) Proceedings of the Conference on Productivity and Conservation in Northern Circumpolar Lands, Edmonton, 1969. IUCN Publ. (new series) No. 16. 344 p.

Andreev, V. N. 1966. Peculiarities of zonal distribution of the aerial and underground phytomass on the East European Far North. Bot. Zh. 51:1410–1411. (in Russian).

Aristovskaya, T. V., and O. M. Parinkina. 1972. Preliminary results of the IBP studies of soil microbiology in tundra, p. 80–92. *In* F. E. Wielgolaski and Th. Rosswall (ed.) Proceedings IV. International Meeting on the Biological Productivity of Tundra, Leningrad, Oct. 1971. Tundra Biome Steering Committee, Stockholm. 320 p.

Bannikov, A. G. 1970. Arctic fox in the USSR, p. 121–130. *In* W. A. Fuller and P. G. Kevan (ed.) Proceedings of the Conference on Productivity and Conservation in Northern Circumpolar Lands, Edmonton, 1969. IUCN Publ. (new series) No. 16. 344 p.

Berg, A. In press. Pigment structure of vascular plants, mosses and Lichens at Hardangervidda, Norway. *In* F. E. Wielgolaski (ed.) Fennoscandian Tundra Ecosystems. Part 1. Plants and Microorganisms. Springer Verlag, Berlin-Heidelberg-New York.

Berg, A., S. Kjelvik, and F. E. Wielgolaski. In press. Measurement of leaf areas and leaf angles of plants at Hardangervidda, Norway. *In* F. E. Wielgolaski (ed.) Fennoscandian Tundra Ecosystems. Part 1. Plants and Microorganisms. Springer Verlag, Berlin–Heidelberg–New York.

Berg, B., L. Karenlampi, and A. K. Veum. In Press. Comparisons of decomposition rates measured by means of cellulose. *In* F. E. Wielgolaski (ed.) Fennoscandian Tundra Ecosystems. Part 1. Plants and Microorganisms. Springer Verlag, Berlin–Heidelberg–New York.

Beschel, R. E. 1970. The diversity of tundra vegetation, p. 85–93. *In* W. A. Fuller and P. G. Kevan (ed.) Proceedings of the Conference on Productivity and Conservation in Northern Circumpolar Lands, Edmonton, 1969. IUCN Publ. (new series) No. 16. 344 p.

Bliss, L. C. 1962. Net primary production of tundra ecosystems, p. 35–46. *In* H. Lieth (ed.) Die Stoffproduktion der Pflanzendecke. Gustav Fischer Verlag. Stuttgart. 156 p.

Bliss, L. C. 1966. Plant productivity in alpine microenvironments on Mt. Washington, New Hampshire. Ecol. Monogr. 36:125–155.

Bliss, L. C. 1970. Primary production within Arctic tundra ecosystems, p. 75–85. *In* W. A. Fuller and P. G. Kevan (ed.) Proceedings of the Conference on Productivity and Conservation in Northern Circumpolar Lands, Edmonton, 1969. IUCN Publ. (new series) No. 16. 344 p.

Bliss, L. C. 1972. Devon Island research 1971, p. 269–275. *In* F. E. Wielgolaski and Th. Rosswall (ed.) Proceedings IV. International Meeting on the Biological Productivity of Tundra, Leningrad, Oct. 1971. Tundra Biome Steering Committee, Stockholm. 320 p.

Bliss, L. C., and F. E. Wielgolaski (ed.) 1973. Primary Production and Production Processes, Tundra Biome. Tundra Biome Steering Committee, Edmonton–Oslo. 256 p.

Bogatschova, I. A. 1972. Leaf-eating insects on willows in tundra biocenoses of Southern Jamal, p. 131–132. *In* F. E. Wielgolaski and Th. Rosswall (ed.) Proceedings IV. International Meeting on the Biological Productivity of Tundra, Leningrad, Oct. 1971. Tundra Biome Steering Committee, Stockholm. 320 p.

Bray, J. R. 1963. Root production and the estimation of net productivity. Canad. J. Bot. 41:65–72.

Brown, J. 1972. Summary of the 1971 U.S. Tundra Biome Program, p. 306–313. *In* F. E. Wielgolaski and Th. Rosswall (ed.) Proceedings IV. International Meeting on the Biological Productivity of Tundra, Leningrad, Oct. 1971. Tundra Biome Steering Committee, Stockholm. 320 p.

Chepurko, N. L. 1972. The biological productivity and the cycle of nitrogen and ash elements in the dwarf shrub tundra ecosystems of the Khibini mountains (Kola Peninsula), p. 236–247. *In* F. E. Wielgolaski and Th. Rosswall (ed.) Proceedings IV. International Meeting on the Biological Productivity of Tundra, Leningrad, Oct. 1971. Tundra Biome Steering Committee, Stockholm. 320 p.

Clarholm, M., V. Lid-Torsvik, and J. H. Baker. In press. Bacterial populations of some Fennoscandian tundra soils. *In* F. E. Wielgolaski (ed.) Fennoscandian Tundra Ecosystems. Part 1. Plants and Microorganisms. Springer Verlag, Berlin–Heidelberg–New York.

Clymo, R. S. 1970. The growth of *Sphagnum:* methods of measurement. J. Ecol. 58:13–49.

Cragg, J. B. 1961. Some aspects of the ecology of moorland animals. J. Anim. Ecol. 31:1–21.

Dahl, E., and A. J. P. Gore (ed.) 1968. Working meeting on analysis of ecosystems: Tundra zone. Ustaoset, Norway. Sept. 1968. 87 p.

Danilov, N. N. 1972. Birds and arthropods in the tundra biogeocenosis, p. 117–121. *In* F. E. Wielgolaski and Th. Rosswall (ed.) Proceedings IV. International Meeting on the Biological Productivity of Tundra, Leningrad, Oct. 1971. Tundra Biome Steering Committee, Stockholm. 320 p.

Dennis, J. G. 1968. Growth of tundra vegetation in relation to

Arctic microenvironments at Barrow, Alaska. Ph.D. dissertation, Duke Univ. 289 p.

Firsova, V. P., S. G. Shijatov, and L. N. Dobriskij (ed.) 1969. Productivity of biogeocenosis in the subarctic. Proceedings of All Union Conference in Sverdlovsk. Sverdlovsk, USSR. 246 p.

Forrest, G. I. 1971. Structure and production of north Pennine blanket bog vegetation. J. Ecol. 59:453–480.

Gaare, E., and T. Skogland. In press. Wild reindeer food habitats and range use at Hardangervidda, Norway. *In* F. E. Wielgolaski (ed.) Fennoscandian Tundra Ecosystems. Part 2. Animals and Systems Analysis. Springer Verlag, Berlin–Heidelberg–New York.

Gorchakovsky, P. L., and N. I. Andreyashkina. 1972. Productivity of some shrub, dwarf shrub and herbaceous communities of forest–tundra, p. 113–116. *In* F. E. Wielgolaski and Th. Rosswall (ed.) Proceedings IV. International Meeting on the Biological Productivity of Tundra, Leningrad, Oct. 1971. Tundra Biome Steering Committee, Stockholm. 320 p.

Gränhall, U., and H. Selander. 1972. Nitrogen fixation studies at the IBP tundra project at Abisko (Stordalen) Sverige, p. 77–78. *In* F. E. Wielgolaski (ed.) IBP i Norden No. 8 (Tundra volume). 111 p.

Gränhall, U., and H. Selander. 1973. Nitrogen fixation in a subarctic mire. Oikos 24:8–15.

Gränhall, U., and V. Lid-Torsvik. In press. Nitrogen fixation by bacteria and free-living blue-green algae in tundra areas. *In* F. E. Wielgolaski (ed.) Fennoscandian Tundra Ecosystems. Part 1. Plants and Microorganisms. Springer Verlag, Berlin–Heidelberg–New York.

Hadley, E. B., and L. C. Bliss. 1964. Energy relationship of alpine plants on Mt. Washington, New Hampshire. Ecol. Monogr. 34:331–357.

Hanssen, J. F., and J. Goksøyr. In press. Biomass and production of soil and litter fungi at Scandinavian tundra sites. *In* F. E. Wielgolaski (ed.) Fennoscandian Tundra Ecosystems. Part 1. Plants and Microorganisms. Springer Verlag, Berlin–Heidelberg–New York.

Heal, O. W. (ed.) 1971. Proceedings of the Tundra Biome Working Meeting on Analysis of Ecosystems, Kevo, Finland. Tundra Biome Steering Committee, London. 297 p.

Heal, O. W. 1972. A brief review of progress in the studies at Moor House, United Kingdom, p. 295–306. *In* F. E. Wielgolaski and Th. Rosswall (ed.) Proceedings IV. International Meeting on the Biological Productivity of Tundra, Leningrad, Oct. 1971. Tundra Biome Steering Committee, Stockholm. 320 p.

Heal, O. W., G. Howson, D. D. French, and J. N. R. Jeffers. 1974. Decomposition of cotton strips in tundra, p. 341–362. *In* A. J. Holding, O. W. Heal, S. F. MacLean, Jr., and P. W. Flanagan (ed.) Soil Organisms and Decomposition in Tundra. Tundra Biome Steering Committee, Stockholm. 398 p.

Hodkinson, I. D. 1971. Studies on the ecology of *Strophingia ericae* (Curtis) (Homoptera:Psylloidea). Ph.D. dissertation, Lancaster Univ.

Holding, A. J., O. W. Heal, S. F. MacLean, Jr., and P. W. Flanagan (ed.) Soil Organisms and Decomposition in Tundra. Tundra Biome Steering Committee, Stockholm. 398 p.

Hubert, B. 1972. Productivity of muskox, p. 272–280. *In* L. C. Bliss (ed.) Devon Island I.B.P. project, high arctic ecosystem. Project report 1970 and 1971. Dept. of Bot., University of Alberta, Canada. 413 p.

Ignatenko, I. V., A. V. Knorre, N. V. Lovelius, and B. N. Norin. 1972. Standing crop in plant communities at the station Ary-Mas, p. 140–149. *In* F. E. Wielgolaski and Th. Rosswall (ed.) Proceedings IV. International Meeting on the Biological Productivity of Tundra, Leningrad, Oct. 1971. Tundra Biome Steering Committee, Stockholm. 320 p.

Jenkin, J. F., and D. H. Ashton. 1970. Productivity studies on Macquarie Island vegetation, p. 851–863. *In* M. W. Holdgate (ed.) Antarctic Ecol. Vol. 2. Academic Press. London and New York. 998 p.

Johnson, P. L., and J. J. Kelley, Jr. 1970. Dynamics of carbon dioxide in an Arctic biosphere. Ecol. 51:73–81.

Kallio, S., and P. Kallio. In press. Nitrogen fixation in lichens at Kevo, North-Finland. *In* F. E. Wielgolaski (ed.) Fennoscandian Tundra Ecosystems. Part 1. Plants and Microorganisms. Springer Verlag, Berlin–Heidelberg–New York.

Kallio, P., and L. Kärenlampi. 1971. A review of the stage reached in the Kevo IBP in 1970, p. 79–91. *In* O. W. Heal (ed.) Proceedings of the Tundra Biome Working Meeting on Analysis of Ecosystems, Kevo, Finland. Tundra Biome Steering Committee, London. 297 p.

Kallio, P., and S. Suhonen. 1972. Lavernas kväveproduksjon, ett nordligt problem (The nitrogen production in lichens, a problem of the North), p. 71–75. *In* F. E. Wielgolaski (ed.) IBP i Norden No. 8 (Tundra volume). 111 p.

Kärenlampi, L. 1970. Distribution of chlorophyll in the lichen *Cladonia alpestris*. Rep. Kevo Subarctic Res. Stat. 7:1–8.

Khodachek, E. A. 1969. The plant matter of tundra phytocenoses in Western Taimyr. Bot. Zh. 54(7):1059–1073. (Translated by P. Kuchar).

Kjelvik, S., and L. Kärenlampi. In press. Plant biomass and primary production of Fennoscandian subarctic and subalpine forests and of alpine willow and heath ecosystems. *In* F. E. Wielgolaski (ed.) Fennoscandian Tundra Ecosystems. Part 1. Plants and Microorganisms. Springer Verlag, Berlin–Heidelberg–New York.

Larcher, W., L. Schmidt, G. Grabherr, and A. Cernusca. 1973. Plant biomass and production of alpine shrub heaths at Mt. Patscherkofel, Austria, p. 65–73. *In* L. C. Bliss and F. E. Wielgolaski (ed.) Primary Production and Production Processes, Tundra Biome. Tundra Biome Steering Committee, Edmonton–Oslo. 256 p.

Makhaeva, L. V. 1959. Winter pasture management in reindeer farming in Murmansk oblast. Probl. Sev. 3:66–77. (Translated from Russian).

Matveyeva, N. V. 1972. The Tareya word model, p. 156–162. *In* F. E. Wielgolaski and Th. Rosswall (ed.) Proceedings IV. International Meeting on the Biological Productivity of Tundra, Leningrad, Oct. 1971. Tundra Biome Steering Committee, Stockholm. 320 p.

Moore, J. J. (ed.) In preparation. Tundra and Related Habitats. Cambridge Univ. Press.

Muc, M. 1973. Primary production of plant communities of the Truelove Lowland, Devon Island, Canada-sedge meadows, p. 3–14. *In* L. C. Bliss and F. E. Wielgolaski (ed.) Primary Production and Production Processes, Tundra Biome. Tundra Biome Steering Committee, Edmonton–Oslo. 256 p.

Novichkova-Ivanova, L. N. 1972. Soil and aerial algae of polar deserts and arctic tundra, p. 261–265. *In* F. E. Wielgolaski and Th. Rosswall (ed.) Proceedings IV. International Meeting on the Biological Productivity of Tundra, Leningrad, Oct. 1971. Tundra Biome Steering Committee, Stockholm. 320 p.

Nygaard, R. T. In press. Acclimatization effect in photosynthesis and respiration. *In* F. E. Wielgolaski (ed.) Fennoscandian Tundra Ecosystems. Part 1. Plants and Microorganisms. Springer Verlag, Berlin–Heidelberg–New York.

Pakarinen, P., and D. H. Vitt. 1973. Primary production of plant communities of the Truelove Lowland, Devon Island, Canada-moss communities, p. 37–46. *In* L. C. Bliss and F. E. Wielgolaski (ed.) Primary Production and Production Processes, Tundra Biome. Tundra Biome Steering Committee, Edmonton–Oslo. 256 p.

Pattie, D. L. 1972. Preliminary bioenergetics and population level studies in high arctic birds, p. 281–292. *In* L. C. Bliss (ed.) Devon Island I.B.P. project, high arctic ecosystem. Project report 1970 and 1971. Dept. of Bot., University of Alberta, Canada. 413 p.

Pospelova, E. B. 1972. Vegetation of the Agapa station and productivity of the main plant communities, p. 204–208. *In* F. E. Wielgolaski and Th. Rosswall (ed.) Proceedings IV. International Meeting on the Biological Productivity of Tundra, Leningrad, Oct. 1971. Tundra Biome Steering Committee, Stockholm. 320 p.

Riewe, R. R. 1972. Preliminary data on mammalian carnivores including man in the Jones Sound region, N.W.T., p. 315–340. *In* L. C. Bliss (ed.) Devon Island I.B.P. project, high arctic ecosystem. Project report 1970 and 1971. Dept. of Bot., University of Alberta, Canada. 413 p.

Rodin, L. E., and N. I Bazilevich. 1966. Production and mineral cycling in terrestrial vegetation. (English translation by G. E. Fogg). Edinburgh. Oliver and Boyd. 288 p.

Rosswall, T. 1972. Progress in the Swedish tundra project 1971, p. 291–294. *In* F. E. Wielgolaski and Th. Rosswall (ed.) Proceedings IV. International Meeting on the Biological Productivity of Tundra, Leningrad, Oct. 1971. Tundra Biome Steering Committee, Stockholm. 320 p.

Rosswall, T., A. K. Veum, and L. Kärenlampi. In press. Plant litter decomposition at Fennoscandian tundra sites. *In* F. E. Wielgolaski (ed.) Fennoscandian Tundra Ecosystems. Part 1. Plants and Microorganisms. Springer Verlag, Berlin–Heidelberg–New York.

Ryan, J. K. 1972. Devon Island invertebrate research, p. 293–314. *In* L. C. Bliss (ed.) Devon Island I.B.P. project, high arctic ecosystem. Project report 1970 and 1971. Dept. of Bot., University of Alberta, Canada. 413 p.

Schultz, A. M. 1964. The nutrient recovery hypothesis for Arctic microtine cycles. II. Ecosystem variables in relation to Arctic microtine cycles, p. 57–68. *In* D. J. Crisp (ed.) Grazing in Terrestrial and Marine Environments. Blackwell Sci. Publ., Oxford. 322 p.

Scott, D., and W. D. Billings. 1964. Effect of environmental factors on standing crop and productivity of an alpine tundra. Ecol. Monogr. 34:243–270.

Shamurin, V. F., T. G. Polozova, and E. A. Khodachek. 1972. Plant biomass of main plant communities at the Tareya station (Taimyr), p. 163–181. *In* F. E. Wielgolaski and Th. Rosswall (ed.) Proceedings IV. International Meeting on the Biological Productivity of Tundra, Leningrad, Oct. 1971. Tundra Biome Steering Committee, Stockholm. 320 p.

Skre, O. In press. CO_2-exchange in Norwegian tundra plants studied by infrared gas analyzer technique. *In* F. W. Wielgolaski (ed.) Fennoscandian Tundra Ecosystems. Part 1. Plants and Microorganisms. Springer Verlag, Berlin–Heidelberg–New York.

Smirnov, V. S., and S. G. Tokmakova. 1972. Influence of consumers on natural phytocenosis production variation, p. 122–127. *In* F. E. Wielgolaski and Th. Rosswall (ed.) Proceedings IV. International Meeting on the Biological Productivity of Tundra, Leningrad, Oct. 1971. Tundra Biome Steering Committee, Stockholm. 320 p.

Speller, S. W. 1972. Biology of *Dicrostonyx groenlandicus* on Truelove Lowland, Devon Island, N.W.T., p. 257–271. *In* L. C. Bliss (ed.) Devon Island I.B.P. project, high arctic ecosystem. Project report 1970 and 1971. Dept. of Bot., University of Alberta, Canada. 413 p.

Stepanova, I. V., and B. Tomilin. 1972. Fungi of basic plant communities in Taimyr tundra, p. 193–198. *In* F. E. Wielgolaski and Th. Rosswall (ed.) Proceedings IV. International Meeting on the Biological Productivity of Tundra, Leningrad, Oct. 1971. Tundra Biome Steering Committee, Stockholm. 320 p.

Stutz, R. C. 1972. Nitrogen fixation studies on Devon Island, p. 252–256. *In* L. C. Bliss (ed.) Devon Island I.B.P. project, high arctic ecosystem. Project report 1970 and 1971. Dept. of Bot., University of Alberta, Canada. 413 p.

Svoboda, J. 1973. Primary production of plant communities of the Truelove Lowland, Devon Island, Canada–beach ridges, p. 15–26. *In* L. C. Bliss and F. E. Wielgolaski (ed.) Primary Production and Production Processes, Tundra Biome. Tundra Biome Steering Committee, Edmonton–Oslo. 256 p.

Thompson, D. Q. 1955. The role of food and cover in population fluctuations of the brown lemming at Point Barrow, Alaska, p. 166–176. *In* J. B. Trefethen (ed.) Transactions of the Twentieth North American Wildlife Conf., March 14–16, 1955. Montreal, Quebec, Canada. Wildlife Management Institute. 683 p.

Tieszen, L. L. 1972. Photosynthesis in relation to primary production, p. 52–62. *In* F. E. Wielgolaski and Th. Rosswall (ed.) Proceedings IV. International Meeting on the Biological Productivity of Tundra, Leningrad, Oct. 1971. Tundra Biome Steering Committee, Stockholm. 320 p.

Tikhomirov, B. A. (ed.) 1971. Biogeocenosis of Taimyr tundra and their productivity. Publ. House "Nauka," Leningrad. 237 p.

Vassiljevskaya, V. D., and L. A. Grishina. 1972. Organic carbon reserves in the conjugate eluvial accumulative landscapes of West Taimyr (Station Agapa), p. 215–218. *In* F. E. Wielgolaski and Th. Rosswall (ed.) Proceedings IV. International Meeting on the Biological Productivity of Tundra, Leningrad, Oct. 1971. Tundra Biome Steering Committee, Stockholm. 320 p.

Vinokurov, A. A., V. A. Orlov, and Yu. V. Okhotsky. 1972. Population and faunal dynamics of vertebrates in tundra biocenoses (Taimyr), p. 187–189. *In* F. E. Wielgolaski and Th. Rosswall (ed.) Proceedings IV. International Meeting on the Biological Productivity of Tundra, Leningrad, Oct. 1971. Tundra Biome Steering Committee, Stockholm. 320 p.

Warren Wilson, J. 1957. Arctic plant growth. Adv. Sci. 13:383–388.

Wielgolaski, F. E. 1972a. Vegetation types and plant biomass in tundra. Arctic and Alpine Res. 4:291–305.

Wielgolaski, F. E. 1972b. Production, energy flow and nutrient cycling through a terrestrial ecosystem at a high altitude area in Norway, p. 283–290. *In* F. E. Wielgolaski and Th. Rosswall (ed.) Proceedings IV. International Meeting on Biological Productivity of Tundra, Leningrad, Oct. 1971. Tundra Biome Steering Committee, Stockholm. 320 p.

Wielgolaski, F. E. (ed.) In press (a). Fennoscandian Tundra Ecosystems. Parts 1 and 2. Springer Verlag, Berlin–Heidelberg–New York.

Wielgolaski, F. E. In press (b). Primary productivity of alpine meadow communities. *In* F. E. Wielgolaski (ed.) Fennoscandian Tundra Ecosystems. Part 1. Plants and Microorganisms. Springer Verlag, Berlin–Heidelberg–New York.

Wielgolaski, F. E., and S. Kjelvik. In press. Plant biomass at the Norwegian IBP sites at Hardangervidda 1969–1972, p. 1–88. *In* R. Vik (ed.) IBP in Norway, Methods and Results. Sections PT-UM Grazing project Hardangervidda, Botanical investigations. Norwegian National IBP Committee, Oslo.

Wielgolaski, F. E., and T. Rosswall (ed.) 1972. Proceedings IV. International Meeting on the Biological Productivity of Tundra, Leningrad, Oct. 1971. Tundra Biome Steering Committee, Stockholm. 320 p.

Zalenskij, O. V., V. M. Shvetsova, and V. L. Voznessenskij. 1972. Photosynthesis in some plants of Western Taimyr, p. 182–186. *In* F. E. Wielgolaski and Th. Rosswall (ed.) Proceedings IV. International Meeting on the Biological Productivity of Tundra, Leningrad, Oct. 1971. Tundra Biome Steering Committee, Stockholm. 320 p.

PRODUCTIVITY OF THE WORLD'S MAIN ECOSYSTEMS

L. E. RODIN, N. I. BAZILEVICH,
and N. N. ROZOV

The total phytomass of 106 terrestrial soil–plant formations, grouped into bioclimatic areas and thermal belts, has been calculated along with estimates of annual productivity. The total phytomass of the land is estimated to be 2.4×10^{12} metric tons dry weight. The bulk of this organic mass is in the tropical zone (56 percent), followed by the boreal (18 percent), subtropical (14 percent), subboreal (12 percent) and polar (1 percent) zones. The majority of phytomass is concentrated in forests (82 percent). Regularities of phytomass distribution in the world's oceans resemble those on the land (alteration of belts with high and low amounts of phytomass; abundance and concentration in the areas of cyclonic cycling of atmosphere and waters), but there exist some pecularities pertinent only to the oceans (maximum accumulation of phytomass in temperate latitudes and the shores). Total phytomass in the world's oceans amounts to 1.7×10^8 metric tons, which is about 15,000 times smaller than that of the land.

The total primary production of the land is estimated to be 1.72×10^{11} metric tons/year. The tropical belt produces 60 percent of this total; subtropical, 20 percent; subboreal, 10 percent; boreal, 9 percent; and polar, 0.8 percent. Forests produce 49 percent of primary production of the land. The total primary production of the oceans is estimated at 4.7 to 7.2×10^{10} metric tons/year (Steemann *et al.*, 1957; Koblents-Mishke, 1968; Bogorov, 1969). Hence, the oceans contribute approximately one-third as much primary production as do terrestrial plant communities. The total primary production of the earth is calculated as being 2.35×10^{11} metric tons/year of dry organic matter (Table 1).

As the most vital of the zones enclosing the earth, the biosphere accumulates and converts the extremely powerful stream of incident solar energy into the chemical energy of organic compounds. This conversion is accomplished by an organic matrix that has constantly reproduced and perfected itself in the process of evolution over the thousands of millions of years of geological history.

Some fifty years ago, V. I. Vernadski (1926) wrote: "Unfortunately, the data currently available [are] still too scarce to pinpoint the exact share of green plants in the earth's total organic matter. For the present, we have to make the best of rather inaccurate figures in an effort to size up the phenomenon in hand." Today the situation has radically changed with the development of plant and soil maps, the accurate calculation of areas within particular soil–vegetation formations, and the determination of biological productivity for many of these formations. It has been shown that the earth's living organic matter (biomass) is dominated (99 percent) by autotrophic and photosynthesizing organisms. As a result of these developments it is now possible to characterize global biotic productivity based primarily upon the primary production of green plants.

Since organic matter is capable of reproduction, growth and accumulation only for a delimited period of time, it is fair to assume that phytomass reserves are correlated with annual increment.* Both parameters may be calculated either on a unit area basis or as total standing crop for the entire area of a given vegetation cover type. The geographi-

* Hereafter figures for phytomass and annual increment including the above- and underground parts of plants will be given as dry weight.

TABLE 1 Areas, Phytomass, and Primary Production of the Earth's Major Vegetation Zones

	Area		Phytomass		Primary Production	
Thermal Zones	10^6 km^2	%	10^9 tn	%	10^9 tn	%
Polar	8.05	1.6	13.77	0.6	1.33	0.6
Boreal	23.20	4.5	439.06	18.3	15.17	6.5
Subboreal	22.53	4.5	278.67	11.5	17.97	7.7
Subtropical	24.26	4.8	323.90	13.5	34.55	14.8
Tropical	55.85	10.8	1,347.10	56.1	102.53	44.2
Land (without glaciers, lakes, and rivers)	133.4	26.2	2,402.5	100	171.54	73.8
Glaciers	13.9	2.7	0	0	0	0
Lakes and rivers	2.0	0.4	0.04	<0.01	1.0	0.4
All continents	149.3	29.3	2,402.54	100	172.54	74.2
Ocean	361.0	70.7	0.17	<0.001	60.0	25.8
The earth	510.3	100	2,402.71	100	232.54	100

TABLE 2 Minimum and Maximum Phytomass Reserve in the Zonal Types of Soil–Vegetation Formations of Different Thermal Belts and Hydrothermal Regions (tn/ha, dry weight)

		Hydrothermal Regions	
Thermal Belt	Type of Soil–Vegetation Formations	Humid	Arid
Polar	Polar desert	5	
	Tundra on tundra soils	28	
Boreal	Needle forest of the northern taiga on gley–podzolic soils	150	
	Needle forest of the middle taiga on podzolic soils	260	
	Needle forest of the southern taiga on turf–podzolic soils	300	
	Broadleaf forest on gray forest soils	370	
Subboreal	Broadleaf forest on brown forest soils	400	
	Semishrub desert on gray–brown desert soils		4.5
Subtropical	Broadleaf forest on red and yellow soils	450	
	Desert on subtropical desert soils		2.0
Tropical	Humid tropical forest on red ferralitic soils	650	
	Desert on tropical desert soils		1.5

cal patterns of phytomass (= plant biomass) distribution per unit area for the major types of terrestrial vegetation already have been summarized and mapped (Rodin and Bazilevich, 1967; Bazilevich and Rodin, 1967).

The data on geographical patterns show that each of the earth's thermal zones is associated with a particular soil-vegetation formation with characteristic biomass. Going from the polar through the boreal, cool temperate or "subboreal," to "subtropic"* to tropic† thermal zones, the range between maximum and minimum biomass increases because of an increase in absolute maximum and a decrease in absolute minimum values (Table 2). This phenomenon is associated, on the one hand, with changes in the efficiencies in energy fixation along pole-to-equator gradients with varying moisture supply and, on the other hand, with genetic properties and life forms of the plant communities. Within the intrazonal soil-vegetation formations these patterns are manifest with the same prominence, yet upon them are superimposed the effects of additional factors. An example is the geochemical sequence of landscapes which involves the redistribution of energy resources as well as water, nutrients, anaerobiosis, salinization, etc. Thus, in semiarid and arid regions, flood-land formations yield large

* i.e., warm temperate in customary English usage.
† Including equatorial.

quantities of biomass, while saline and alkaline areas yield very small quantities. In calculating the total global reserves of biomass, surface areas were categorized according to suitable types of soil-vegetation formations using the soil and continental vegetation maps from the Physico-Geographical Atlas of the World (Senderova, 1964). These data were synthesized into 106 soil-vegetation formations which, in turn, were classed with thermal zones and hydrothermal (bioclimatic) subzones (Table 3).

Since our primary objective is to estimate the earth's biological potential, calculations of biomass and annual increments assumed that the vegetation cover existed in its precultivated or natural state (not exceeding 15 percent of the total dryland area) or prelogging status. Besides the materials furnished by L. E. Rodin and N. I. Bazilevich (1967), the authors used extensive new data on productivity of vegetation cover obtained during the International Biological Programme (Molchanov, 1964; Bazilevich, 1967;

TABLE 3 Phytomass and Annual Primary Production of the Earth's Land Areas

	Phytomass		Production	
Thermal Belts, Bioclimatic Regions, and Soil–Vegetation Formations	t/ha	10^6 t	t/ha	10^6 t
Polar belt, humid and semihumid regions				
Polar deserts (Arctic) on polygonal and other Arctic soils	5	353	1.0	70.6
Tundras on tundra gley soils	28	10,517	2.5	939.0
Bogs (polar) on bog permafrost soils	25	470	2.2	41.4
Floodplain formations	10	12	1.7	2.04
Mountainous polar desert formations on Arctic mountain soils	8	352	1.5	66.0
Mountainous tundra formations on mountain-tundra soils	7	2,062	0.7	206.22
TOTAL	17.1	13,766	1.6	1,325.26
Boreal belt, humid and semihumid regions				
Maritime herbaceous-forest formations on volcanic ash	100	1,350	10.0	135.0
Open forest-tundra woodland and northern taiga forest on gley-podzolic soils	125	11,000	5.0	440.0
Same, on gley-permafrost taiga soils	100	12,190	4.0	487.6
Middle taiga forest on podzolic soils	260	92,846	7.0	2,499.7
Same, on permafrost taiga soils	200	49,260	6.0	1,477.8
Southern taiga and mixed broadleaf and needle forest on turf-podzolic soils	300	95,460	7.5	2,386.5
Same, on yellowish-podzolic soils	350	13,755	10.0	393.0
Same, on turf-calcareous and turf-gley soils	350	10,535	10.0	301.0
Same, on taiga bog soils with bogs	80	6,824	4.0	341.2
Small-leaf forest on gray and gray solodized forest soils	200	4,540	8.0	181.6
Broadleaf forest on gray forest soils	370	26,307	8.0	568.8
Bogs	35	3,892	3.5	389.2
Floodplain formations	60	3,558	6.0	355.8
Mountain-taiga forests on mountain podzolic soils	170	45,832	6.0	1,617.6
Same, on mountain permafrost soils	160	48,640	5.0	1,520.0
Mountain meadows on gray mountain forest soils	300	7,560	7.5	189.0
Mountain meadows on mountain meadow soils	35	5,512.5	12.0	1,890.0
TOTAL	189.2	439,061.5	6.54	15,173.8
Subboreal belt, humid regions				
Broadleaf forests on brown forest soils	400	99,400	13	3,230.5
Same, on rendzinas	370	999	12	32.4
Herbaceous prairie on meadow chernozemlike soils (brunizems)	35	1,984.5	15	850.5
Broadleaf forests, swampy, with small bog areas	300	9,900	13	429.0
Bogs	40	64	25	40.0
Floodplain formations	90	1,782	12	237.6
Mountain forest on brown mountain forest soils	370	139,453	12	4,522.8
TOTAL	342	253,582.5	12.6	9,342.8
Subboreal belt, semiarid regions				
Steppe on typical and leached chernozems	25	1,607.5	13	1,355.9
Same, on ordinary and southern chernozems	20	2,942.0	8	1,176.8
Same, on solonets chernozems	20	500.0	8	200.0
Steppified formations on solonets	16	104.0	5	32.5
Halophytic formations on solonchak (in steppe)	12	7.2	4	2.4
Psammophytic formations on sand (in steppe)	18	108.0	8	48.0
Dry steppe on dark chestnut soils	20	1,480.0	9	666.0
Desert steppe on light chestnut soils	13	1,574.3	5	605.5
Dry and desert steppe on chestnut and solonets complexes	14	938.0	5	335.0
Same, on solonets	14	296.8	5	106.0

TABLE 3 (continued)

Thermal Belts, Bioclimatic Regions, and Soil-Vegetation Formations	Phytomass		Production	
	t/ha	10^6 t	t/ha	10^6 t
Halophytic formations on solonchak (in dry and desert steppe)	2	15.2	0.7	5.3
Psammophytic formations on sand (in dry and desert steppe)	15	213.0	6	85.2
Herbaceous bog on meadow-bog soils	15	85.5	7	39.9
Floodplain formations	80	2,280.0	12	342.0
Mountain dry steppe on mountain chestnut soils	15	1,245.0	7	581.0
Mountain steppe on mountain chernozems	25	572.5	10	229.0
Mountain meadow steppe on subalpine mountain meadow steppe soils	25	1,885.0	11	829.4
TOTAL	20.8	16,854.0	8.2	6,639.9
Subboreal belt, arid regions				
Steppified desert on brown semidesert soils	12	1,696.8	4.0	565.6
Same, on brown-soil and solonets complexes	10	407.0	3.5	142.5
Same, on solonets	9	126.0	3.2	44.8
Desert on gray-brown desert soils	4.5	617.9	1.5	205.9
Psammophytic formations on sand (in desert)	30	2,484.0	5.0	414.0
Desert on takyr soils and takyrs	3	36.3	1.0	12.1
Halophytic formations on solonchak (in desert)	1.5	32.7	0.5	10.9
Floodplain formations	80	1,088.0	13	176.8
Mountain desert on brown mountain semidesert soils	9	316.8	3	105.6
Same, on desert highland soils	7	1,437.8	1.5	308.1
TOTAL	11.7	8,234.3	2.8	1,986.3
SUBBOREAL BELT TOTAL	123.6	278,679.8	7.9	17,969.0
Subtropical belt, humid regions				
Broadleaf forest on red soils and yellow soils	450	88,965	20	3,954.0
Same, on red-colored rendzinas and terra rossa	380	6,536	16	275.2
Herbaceous prairie on reddish black soils and rubrozems	30	1,428	13	618.8
Broadleaf forest, swampy, with small bog areas	400	5,040	22	277.2
Meadow-bog and bog formations	200	5,380	130	3,497.0
Floodplain formations	250	17,050	40	2,728.0
Mountain broadleaf forest on mountain yellow soils and red soils	410	104,017	18	4,566.6
TOTAL	366.1	228,416	25.5	15,916.8
Subtropical belt, semiarid regions				
Xerophytic forest on brown soils	170	27,081	16	2,548.8
Shrub-steppe formations on gray-brown soils	35	9,331	10	2,666
Same, on gray-brown solonets soils with small solonets areas	20	150	6	45
Same, on subtropical chernozemlike and coalesced soils	25	1,187.5	8	380
Psammophytic formations on sandy soils and sand	20	102	5	25.5
Halophytic formations on solonchak soils and solonchak	1.5	10.95	0.5	3.65
Floodplain formations	250	15,725	40	2,516
Mountain xerophytic forest on brown mountain soils	120	26,832	13	2,906.8
Mountain shrub-steppe formations on gray-brown mountain soils	30	1,479	8	394.4
TOTAL	98.7	81,898.45	13.8	11,486.15
Subtropical belt, arid regions				
Steppified desert on serozems and meadow-serozem soils	12	2,376	10	1,980.0
Desert on subtropical desert soils	2	857.8	1	428.9
Psammophytic formations on sand	3	562.5	0.1	18.75
Desert on takyr soils and takyrs	1	14.7	0.5	7.35
Halophytic formations on solonchak	1	21.0	0.2	4.2
Floodplain formations	200	8,780.0	90	3,951.0
Mountain desert on mountain serozems	15	915.0	12	732.0
Same, on subtropical mountain desert soils	3	54	1	18.0
TOTAL	13.9	13,581	7.3	7,140.2
SUBTROPICAL BELT TOTAL	133.5	323,895.4	14.2	34,543.15
Tropical belt, humid regions				
Humid evergreen forest on red-yellow ferralitic soils	650	560,950	30	25,890
Same, on dark-red soils	600	11,820	27	531.9
Seasonally humid evergreen forest and secondary tall-grass savanna on red ferralitic soils	200	175,720	16	14,057.6
Same, on black tropical soils	80	1,128	15	211.5
Humid evergreen swampy forest on ferralitic gley soils	500	112,200	25	5,610
Bog formations	300	19,920	150	9,960

TABLE 3 (continued)

Thermal Belts, Bioclimatic Regions, and Soil–Vegetation Formations	Phytomass		Production	
	t/ha	10^6 t	t/ha	10^6 t
Floodplain formations	250	29,375	70	8,225
Mangrove forest	130	6,214	10	478
Humid tropical mountain forest on red-yellow ferralitic mountain soils	700	169,330	35	8,466.5
Seasonally humid tropical mountain forest on red ferralitic mountain soils	450	79,515	22	3,887.4
TOTAL	440.4	1,166,172	29.2	77,317.9
Tropical belt, semiarid regions				
Xerophytic forest on ferralitized brownish-red soils	250	115,675	17	7,865.9
Grass and shrub savanna on ferralitic red-brown soils	40	26,188	12	7,856.4
Same, on tropical black soils	30	6,150	11	2,255.0
Same, on tropical solonets soils	20	470	7	164.5
Meadow and swamp savanna on ferralitized red and meadow soils	60	5,364	14	1,251.6
Floodplain formations	200	4,420	60	1,326.0
Xerophytic mountain forest on brownish-red mountain soils	200	9,940	15	745.5
Mountain savanna on red-brown mountain soils	40	3,752	12	1,125.6
TOTAL	107.4	171,959	14.1	22,590.5
Tropical belt, arid regions				
Desertlike savanna on reddish-brown soils	15	6,435	4	1,716.0
Desert on tropical desert soils	1.5	700.95	1	467.3
Psammophytic formations on sand	1.0	282	0.1	28.2
Desert on tropical coalesced soils	1.0	21.6	0.2	4.32
Halophytic formations on solonchak	1.0	11.5	0.1	1.15
Floodplain formations	150	1,500	40.0	400.0
Mountain desert on tropical mountain desert soils	1	62.6	0.1	6.26
TOTAL	7.0	9,013.65	2.0	2,623.23
TROPICAL BELT TOTAL	243.3	1,347,144.65	18.5	102,531.63
Earth's total land area (without glaciers, streams, lakes)	180.1	2,402,547.4	12.8	171,542.86
Glaciers	0	0	0	0
Lakes and streams	0.2	40.0	5	1,000
TOTAL FOR ALL CONTINENTS	160.9	2,402,587.4	11.5	172,542.86

Pyavchenko, 1967; Ghilarov *et al.*, 1968; Dilmy, 1969; Golley *et al.*, 1969; Jenik, 1969; Kira and Ogawa, 1969; Whittaker, 1971; Bazilevich and Rodin, 1971; Soil regimes and processes, 1970; Biological productivity and regimes of soils, 1971).

The total land phytomass reserves of the earth are estimated as 2.4×10^{12} metric tons. The bulk of this biomass occurs in the tropical zone—1.35×10^{12} metric tons or over 56 percent of the total continental phytomass (minus rivers, lakes and glaciers).* This phenomenon is not unexpected since the area of the tropical zone makes up almost 42 percent of the earth's total land area and almost half of this zone is vegetated by highly productive moist tropical forests (Table 4).

The boreal zone (18 percent of the global reserves) is second in biomass reserves followed by the subtropical (approximately 14 percent), subboreal (approximately 12 percent), and the polar zone (less than 1 percent). It is noteworthy that the boreal, subboreal and subtropical zones are approximately equal in area. The differences in phytomass are primarily determined by the degree to which the landscape is covered with forests (which is greatest for the boreal zone).

Within each of the thermal zones, a precipitous decrease occurs in total phytomass and average figures per unit area (biomass) from humid to semiarid and arid bioclimatic regions, even though the latter may be more extensive in area. However, biomass values (phytomass per area unit) for humid areas only (dominated by forest communities) increases from north to south. This is a crucial factor responsible for the progressive increase in the biomass estimates throughout the thermal zones, moving similarly from subboreal to subtropical zones.

Thus, the geographic distribution of phytomass reserves over the earth's land is determined by the forest types of the different soil-vegetation formations. Indeed, the total phytomass reserves of the world's forests are 1.96×10^{12} metric tons, i.e., almost 82 percent of the entire terrestrial phytomass, with the total under forest cover amounting to

* The same holds for the total continental area of the entire earth since the phytomass reserves of inland reservoirs (rivers and lakes) as well as the global oceans are much lower than in vegetation communities on land.

TABLE 4 Areas, Phytomass, and Annual Production of the Earth (dry weight)

Thermal Belts and Bioclimatic Regions	Area				Phytomass					Annual Production				
		% of Area of					% of Area in					% of Area in		
	10^6 km^2	Land[a]	Conti-nents	Earth	Mean (t/ha)	Total (10^9 t)	Land[a]	Conti-nents	Earth	Mean (t/ha)	Total (10^9 t)	Land[a]	Conti-nents	Earth
Polar humid and semihumid	8.05	6.0	5.4	1.6	17.1	13.77	0.6	0.6	0.6	1.6	1.33	0.8	0.8	0.6
Boreal humid and semihumid	23.20	17.4	15.5	4.5	189.2	439.06	18.3	18.3	18.3	6.5	15.17	8.8	8.8	6.5
Subboreal:														
humid	7.39	5.5	4.9	1.5	342.0	253.58	10.5	10.5	10.5	12.6	9.34	5.4	5.4	4.0
semiarid	8.10	6.1	5.4	1.6	20.8	16.85	0.7	0.7	0.7	8.2	6.64	3.9	3.8	2.8
arid	7.04	5.3	4.7	1.4	11.7	8.24	0.3	0.3	0.3	2.8	1.99	1.2	1.2	0.9
TOTAL	22.53	16.9	15.0	4.5	123.6	278.67	11.5	11.5	11.5	7.9	17.97	10.5	10.4	7.7
Subtropical:														
humid	6.24	4.7	4.2	1.2	366.1	228.42	9.5	9.5	9.5	25.5	15.92	9.3	9.3	6.8
semiarid	8.29	6.2	5.0	1.6	98.7	81.90	3.4	3.4	3.4	13.8	11.49	6.7	6.7	4.9
arid	9.73	7.3	6.5	2.0	13.9	13.58	0.6	0.6	0.6	7.3	7.14	4.1	4.1	3.1
TOTAL	24.26	18.2	16.3	4.8	133.5	323.90	13.5	13.5	13.5	14.2	34.55	20.1	20.1	14.8
Tropical:														
humid	26.50	19.9	17.7	5.2	440.4	1,166.17	48.6	48.6	48.6	29.2	77.32	44.9	44.6	33.5
semiarid	16.01	12.0	10.8	3.1	107.4	171.96	7.1	7.1	7.1	14.1	22.59	13.2	13.0	9.6
arid	12.84	9.6	8.6	2.5	7.0	9.01	0.4	0.4	0.4	2.0	2.62	1.7	1.7	1.1
TOTAL	55.35	41.5	37.1	10.8	243.3	1,347.1	56.1	56.1	56.1	18.5	102.53	59.8	59.3	44.2
Land (without glaciers, streams, lakes)	133.4	100.0	89.3	26.2	180.1	2,402.5	100.0	100.0	100.0	12.8	171.54	100.0	99.4	73.8
Glaciers	13.9	–	9.3	2.7	0	0	–	0	0	0	0	–	0	0
Streams and lakes	2.0	–	1.4	0.4	0.2	0.04	–	<0.1	<0.01	5.0	1.0	–	0.6	0.4
Continent total	149.3	–	100.0	29.3	160.9	2,402.54	–	100.0	100.0	11.5	172.54	–	100.0	74.2
Oceans[b]	361.0	–	–	70.7	0.005	0.17	–	–	<0.01	1.7	60.0	–	–	25.8
Earth as a whole	510.3	–	–	100.0	47.1	2,402.71	–	–	100.0	4.5	232.54	–	–	100.0

[a] Without glaciers, lakes, streams.
[b] After Bogorov, 1969, and Koblents-Mishke *et al.*, 1968.

5×10^7 km^2 or 39 percent* of the area of the earth's surface. Tropical zone forests make up half of this total (1.03×10^{12} metric tons), boreal zone forests account for some 20 percent (0.4×10^{12} metric tons) and subboreal and subtropical forests an additional 15 percent each. It is significant that the phytomass reserves of desert types of soil-vegetation formations, which occupy a total area of 2.89×10^7 km^2 (22 percent of the earth's surface) make up only 0.02×10^{12} metric tons or 0.8 percent of total land phytomass.†

The distribution of oceanic phytomass generally follows the principles established for terrestrial environments, and yet there are some differences. The concept of the marine zonation was proposed by L. A. Zenkevich (1948). Underlying this concept of natural oceanic zones is a combination of the same five factors that influence terrestrial productivity: light, temperature, nutrients, substrate and the interrelationships of organisms (Bogorov, 1969). Also important in the geography of terrestrial and oceanic phytomass is the effect of vertical (for land) and abyssal (for ocean) zonality. Similar to the land, the ocean is characterized by the alternation of zones high and low in phytomass (as well as in zoomass). The zones of land and ocean which are relatively rich in living matter are characterized by the cyclonic regime of air and/or water circulation. Such regimes in the ocean cause mixing of water zones and upwelling of nutrients to surface phytoplankton. In areas with an anticyclonic regime, the surface waters sink and thus deprive the upper layers of nutrients. The areas poor in living matter are confined to anticyclonic regions in the center of the Pacific and Atlantic Oceans and in the southern Indian Ocean. The areas high in living matter occupy less than a quarter of the world's oceans (Bogorov *et al.*, 1968; Koblenz-Mishke *et al.*, 1968).

And yet, in contrast to the land, the highest marine accumulation of living organic matter occurs at moderate temperate rather than tropical latitudes. This results not only from the cyclonic water currents and divergence, but also from the more intensive mixing of water layers in temperate latitudes under the influence of winter-autumn temperature fluctuations. Additional factors in the high biomass of the littoral zone are sea-to-shore winds and nutrient inflow from river discharge, abrasion effects, discharge of underground waters, etc. The low amounts of biomass in high latitude and Antarctic waters are explained by the shorter growing season and lower water temperatures. The absolute phytomass reserves in the world ocean are not high, 0.15×10^9 metric tons of phytoplankton and 0.02×10^9 metric tons of phytobenthos for a total of 0.17×10^9 metric tons. Thus, the phytomass reserves on land exceed almost 15,000-fold those in the ocean (Bogorov, 1969). The earth's total phytomass (including 0.04×10^9 metric tons occurring in rivers and lakes as calculated by Whittaker (1971) amounts to 2.4×10^{12} metric tons.

The annual production of phytomass also may be expressed in dry weight per area unit by considering the areas of different soil-vegetation formations. These data have previously been published and plotted as schematic maps of the earth's terrestrial productivity (Rodin and Bazilevich, 1967; Bazilevich and Rodin, 1967). These data reveal the same patterns obtained in examining the distribution of phytomass over the earth's land. Plant biomass production increases sharply from the pole to the equator within humid cyclonic regions and just as sharply drop to a minimum in arid anticyclonic regions (Figure 1). Phytomass production per area unit in humid regions gorws from less than 2 metric tons per hectare on the average in the polar zone to between 6 and 13 metric tons per hectare in the boreal and subboreal zones and between 26 and 29 metric tons per hectare in subtropical and tropical zones. At the same time, the annual increment in arid regions follows a different pattern, increasing from 3 to 7 metric tons per hectare going from the subboreal to subtropic regions and dropping sharply in the arid tropics to an average of 2 tons per hectare. This is due to the wide occurrence in the arid subtropics of plant communities dominated by ephemers and ephemeroids yielding abundant phytomass.

These geographic regularities of distribution of annual increment (and also phytomass reserves) correlate very well with climatic factors. This correlation is elucidated in the literature (Budyko, 1956; Grigoriev and Budyko, 1965; Budyko and Efimova, 1968; Bazilevich, Drozdov and Rodin, 1968). In another instance Bazilevich and Rodin (1971) used the most recent data on phytomass production in conjunction with data on the environmental factors affecting production, including the influx of heat (R, in kcal/cm^2 per year) and the degree of moistening (determined by the dryness factor, R/Lr, where r is precipitation rate and L is the latent heat of phase conversion). An increase in heat reserves results in an especially large increase in production when $R < 35$–40 kcal/cm^2/year (regions north of the middle-taiga subzone). When R is > 35–40 kcal/cm^2/year, moisture was the key limiting factor. The annual increment of zonal soil vegetation formations was the highest in regions with $R < 35$–40 kcal/cm^2/year and $R/Lr > 1$ (1.2 for northern taiga and 1.5 for tundra) and not in regions with $R > 35$–40 kcal/cm^2/year and $R/Lr < 1$ (0.7 for subtropics and 0.5 for tropics). Hence, conclusions drawn by previous researchers with respect to climatic control over biotic production have not been entirely accurate (Figure 2).

Thus, if heat reserves are sufficient, excess precipitation will result in higher production; even if moisture is ample, if heat reserves are insufficient, lower production will result. A case in point is the production of flood-land vege-

* According to FAO data, forested area is somewhat smaller (some 40,000,000 km^2) since FAO considered only existing and not potential forested areas.

† According to FAO data, the desert areas are much more extensive, arid steppes are partly classed with deserts.

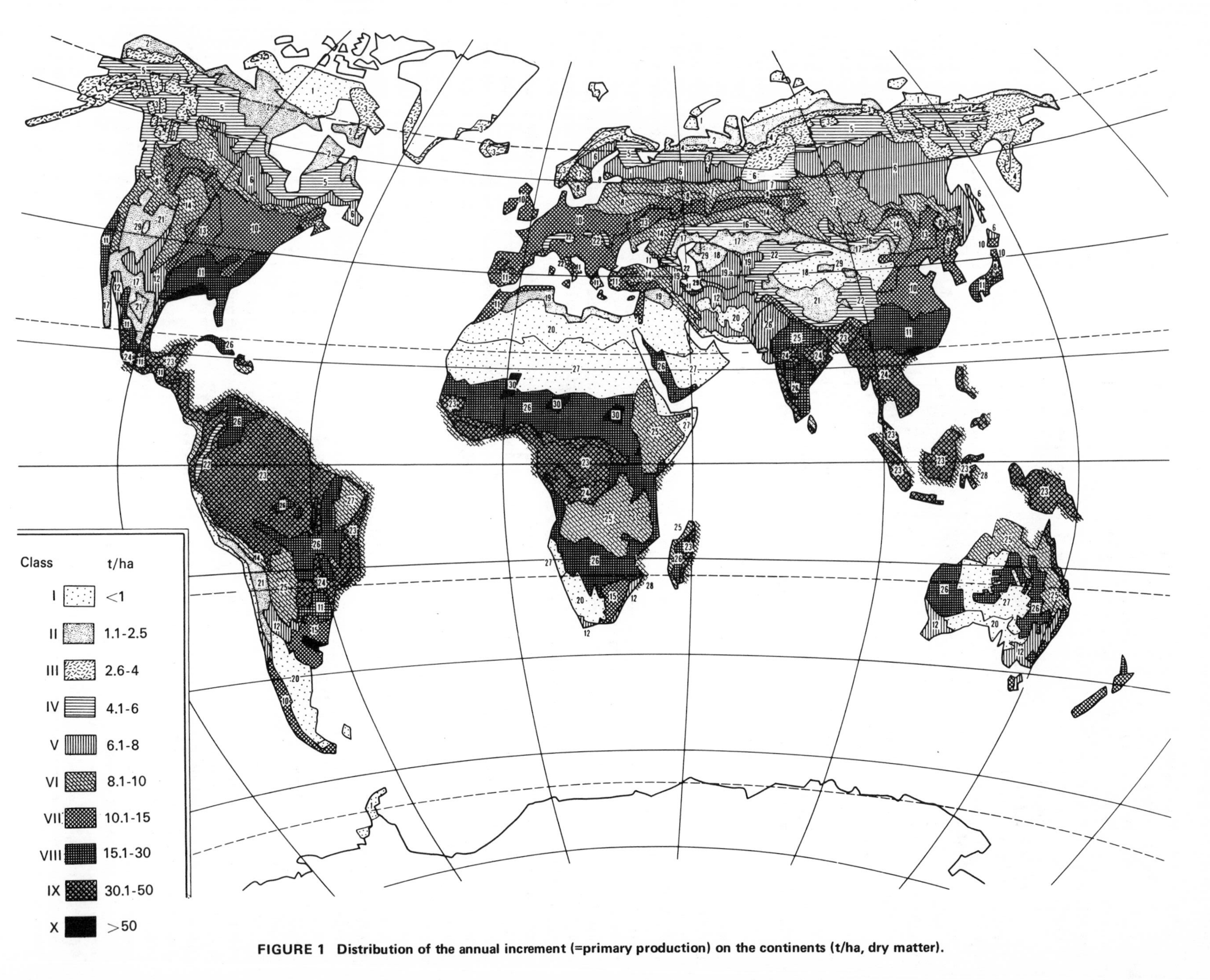

FIGURE 1 Distribution of the annual increment (=primary production) on the continents (t/ha, dry matter).

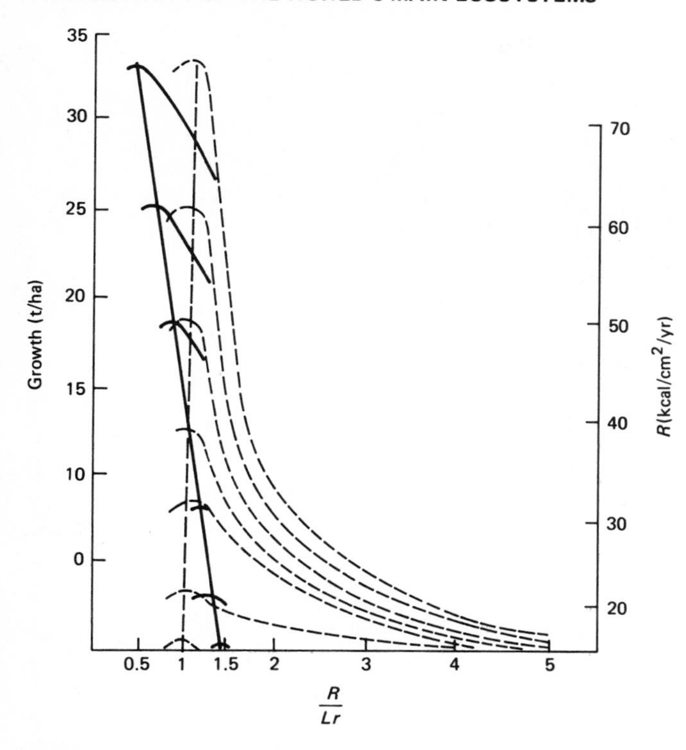

FIGURE 2 Relation between annual production (plant growth) of soil–vegetation formations and total radiation (*R*) and dryness index (*R/Lr*). The relationship established by Budyko and Yefimova, 1968, is shown in dotted lines.

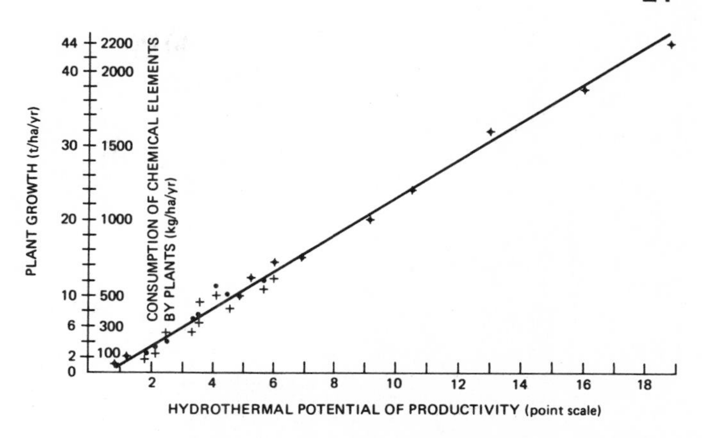

FIGURE 3 Correlation between the hydrothermal potential of plant productivity and the actual plant growth (dots) and the consumption of chemical elements by plants (crosses). After A. M. Ryabchikov, 1968.

tation in subtropics and tropics with an annual increment in excess of 90 metric tons per hectare. According to M. I. Budyko (1956), the flood-lands and especially deltas of rivers in these zones should be classed with the warmest and best moistened areas on land.* Yet, a simple temperature : moisture ratio fails to account for all possible patterns of distribution of primary productivity. It is also important to consider the duration of the growing season and effective precipitation (precipitation minus surface discharge). A. M. Ryahchikov (1968) has proposed a hydrothermal productivity potential (Kp) calculated as

$$Kp = \frac{WT_v}{36R},$$

where W is the average annual effective precipitation (mm), T_v is the duration of the growing season in terms of 10-day periods (36 10-day periods in a year); and R is average annual radiation balance in kcal/cm^2. The hydrothermal productivity potential correlates extremely well with biomass production (Figure 3).

The annual total phytomass increment of the terrestrial vegetation (Table 4) is put at 1.72×10^{11} metric tons (7 percent of total phytomass reserves). The bulk of the phytomass of 1.03×10^{11} metric tons (60 percent of total phytomass) is produced in the tropical zone, while the soil-vegetation formations of humid regions are responsible for 7.73×10^{10} metric tons or 45 percent of the total increment of the earth's land. Second to the tropic is the subtropical zone (34.6×10^9 ton, or 20 percent). Being roughly equal in area, the plant formations of the subboreal and boreal zones produce much less in organic matter—18×10^9 ton (10 percent) and 15×10^9 ton (9 percent), respectively. The annual increment in polar regions is the lowest: a little more than 1×10^9 (0.8 percent).

The same as for phytomass distribution, there is a sharp decline in annual increment within each of the thermal zones in the direction from humid to semiarid and especially arid regions. This pattern is the least pronounced in the subtropical zone and the most pronounced in the tropical one, which is associated with the wide occurrence of arid types of vegetation largely dominated by ephemere in the former zone and vast areas practically devoid of vegetation cover in the latter. The annual total increment of desert formations fails to exceed 7.22×10^9 ton or some 4 percent of the total land increment (the earth's area under desert makes up some 22 percent).

Making up some 39 percent in area, forests produce almost half (49 percent, or 84.1×10^9 ton) the annual total land increment.

It is also significant that, while occupying a very small area of some 3 percent, the soil–vegetation formations of deltas and flood-lands produce over 20×10^9 ton (12 percent) of organic matter to additionally confirm their special biogeochemical character.

The annual increment to oceanic phytomass according to the latest samplings at home and abroad is put at 1.3–

* Consideration should also be given to nutrient abundance in flood-lands and delta formations resulting from the positive geochemical balance of the landscapes.

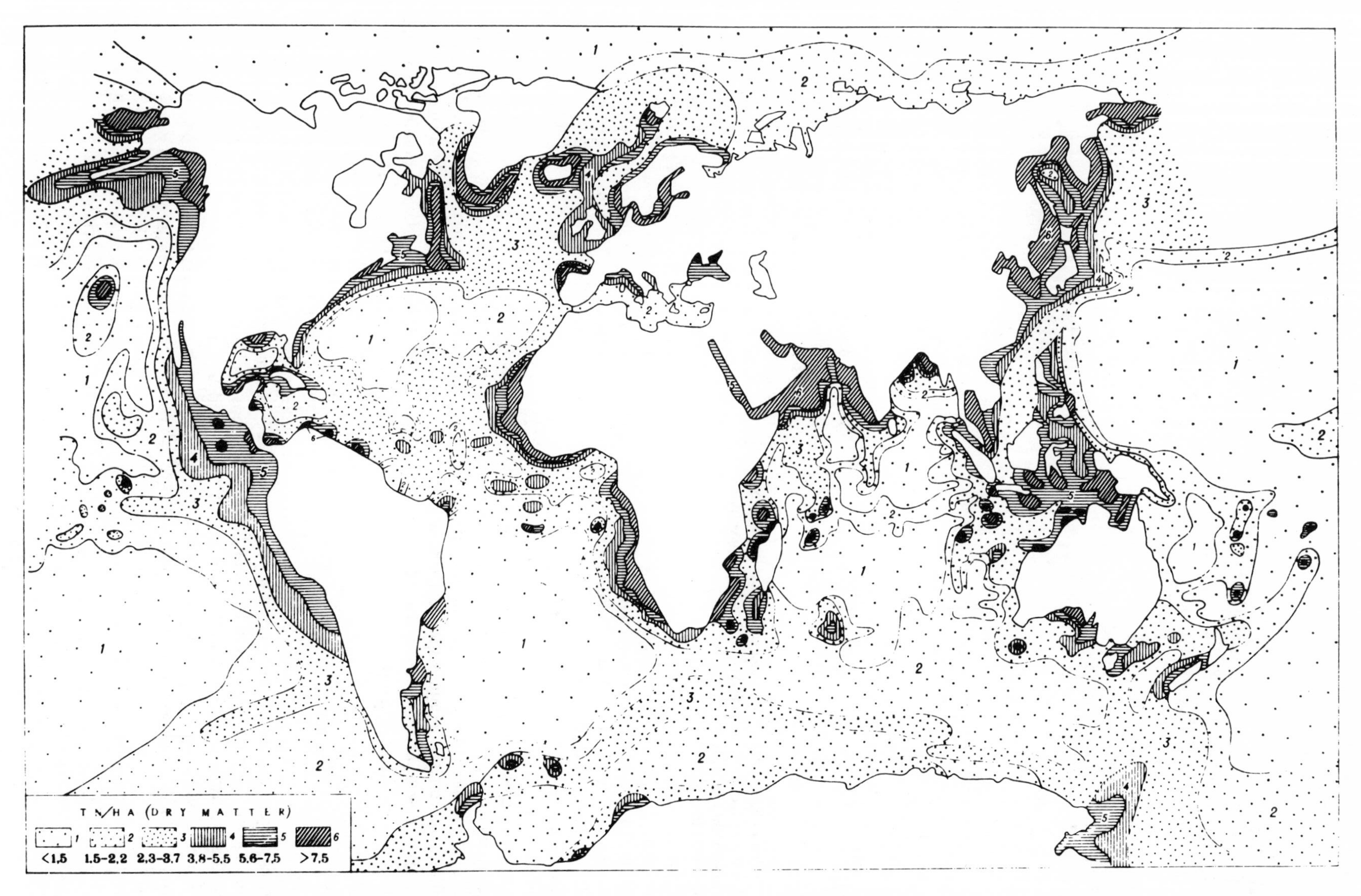

FIGURE 4 Distribution of the annual increment (=primary production) in the World Ocean (t/ha, dry matter).

2.0 ton per hectare or, making a total at 47–72 $\times 10^9$ ton (Steemann *et al.*, 1957, quoted by Lieth, 1964–65; Gessner, 1959; Koblenz-Mishke *et al.*, 1968; Whittaker, 1971). Thus, the total annual increment of the world ocean exceeds 30,000 percent of its total phytomass reserve.* This is all too natural, since the ocean is dominated by unicellular plants with extremely rapid reproductive potentials. And yet, the annual increment of terrestrial communities is three times greater than the world's oceans (including the increment of rivers and lakes which, according to Whittaker, 1971) produce 1×10^9 metric tons/year. It is significant that even though the ocean covers 180 times the area of rivers and lakes, its phytomass reserves are only four times as large. The geographical pattern of productive and low-productive aquatoria is correctly described in the previous studies of primary production distribution over the world's oceans (Lieth, 1964–65; Koblenz-Mishke *et al.*, 1968) (Figure 4).

The data now available make it possible to estimate the primary production of the entire earth at 2.33×10^{11} metric tons/year of organic matter, with some 74 percent being produced by plant communities on land. The scanty data available in the literature on phytomass reserves both on land and in the ocean are summarized in Table 5. This table also includes data on the mass of consumers and reducers for the purpose of assessing the total quantity of living matter of the entire planet. It is obvious from the table that wide gaps between data from different sources still exist in some cases. Since phytomass data are limited, especially those from the beginning of this century, many researchers have attempted to estimate both phytomass reserves and total organic matter even though they based their conclusions on indirect information correlated only to the above-ground parts of plants.

The productivity data presented in Table 5 reject the once prevalent opinion that most of the living matter is concentrated in the ocean. Recent studies have repudiated this past misconception (Koblenz-Mishke *et al.*, 1968; Bogorov *et al.*, 1968; Olson, 1970; Whittaker, 1971).

According to our estimates,† the earth's total living matter is equal to 2.42×10^{12} metric tons. The living matter of the land thus exceeds by some 750 times that of the world's oceans.

The figures of phytomass primary production in literature vary widely from author to author (Table 6). This is especially true of the estimated production on the continents. According to our calculations, production by the terrestrial landscape far exceeds that published by most other authors with the exception of Deevey (1960), the American workers as quoted by Duvigneaud and Tanghe (1968), and the most recent data of Olson (1970), Whittaker (1971) and Lieth (1972). Such large differences may be most rationally explained by the fact that (i) many authors have relied on second-hand information and have extrapolated widely in estimating primary production, (ii) many have failed to take notice of the considerable annual increment produced by the underground parts of plants and (iii) there has been a paucity of data of recent origin based upon modern analytical techniques.

Generalizations of potential primary productivity may best be estimated by evaluating the ratio between annual increment and phytomass of the continents. To calculate this ratio incorporating the data of those authors who fail to account for the extent phytomass reserves, it is advisable to use the value 2.4×10^{12} metric tons which we have found to be the average phytomass value cited in most recent publications on the subject. These publications give ratios of increment to phytomass from 0.15 to 1.9 percent with the exception of Deevey, Lieth, Whittaker, and the authors of this paper. Yet, actual sampling of different types of terrestrial vegetation shows that there are few plant communities which yield such low ratios. Thus, the ratio between annual increment and phytomass reserves for tundra is between 10 and 20 percent, for boreal and sub-boreal forests 2 to 5 percent, for subtropical forests 5 to 6 percent, for tropical forests 8 to 10 percent, for the grass communities of steppes, prairies and savannas 20 to 55 percent, for desert communities 30 to 75 percent, and for communities of annual field crops it is 100 percent (Rodin and Bazilevich, 1967). Thus, the absolute value of 1.09 to 1.72×10^{11} metric tons and the relative (6 to 7 percent) values for annual increase in phytomass on land seem to be consistent with those of Whittaker (1971).

Estimates of the annual increments and reserves of land phytomass currently under discussion have been based on actual determinations. The patterns of distribution of living plant organic matter were established by a detailed calculation of the areas occupied by various soil-vegetation formations. This approach enabled us to calculate with greater precision the phytomass reserves and the relationships with the annual increment for the land. These results have been compared with our calculations of the corresponding values for the world's oceans, which illustrates the role of mineral nutrition in primary productivity. Phytomass production has been shown to be highest in the zone in which physiographic processes are most intensive due to a favorable combination of warmth and moisture, such as the tropical zone. Geographical regularities of phytomass distribution per unit area (= biomass) for basic types of the earth's vegetation (Rodin and Bazilevich, 1967; Bazilevich and Rodin, 1967) indicate that soil-vegetation with both high and low values are characteristic

* Some 1,700 percent according to Whittaker (1971) since he put oceanic phytomass reserves at 3.3×10^9; however, we relied upon a value of 0.17×10^9 ton in dealing with data from literature.
† The sequence of figures for consumers and reducers adopted in this paper is based upon the data of Duvigneaud and Tanghe (1968).

TABLE 5 Mass of Living Matter on Earth (tons, dry weight)

Authors	Continents: Plant Mass	Continents: Consumers, Reducers	Continents: Total	Oceans: Plant Mass	Oceans: Consumers, Reducers	Oceans: Total	Total
Le Chatelier, 1908	–	–	$22 \cdot 10^9$	–	–	–	–
Vernadski, 1926	–	–	–	–	–	–	$n \cdot 10^{15}$
Vernadski, 1934–40	–	–	$n \cdot 10^{11}$	–	–	–	$n \cdot 10^{13}$–10^{14}
Titlyanova, 1967	–	–	–	–	–	–	$n \cdot 10^{13}$
Kovda, 1969	$3.1 \cdot 10^{12}$–$1 \cdot 10^{13}$	–	$3.1 \cdot 10^{12}$–$1 \cdot 10^{13}$	–	–	–	–
Bogorov, 1969	–	–	–	$0.17 \cdot 10^9$	$3.3 \cdot 10^9$	$3.47 \cdot 10^9$	–
Whittaker, 1971	$1.85 \cdot 10^{12}$	–	–	$3.3 \cdot 10^9$	–	–	–
Our own data, 1970	$2.4 \cdot 10^{12}$	$0.02 \cdot 10^{12}$	$2.42 \cdot 10^{12}$	$0.0002 \cdot 10^{12}$[a]	$0.003 \cdot 10^{12}$[a]	$0.0032 \cdot 10^{12}$	$2.423 \cdot 10^{12}$ $2.42 \cdot 10^{12}$

[a]Taken from Bogorov (1969).

TABLE 6 Primary Production (10^9 metric tons a year, dry weight)

	Continents		Oceans		Total	
Authors	Production	Percent of Plant Mass	Production	Percent of Plant Mass	Production	Percent of Plant Mass
Schröder, 1919[a]	36.2	1.5	–	–	–	–
Ereny, 1920[b]	33.0	1.3	–	–	–	–
Noddak, 1937[a]	33.5	1.4	113	59,000	146.5	6.1
Riley, 1944	–	–	176–815	95,500–425,000	–	–
Steemann *et al.*, 1957	–	–	60	31,000	–	–
Fogg, 1958	46.0	1.9	128	66,700	174	7.2
Gessner, 1959	–	–	47–60	25,000–31,000	–	–
Müller, 1960	22.8	0.9	–	–	–	–
Deevey, 1960	182.0	7.6	–	–	–	–
Duvigneaud, 1962	34.3	1.4	–	–	–	–
Lieth, 1964–65	46.6	1.9	–	–	–	–
Lieth, 1972	100.2	4.2	55	35,000	155.2	6.4
Mattson, 1965[a]	16.0	0.7	–	–	–	–
Duvigneaud and Tangue, 1968	53.0	2.2	30	17,600	83	3.4
Nichiporovich, 1968	50.0	2.0	50	29,000	100	7.2
Bykhovski and Bannikov, 1968	–	–	–	–	92	3.8
Koblents-Mishke *et al.*, 1968	–	–	60–72	35,290	–	–
U.S. authors (cited by Duvigneaud and Tangue, 1968)	–	–	–	–	70–280	2.9–11.5
Bogorov, 1969	–	–	55	32,350	–	–
Kovda, 1969	15–55(16)	0.15–1.8	–	–	–	–
Whittaker, 1971	109.0	6.0	55	1,666	164	8.8
Our own data, 1970	172.5	7.0	60[c]	35,290	232.5	9.6

[a]Cited by Lieth, 1964–65.
[b]Cited by Vernadski, 1934.
[c]Taken from Koblents-Mishke, 1968.

of every thermal belt. This is explained by a nonuniform latitudinal distribution of energy resources, different conditions of moisture supply, and by the genetic properties of primary producers.

REFERENCES

Bazilevich, N. I. 1967. Productivity and the biological cycle in moss bogs of the southern Vasyugan'ye. Rastitel'nyye resursy 3(4): 576–588.

Bazilevich, N. I., A. V. Drozdov, and L. E. Rodin. 1968. Productivity of the earth's plant cover, general regularities of distribution and its relationship to climate. J. Gen. Biol. Moscow 29(3): 261–271.

Bazilevich, N. I., and L. E. Rodin. 1967. Maps of productivity and the biological cycle in the earth's principal terrestrial vegetation types. Izvestiya Geographicheskogo Obshchestva, Leningrad 99(3):190–194.

Bazilevich, N. I., and L. E. Rodin. 1971. Productivity and the circulation of elements in natural and cultivated plant communities (with particular reference to the USSR), p. 5–32. *In* Biological productivity and mineral cycling in the terrestrial plant communities. Nauka, Leningrad.

Bazilevich, N. I., L. E. Rodin, and N. N. Rozov. 1971. Geographical aspects of biological productivity. Soviet Geography: review and translation. May:293–317.

Biological productivity and regimes of soils. 1971. Transactions of Estonian Agricultural Academy. Tartu. 71. 342 p.

Bogorov, V. G. 1969. Life of the Oceans. Znaniye. Biology Series, Moscow 6:3–5.

Bogorov, V. G., M. E. Vinogradov, N. M. Voronina, J. N. Kanaeva, and J. A. Suetova. 1968. Distribution of the biomass of zooplankton in the surface layers of oceans. Doklady Akad. Nauk, USSR, Ser. Biol. 182(5):1205–1207.

Budyko, M. I. 1956. The heat balance of the earth's surface. Translated by Nina A. Stepanova. U.S. Department of Commerce, PB316992.

Budyko, M. I., and N. A. Yefimova. 1968. The use of solar energy by the natural plant cover in the USSR. Bot. J., Leningrad 53(10):1384–1389.

Bykhovski, B. E., and A. G. Bannikov. 1968. The International Biological Program. Znanie. Biology Series, Moscow 12:3–15.

Deevey, E. S., Jr. 1960. The human population. Scientific American 203(3):195–204.

Dilmy, A. 1971. The primary productivity of equatorial tropical forests in Indonesia, p. 333–337. *In* P. Duvigneaud (ed.) Productivity of forest ecosystems. Proc. Brussels Symp., 27–31 October 1969. UNESCO.

Duvigneaud, P., and M. Tanghe. 1968. Biosfera i mesto v neĭ cheloveka. Progress. Moscow. (Translated from: Ecosystems et Biosphere. Brussels, 1967). 253 p.

Gessner, F. 1959. Hydrobotanik, II. Berlin. 701 p.

Ghilarov, M. S. (ed.) 1968. Methods of productivity studies in root systems and rhizosphere organisms. International Symposium, USSR. Nauka, Leningrad. 240 p.

Golley, F. B., J. T. McGinnis, R. G. Clements, G. I. Child, and M. J. Duever. 1969. The structure of tropical forests in Panama and Columbia. Bioscience 19(8):693–696.

Grigorjev, A. A., and M. I. Budyko. 1956. The periodicity law of geographic zonality. Doklady Akad. Nauk, USSR 110(1):129–132.

Grigorjev, A. A., and M. I. Budyko. 1965. The relationship between heat and water balances and the intensity of geographical processes. Doklady Akad. Nauk, USSR 162(1):151–154.

Jenik, J. 1971. Root structure and underground biomass in equatorial forests, p. 323–331. *In* P. Duvigneaud (ed.) Productivity of forest ecosystems. Proc. Brussels Symp., 27–31 October 1969. UNESCO.

Kira, T., and H. Ogawa. 1971. Assessment of primary production in tropical and equatorial forests, p. 309–321. *In* P. Duvigneaud (ed.) Productivity of forest ecosystems. Proc. Brussels Symp., 27–31 October 1969. UNESCO.

Koblenz-Mishke, O. I., V. V. Volkovinski, and Yu. G. Kabanova. 1968. New data on a magnitude of primary production in the oceans. Doklady Akad. Nauk, USSR, Ser. Biol. 183(5):1189–1192.

Kovda, V. A. 1969. The problem of biological and economic productivity of the earth's land areas, p. 8–24. *In* Basic problems of biological productivity. Nauka, Leningrad. (Translated in Soviet Geography, January 1971).

Le Chatelier, H. 1908. Leçons sur le carbone. Paris.

Lieth, H. 1964–65. A map of plant productivity of the World. Geographisches Taschenbuch, Wiesbaden, Frank Steiner. p. 72–80.

Lieth, H. 1972. Construction de la productivité primaire du globe. Nature et Resources, UNESCO, Paris 8(2):6–11.

Molchanov, A. A. 1964. Scientific fundamentals of farming in the oak groves of the wooded steppe. Nauka, Moscow. 255 p.

Müller, D. 1960. The circulation of carbon. Handbuch der Pflanzenphysiologie, Berlin 12(2):934–1254.

Nichiporovich, A. A. 1968. The International Biological Program and the processes of formation of primary biological production on the earth. Znanie. Biol. Ser., Moscow 12:22–48.

Olson, J. S. 1970. Geographic index of world ecosystems, p. 297–304. *In* D. E. Reichle (ed.) Analysis of temperate forest ecosystems. Ecological Studies 1. Springer Verlag, New York.

P'yavchenko, N. I. 1967. Some results of field station study of forest–bog relationships in Western Siberia, p. 7–43. *In* Vzaimootnosheniya lesa i bolota. Nauka, Moscow.

Riley, A. G. 1944. The carbon metabolism and photosynthetic efficiency of the earth as a whole. American Scientist 32: 128–134.

Rodin, L. E., and N. I. Bazilevich. 1968. Production and mineral cycling in terrestrial vegetation. (English translation by G. E. Fogg). Edinburgh. Oliver and Boyd. 288 p.

Ryabchikov, A. M. 1968. Hydrothermal conditions and the productivity of plant mass in the principal landscape zones. Vestnik MGU, Geogr., Moscow 5:41–48.

Senderova, G. M. (ed.) 1964. Physico-geographical Atlas of the World. USSR Acad. Sci. (Legend translated in Soviet Geography, May–June 1965).

Soil regimes and processes. 1970. Transactions of Estonian Agricultural Academy, 65, Tartu. 342 p.

Steemann, N., and E. A. Jensen. 1957. Primary oceanic production. The autotrophic production of organic matter in the oceans. Sci. Rept. Danish Deep-Sea Exped., 1950–52. Galathea Rept. 1:49–135.

Titlyanova, A. A. 1967. Lectures on the Biosphere. Novosibirsk Univ. 131 p.

Vernadski, V. I. 1926. Biosfera. Leningrad. 146 p.

Vernadski, V. I. 1934. Essays in Geochemistry. Russian edition. Leningrad. 224 p.

Vernadski, V. I. 1940. Biogeochemical Essays. Moscow-Leningrad. 250 p.

Whittaker, R. H. 1971. Communities and ecosystems. Third Printing. Macmillan Company, New York. 150 p.

Zenkevich, L. A. 1948. The biological structure of the oceans. Zool. Journ., Leningrad 27(2):113–124.

PRODUCTIVITY OF MARINE ECOSYSTEMS

M. J. DUNBAR

ABSTRACT

The present knowledge of the levels of biological production in the world ocean is briefly reviewed, together with a discussion of variation in ecosystem patterns with reference to the evolution of the systems in different marine climates. Emphasis is given to (1) energy supply and the entrapment of energy within the production systems; (2) nutrient supply and nutrient capital; (3) metabolic regulation with respect to temperature; and (4) the importance of seasonality in the building up of commercially exploitable natural stocks. Recent Canadian work on the effects of freshwater run off on marine productivity is discussed, and attention is drawn to the need for more intensive study of climatic cycles and their effect in causing geographic shifts in maximum production zones.

INTRODUCTION

The biological productivity of the seas is of very great practical significance to man. As a consequence, the literature on this subject has been growing logarithmically for a century, with many notable summaries in book form. In spite of this it is unfortunately still true [and I quote here from Riley (1972)] that "we still know very little about marine productivity." The higher trophic levels, particularly those of commercial interest, are better known and easier to measure than producer trophic levels. In fact, the relative richness of different parts of the world oceans can be measured by sustained commercial take as well as, or better than, in terms of primary production.

Presenting a coherent general account of our present knowledge of marine production is therefore not simple. It is unnecessarily rendered even more difficult by the lack of standardization of terms and units, a matter that editors of the IBP Synthesis Volumes should vigorously address. "There are several different ways of measuring ^{14}C fixation, and at least three other techniques of investigating primary production" (Riley, 1972). Secondary production is far more difficult to measure, as was emphasized most recently at the IBP/PM Working Conference in Rome last year. No mathematical genius is required to convert milligrams to grams, or even saturation values of oxygen concentration to milliliters per liter, but to convert milligrams of carbon fixation per square meter per day to milligrams per cubic meter per year is impossible without other information which is often not supplied. Biomass values expressed in units per volume or per area per day give totally different sorts of information from those conveyed by average values per year, etc.

ECOSYSTEM PROCESSES

For a scholarly review of our understanding of marine ecosystem processes, and for detailed comparison of the Sargasso Sea, the Eastern Tropical Atlantic, and Long Island Sound, as examples of marine areas, I refer you to Riley (1972). Although this paper emphasizes interrelations between elements within the ecosystem rather than comparisons of ecosystems, it is particularly valuable for its insights into production processes. I will begin here by mentioning some of Riley's points:

1. The total range of productivity in marine ecosystems is about the same as that of terrestrial ecosystems.
2. Apart from extremes of production, the richest areas

of the sea are only some three to five times as productive as the poorest areas. The extremes of the productivity spectrum are represented by coral reefs and *Thalassia* communities on the one hand and the Arctic Ocean on the other; total annual production per unit area in the Arctic Ocean is less than one day's production on a coral reef.

3. Primary production is channeled into different secondary and tertiary trophic levels in different regions, i.e., toward production of large carnivores (whales or fishes—and it should be noted that there appears to be an ecological choice within the carnivore group) or bacteria and detritus feeders or herbivores.

4. Open ocean ecosystems are most nearly analogous to grassland communities. Both systems are of intermediate productivity; rates of production are high but the plant community is small in terms of total biomass due to constant cropping by consumers. Both ecosystems support a diversified system of herbivores and carnivores. The grassland ecosystem produces a residuum of humus which is worked over by soil fauna and flora. The marine equivalent is a residuum of particulate and dissolved organic matter in the water column and in bottom sediments. Organic detritus is two orders of magnitude larger than that of the living biomass; and its accumulation obviously is due to its refractory nature, yet, in time it is utilized and contributes to ecosystem productivity, as evidenced by the relatively small organic content that is left in deep sea sediments (Riley, 1972).

5. Estimates for net primary production in different latitudes of the world ocean are in mgC/m^2/day: Gulf of Guinea, 365 gC/m^2/year; Long Island Sound, 190 gC/m^2/year; Sargasso Sea, 70 to 145 gC/m^2/year. The normal range of oceanic primary production is estimated by Riley (1970) to be between 50 and 150 gC/m^2/year (150 to 300 g dry weight/m^2/year) which allows for seasonal variation. Production in inshore areas and areas of upwelling can be much higher, up to ten times greater than these oceanic values.

These observations stimulate three comments. First, coral reefs are no longer officially included among the themes under IBP/PM study, which is a pity; they are oases of extremely high production in oceanic tropical deserts, or near-deserts, and as such pose unresolved questions on the mechanisms by which the essential nutrient capital is retained and recycled by biota. Secondly, turtle grass (*Thalassia*) beds are similar localized areas of high production in tropical oceans, and Patriquin (1971) has shown that "nitrogen for the growth of *Thalassia* is derived exclusively from gaseous nitrogen fixed by anaerobic bacteria in the rhizosphere. It had previously been unsuspected that such a phenomenon might be important in the ecology of aquatic plants." This is an important discovery and doubtless a major stimulant for further research. Thirdly, the differences in the end products of the food chains are crucial and have received little attention by marine ecologists even as late as the food chain symposium in Denmark in 1968. Why large whales as opposed to large fishes or herbivores as opposed to detritus eaters form the final or dominant link in the food chain in different areas may well be explained by environmental differences. For example, the depth of the euphotic zone in the Sargasso Sea causes a thin dispersion of food and hence discourages the development of predator populations. Selection of the final link may, however, also have much to do with evolutionary chance or with the particular advantages or disadvantages of different body organizations and metabolic systems under different environmental conditions. Resolution of these questions could explain for instance why there are no large exploitable fish populations in Antarctic waters. These evolutionary aspects of marine ecosystem productivity form part of my concerns in this paper.

ECOSYSTEM PRODUCTION

One recent publication comparing productivity of different parts of the sea is the collection entitled *Fertility of the Sea* (Costlow, 1971), which places some emphasis on the importance of organic substances in the development of marine plant populations, both planktonic and benthic. Although pioneer work on external metabolites and their ecological significance dates back at least to 1947 (Lucas, 1947), we still tend to think only in terms of inorganic nutrients when considering primary production. Provasoli (1963) discussed the organic regulation of phytoplankton fertility, Prakash (1971) summarized the importance of terrigenous organic matter in coastal primary production—particularly the part that low molecular weight fractions of humic matter apparently play. Prakash (1971) differentiates sharply between coastal waters and open ocean as being two distinct environments, not only in terms of production itself, but also in terms of chemical constituents (especially the role of humic substances in the coastal zone). Although the marine section of IBP has been concerned mainly with coastal waters, the organic enrichment of coastal water has not received much attention.

IBP research has made clear the importance of forms of primary production other than the purely phytoplanktonic. Canadian work in St. Margaret's Bay, Nova Scotia, has shown that the annual seaweed production in coastal enclosed waters can be three times as large as the primary planktonic production. Similar results have appeared in IBP studies in the Philippines and in Japan, and work in the Japanese, Romanian, Netherlands, and South African IBP programs has demonstrated the high production of microphytobenthos in shallow water. Much of this research is only now beginning to reach the open literature. Rather than attempting a review of the entire IBP/PM research on productivity I shall consider the general aspects of marine productivity and, in particular, the patterns of ecosystem cycling found in the sea. This is best introduced with some of the principles essential for modeling the marine production cycle.

ENERGY SUPPLY (LIGHT AND HEAT) FROM THE SUN

I wish I had the courage or the brashness of physicists and engineers, who claim to know what energy is. Most biologists make no such claim. What energy *does* is simpler to understand, but I think that when engineers, and perhaps some of the physicists, too, talk of energy they really mean difference in energy level, which is quite a different thing. Energy, on its way through the ecosystem, becomes incorporated into various forms of organic matter for varying periods of time. This is not textbook ecology, but the best description of how an ecosystem, or an individual, functions. Whatever energy is, matter is one of its forms, and it is time ecological theory came to terms with early twentieth-century physics.

NUTRIENT SUPPLY AND NUTRIENT CAPITAL

If a word other than "energy" is required here, let us coin something like "biopotential" for the nutrients without which production is not possible. The availability of nutrients is probably the limiting factor in most marine systems; e.g., although the polar winter obviously renders photosynthesis impossible, there is an abundance of light in the polar summer, and it is not shortage of light that renders the Arctic Ocean so low in productivity, but the scarcity of nutrients in the euphotic zone—a result of intense vertical density stratification.

TEMPERATURE

Life has adapted to the physical variables of the environment, both in the course of evolution time and in the functional dynamics of the system, so that environmental temperature becomes a basic variable of the system to which organic responses are made at both the proximate and the ultimate levels. Regulation of the relation between temperature and metabolic rate is normal, so that in colder environments the temperature-metabolism curve is simply shifted to the left, toward the lower temperature end of the horizontal axis. This has an obvious selective advantage; in fact, according to the Q_{10} relationships extrapolated from temperate climates, life for poikilotherms would be impossible in Arctic and Antarctic seas; but there is plenty of life in both. Growth rates, also, can be compensated at low temperatures if the evolutionary (adaptive) advantage is necessary for survival—as in the larval stages of certain invertebrates (for a summary of these phenomena, see Dunbar, 1968). Temperature, therefore, becomes in ecological theory a less limiting variable than it was formerly thought to be.

VERTICAL STABILITY OF THE WATER COLUMN

The supply of nutrients to the euphotic upper layer is entirely dependent upon instability of the water column, except where direct outflow of nutrients from the land is concerned, which is usually a local effect. Analysis of the world map of marine production will show that it is this factor of vertical instability, and not temperature or latitude (light), which controls the pattern of productivity. Instability in the water column is achieved in various ways: upwelling caused by wind or Coriolis, or both; vertical exchange in winter in temperate and subarctic regions; mixing of water masses; storm turbulence; tidal mixing.

SEASONALITY

The gradient between the weak seasonality of the tropics and the extreme winter-summer oscillation of polar regions has not been fully recognized, probably because its importance depends far less upon temperature than upon the cycling of nutrients and of plant materials which form the food supply for the secondary producers. Where winter dictates a seasonal pause (a very long pause indeed in high northern latitudes) in primary production, the life cycles of the secondary producers are long. The lifetime of the larger copepods in the Arctic and much of the subarctic is one year at least, sometimes two or even three years. The one-year minimum has most probably evolved not as a necessary response to low temperature but in response to the need for delaying spawning periods until the next bloom of the phytoplankton assures a food supply for the next generation. This is the basic reason for the high standing crops of zooplankton in temperature and subpolar waters and, conversely, for the low standing crops in the stable tropics. The high standing crops in mid-latitudes have much to do with the support of commercially exploitable stocks at the tertiary or higher food chain levels, i.e., fish and mammals.

The effect of this seasonality is thus to cause energy storage in the system, to delay the flow of energy through it. This is to the economic advantage of mankind, for it is largely the lack of this storage factor in the stable tropics that renders those regions so poor in exploitable populations, together with the low nutrient capital and the deep euphotic zone. Where upwelling occurs in tropical latitudes, on the other hand, bringing a constant and large supply of nutrients into the system, high standing crops of zooplankton and of higher trophic levels become possible; energy is caught in its rapid flight and stored in exploitable stocks.

Seasonality thus controls to a large extent the type of cycling in the system. In mid and high latitudes the phytoplankton blooms, which are seasonal, support long-lived zooplankton which carry the populations through to the next time of phytoplankton production. In the stable tropics the lack of seasonality, coupled with low nutrient capital, results in rapid use of both nutrients and phytoplankton as they become available. The cycling is rapid and the standing crops are low. This has been supposed to result in constant concentrations of primary and secondary producers, but recent work, for instance that of Steven and Glombitza (1972) in Barbados waters, has shown that there can be well-marked oscillations in both.

There is much still to be done on these planktonic cycles. T. R. Parsons pointed out at the Rome Working Conference (1971) that different types of plankton cycles are found even in different regions where the general conditions might be considered to be much the same on first examination. The classic pattern based on work in the North Sea shows a major peak of zooplankton following the spring phytoplankton bloom and a minor zooplankton peak following an autumnal bloom. In the North Pacific it seems that the zooplankton peak occurs in summer at the same time as the maximum of phytoplankton production, and the phytoplankton peak is far less steep. These differences may involve differences in the food-chains and in physical conditions such as the behaviour of the thermocline and the critical depth.

The Rome IBP Conference in 1971, in fact, recommended detailed study of these differences using data already available. Associated with these phenomena is the successional dominance of planktonic species, a pattern which also differs from region to region. Voronina (1970) drew attention to the seasonal cycles of three common copepod species in the Antarctic: "Different timing in the summer biomass maximum in these copepods provides a mechanism leading to the spatial isolation of their maxima. As a rule the maximum biomass of *Calanoides acutus* is developed in a more southern position from that of *Calanus propinquus* or in a deeper layer, while *Rhincalanus gigas* has the northernmost maxima. In the Antarctic convergence zone, where there is a mechanical concentration of plankton, the prevailing species succeed one another in the same order. The biological importances of all these relationships is obvious. The sequence in appearance in the plankton of the numerous herbivorous species increases the intensive grazing period and the degree of phytoplankton utilization. The spatial differences in the maxima of different species decrease the competition between them." Again, the importance of evolutionary considerations in the study of present day marine production is illustrated. The production : biomass ratio is, of course, intimately involved in all these cyclical patterns. Several widely dispersed IBP projects have been engaged in the study of production : biomass ratios, and their synthesis and comparison will constitute an important advance. Perhaps we should introduce a new concept of "production per cycle per square meter," in addition to production rates per unit time. Low production per cycle would mean a low standing crop and a high production : biomass ratio.

FRESHWATER RUNOFF FROM THE LAND

In estuarine and enclosed coastal regions freshwater runoff is often vitally important in establishing and maintaining fertility by virtue of the entraining effect, which brings water from deeper layers to the surface. This is well known in eastern Canada; for example, in the Gulf of St. Lawrence. Sutcliffe (1972) has recently shown that annual variations in the land drainage inflow to the Gulf of St. Lawrence can be correlated positively with the commercial catches of several species of fish and invertebrates. There is a time lag appropriate for each species, namely the number of years between spawning and recruitment into the commercial stock. It may not be generally recognized, however, that hydroelectric development in many parts of the world has altered the seasonal balance of the runoff profoundly, and that this must have serious effects on productivity. It is the high natural spring runoff that is important here. Hans Neu, of the Bedford Institute of Oceanography (personal communication), estimates that the natural ratio of spring to autumn inflow at the beginning of this century into the Gulf of St. Lawrence was approximately 3 : 1. The present ratio, following hydroelectric development and the holding back of the spring inflow, is 1.6 : 1. The "ideal" ratio conceived by the power company would be 1 : 1. Here is an obvious conflict of legitimate commercial interests and ecological principles which must somehow be adjusted.

SUMMARY

I have put some emphasis on the evolutionary aspects of the study of marine productivity; a more detailed discussion would require more space than is available here. But it is important to mention the impermanence of particular geographic patterns in marine production. Nothing is so certain as change itself. Marine (or hydrospheric, subsurface) climates, like the atmospheric climates with which they are linked, change cyclically with various periodicities and amplitudes. Since the pattern of marine productivity is in part climatically determined, the pattern must be expected to vary. For example, paleoclimatic studies involving deep-sea cores make clear that the productivity of surface waters is quite changeable over a long time scale. The history of the last century, moreover, shows that marine climates are also changeable in large amplitude over a much smaller time scale as well. The shift in the cod and halibut fisheries in the North Atlantic and subarctic serves as an impressive example. The growth of sea-going salmon fisheries in West Greenland and northern Norway during the past 10 years also may be related to climatic change.

It is probably within our power to predict such marine climatic changes, if we mobilize our international resources to attack the problem. The changes are of immense economic significance. The international coordination of scientific resources toward specific ends has been a concern of both IGY and IBP; and their coordination has been a success. The problems of prediction of hydrospheric climatic change offer a great opportunity for the geophysicists and the biologists to pool their resources and their skills.

REFERENCES

Costlow, J. D. (ed.) 1971. Fertility of the sea. Gordon and Breach Science Publishers, New York, London, Paris. 2 volumes.

Dunbar, M. J. 1968. Ecological development in polar regions. Prentice-Hall, Englewood Cliffs, N.J. 119 p.

Lucas, C. E. 1947. The ecological effects of external metabolites. Biol. Rev. 20:270–295.

Patriquin, D. G. 1971. The origin of nitrogen and phsophorus for growth of the marine angiosperm *Thalassia testudinum* König. M.S. thesis, McGill Univ. 193 p.

Prakash, A. 1971. Terrigenous organic matter and coastal phytoplankton fertility, p. 351–368. *In* J. D. Costlow (ed.) Fertility of the Sea. Vol. 2. Gordon and Breach Science Publishers, New York, London, Paris.

Provasoli, L. 1963. Organic regulation of phytoplankton fertility, p. 165–219. *In* M. N. Hill (ed.) The Sea. Vol. 2. Interscience, London.

Riley, G. A. 1970. Particulate organic matter in sea water, p. 1–118. *In* Russell and Yonge (ed.) Adv. Mar. Biol. 8.

Riley, G. A. 1972. Patterns of production in marine ecosystems, p. 91–112. *In* J. A. Wiens (ed.) Ecosystem Structure and Function. Oregon State Univ. Press.

Steven, D. M., and R. Glombitza. 1972. Oscillating variations of a phytoplankton population in a tropical ocean. Nature 237(5350): 105–107.

Sutcliffe, W. H., Jr. 1972. Some relations of land drainage, nutrients, particulate material, and fish catch in two eastern Canadian bays. J. Fish. Res. Bd. Can. 29:357–362.

Voronina, N. M. 1970. Seasonal cycles of some common Antarctic copepod species, p. 162–172. *In* M. W. Holdgate (ed.) Antarctic Ecology. Vol. 1. Academic Press, London, New York. 604 p.

AN ANALYSIS OF FACTORS GOVERNING PRODUCTIVITY IN LAKES AND RESERVOIRS*

M. BRYLINSKY and K. H. MANN

ABSTRACT

Data collected as part of the International Biological Program from 43 lakes and 12 reservoirs, distributed from the tropics to the Arctic, were subjected to statistical analysis to establish which factors are important in controlling production and how they are related. In the whole body of data, variables related to solar energy input have a greater influence on production than variables related to nutrient concentration; in lakes within a narrow range of latitude, nutrient-related variables assume greater importance. Morphological factors have little influence on productivity per unit area in either case. Chlorophyll *a* concentration is a good indicator of nutrient conditions and when combined with an energy-related variable constitutes a good estimator of primary production.

* This paper is published in full in the January 1973 issue of Limnology and Oceanography (Vol. 18, No. 1).

PRODUCTIVITY OF FOREST ECOSYSTEMS*

JERRY S. OLSON

INTRODUCTION

This paper is but one step toward IBP's synthesis of understanding the biological basis of productivity and human welfare. My *primary* objective is to assess major components of total production estimates for wooded ecosystems: forests and other stands typified by numerous, usually large, more or less long-lived trees. The Terrestrial Productivity (PT) Woodlands Working Group's workshop held in Oak Ridge, August 13-26, 1972, served to highlight progress and problems, but its numerous data and interpretations will emerge mainly in later synthesis.† Workshops of IBP/PT Grassland and Tundra Working Groups, and on Arid Lands and Wetlands research (being integrated elsewhere), each have their own themes on major biome types of the world. All these five groups are expected to contribute to the PT Section theme 6: "Analysis of Ecosystems." Yet the ecosystem analysis theme has had no working group, and only informal modes of working between other groups.

Two working hypotheses are that woodlands have (i) a predominant part of the whole biosphere's live biomass, and probably also (ii) a higher biological production than other biomes. A *second* objective of this chapter and some others (especially Rodin *et al.*, pp. 13-26) is therefore to marshal tabular estimates comparing terrestrial biome regions. In brief, both hypotheses seem amply confirmed by my review and by related summaries (cf. Olson, 1970a, 1974; Whittaker and Likens, 1973; Reiners *et al.*, 1973).

Present data on mass and production per unit area and on biome areas are obviously preliminary and will need improvement after IBP. Further contributions also will be needed for a longer term goal: the integration of knowledge about the main terrestrial ecosystems with that of the freshwater and marine systems (PF and PM sections) into a better global perspective. We can settle for no smaller scale than the whole Earth in treating problems like the biosphere's carbon exchange with the atmosphere (cf. Olson, 1970a; Olson *et al.*, 1970; Whittaker and Woodwell, 1971; Whittaker and Likens, 1973).

BACKGROUND

To review the ecosystem-oriented research on forests even briefly would have been an impossibly large task except for three circumstances. First, Ovington's (1962, 1965) reviews,* many surveys of Rodin and Bazilevich (pp. 13-26)

* Research supported by the Eastern Deciduous Forest Biome, US-IBP (Contribution No. 148), funded by the National Science Foundation under Interagency Agreement AG-199, 40-193-69, and by the Oak Ridge National Laboratory, which is operated by the Union Carbide Corporation under contract for the U.S. Atomic Energy Commission.

† Reports and data banks brought to that workshop or developed during it have been collated for participant editing by Reichle *et al.* (1973c). I thank those participants and many authors cited in my bibliography for contributing to a global perspective that could hardly have been feasible without IBP.

* Mr. R. G. Fontaine of the Forestry Department of the Food and Agriculture Organization of the United Nations suggested that Ovington's paper on tropical forests be presented by title at the World Woodlands Workshop at Oak Ridge. Copies are available from FAO, Via della Terme di Caracella, Rome, which commissioned Ovington's study. My oral presentation was to have been restricted to temperate (and boreal) forests, but a few comparisons with tropical systems (see also Golley, pp. 106-115) and other biomes will be noted for comparisons.

and symposia edited by Young (1968, 1971), Reichle (1970), Duvigneaud (1971), Andersson (1972), and Woodwell and Pecan (1973) provide many results and references which can be consulted for details. Second, many specific newer contributions of the IBP are to be presented later in this symposium (Harris *et al.*, pp. 116–122; Golley, pp. 106–115). Third, modeling research now provides an improving dynamic framework to help sharpen our focus on *both* similarities and the differences among ecosystems of contrasting types and regions.

Models are simplified for particular purposes and local conditions and often need improvement. One direction of improvement can be better injection of basic knowledge into practical models that are still oriented mainly for professional applications (e.g., forestry, wildlife, grazing). Another improvement would be for models of ecosystem structure and function to anticipate whole classes of management questions that might arise without restriction to conventional or *ad hoc* models. In improving management of our resources, modeling cannot serve all roles or purposes at every stage, although the most interesting models will be those having values that *reach beyond the specific objectives* of any particular modeling exercise. The present object and that of Reichle *et al.* (1973c) is toward *descriptive* models: summarizing pool sizes and rates of income and loss.

PRIMARY PRODUCTIVITY

Some terms in the ecosystem's production budget are inherently more difficult than others to measure or derive with confidence. Using gas exchange methods to estimate gross primary production, where

$$\text{Gross primary production (GPP)} = \text{Gross photosynthesis} = \text{Surplus production (SP)} + \text{Respiration}_{\text{green plant parts}} \quad (1)$$

involves difficulties in controlling temperature, ventilation, light and appropriate carbon dioxide levels in the chamber of measurement (cuvette). Controlling conditions approximating those sensed outside the cuvette in natural canopies have been improved during IBP (Walker, pp. 60–63). Yet the extension of results measured at a few selected heights to the natural canopy strata poses several questions: How well can quick leaf responses be related to the sunfleck pattern of (i) truly direct sunbeams; (ii) more or less diffused and fluttering flecks; and (iii) the "green" light transmitted through the leaves? Do changes in the aerodynamic boundary layers around individual leaves as well as sunlit or shaded layers (or sides) of tree crowns distort the interpretation of what takes place in a whole stratum or canopy from measurements of individual leaves? For days of differing weather type, how well can the hourly patterns of stomatal opening, transpiration and photosynthesis be related to the cumulative input of photosynthate to the green plant parts? How does the rising curve of cumulative input, the integrated form of equation (1), vary from year to year? From place to place? Are most of the variations predictable from a fairly small number of ecosystem parameters (like leaf area index, LAI) and environmental variables?

Such questions thus involve time responses ranging from seconds or days to months and years. The Production Process (PP) section of IBP has tended to focus physiological interest, especially toward the faster processes, and all the *rates* of change expressed in equation (1). Each biome theme group in Terrestrial Productivity (PT) is interested in the *integral* of these changes over whole seasons, and its ecological prediction. One approach to prediction is through models with fine resolution and insight (but also exacting requirements for input data), to test how well all the component processes forecast their total result. Yet because such tests are feasible in very few places, complementary models calling for rather few predictors (like standard climatic variables) are still needed. These are being improved and calibrated empirically to rationalize the regional patterns of the world.

Models of *either* type can provide a scale against which to compare estimates of total ecosystem production and of that fairly small fraction of production which is currently used or usable by man. Both types are desirable.

Respiration of the green parts (Eq. 1) limits the magnitude of surplus production—organic material or energy exportable for building and maintaining nongreen plant parts,

$$\text{Surplus production (SP)} = \text{Export from leaves (and twigs?)} = \text{Net primary production (NPP)} + R_{\text{nongreen parts}} \quad (2)$$

The last term, respiration of nongreen parts of green plants, is an additional tax limiting net primary production. It is still not clear how much both respiratory taxes in (1) and (2) increase as a function of income rate (i.e., as a kind of "income tax" on photosynthate, *sensu* Olson, 1964, p. 107) or of biomass and condition of the respiring tissues (as a "property tax"). Perhaps photorespiration partakes most of the first aspect, and dark respiration of the latter, but that correlation need not be a sharp one (cf. Richardson *et al.*, 1973). Relations between respiration of green and nongreen parts, light and dark times, and "income" vs. "property" tax models hopefully could be aided by radiocarbon or organic tritium tracer work to clarify the translocation as well as the prompt turnover of plants' labile and nonlabile pools of organic material (cf. Harris *et al.*, pp. 116–122).

In principle, *net primary production* should be measurable, by definition, from photosynthesis minus plant respiration (3a).

$$\text{NPP} = \text{GPP} - R_{\text{green plant parts}} - R_{\text{other parts of green plants}} \quad (3a)$$

$$= \Delta\ \text{Biomass}_{\text{plant}} + L \quad (3b)$$

TABLE 1 Preliminary Forest Productivity Estimates, Based on 1972 IBP Woodland Project Reports, First-Order Models and Seasonal Models (From Reichle *et al.*, 1973)

	g Dry Matter/m^2/yr		
Major Forest and Woodland Ecosystem Group	Net Primary Production	Surplus Production	Gross Primary Production
Cool temperate to boreal conifer forest	480–1,280	1,508–2,500	2,250–4,880
Temperate deciduous forests:			
Beech	1,440–1,780	2,100–3,680	2,360–4,250
Oak	1,400–1,944	2,520–4,210	2,890–4,660
Other	1,490–2,100	2,000–3,030	3,360–3,530
Warm temperate broadleaved evergreen forests	830–1,500	2,393–3,100	2,430–5,500
Tropical evergreen forest and deciduous woodlands	743–2,100	1,423–7,370	3,400–12,730

In practice widespread data are becoming available, but only by a complementary approach (note 3b). This is from careful measurements of plant biomass change (ΔB) *plus* allowance for the appropriate losses (L) of produced material in the time interval, Δt. In almost exact parallel with Coupland's critique for grasslands (pp. 44–49), it seems fair to say that upward corrections of L are still needed. Some components of litterfall may be missed (e.g., coarse branches), while others (leaves, flowers, fruits) are underestimated due to decay or consumption. Missing or preliminary terms are being corrected by the working group, but general magnitudes for production estimates of the four forest groups in the Woodlands Workshop are given in Table 1 (after Reichle *et al.*, 1973c).

Leaf Production

In seasonal forests, some early leaves, stipules, and reproductive parts are typically shed before annual foliage reaches its maximum value. Year-round litterfall collections provide a minimum correction for such loss, but the easy decomposability and leachability of some of these soft materials (before falling or even within the litter traps) calls for augmenting such a minimum correction.*

Production of understories and groundcover may be underestimated because of asynchrony of harvest times and growth phases, and other problems discussed in chapters on grass and shrub systems. Lower strata (and uncertainties about their productivity) might be relatively unimportant in large, dense forests compared with problems of estimating primary and secondary production for the main tree strata. The simplification of past studies could become increasingly misleading as ecosystem studies move to open woodlands and savannas in post-IBP years.

* In rainforest, as at the Pasoh project in Malaysia, I was shown appreciable masses of *Dipterocarpus* and other leaves fallen from the overstory but "hung up" (decomposing somewhat en route) in the understory canopy layers, before rains and winds finally brought the partly decayed debris down to the litter traps.

Aboveground Support Structures

Stem biomass, and hence production rates, has been subdivided somewhat differently among IBP projects. Current twigs have too seldom been distinguished from and related to other stem material and (leafy) green parts.

Reproductive "support" structures (flowers, fruits, etc.) are recognized as highly variable but ecologically important fractions of the plants' allocated production. They are seldom a large part of the total production (for trees), but estimates need to be filled in where missing, or desegregated in cases where they have been pooled with stems or with foliage, in order for results from different projects to have comparable meaning. Aggregation of unlike components can often be avoided for a major species or group (and for a "target" group of interest for management even if it is not the most abundant one), but in descriptive modeling groups that are less abundant or lower in stature often tend to get lumped together.

For trees of diverse size, the fundamental biological problem of relating dimensions (allometry) has taken on very practical importance: e.g., using easily measured diameters (and perhaps height) for estimating more difficult measures like biomass. It is natural for these statistical considerations to be applied first to biomass, but there is some progress in making application to production rates (Hozumi *et al.*, 1968, 1969). The common tendency to underestimate aboveground biomass is still one contributor toward *under*estimation of some productivity *rates.*

BELOWGROUND PRODUCTION

As with other ecosystems, woodland studies have had further errors (again usually estimating on the low side) for production *rates* of roots and other belowground parts.

Root mass poses formidable problems. The butt root or underground portion of stump and the immediately adhering lateral bases (or buttresses in the tropics) can and should be treated by further extension of regression methods already noted for aboveground supporting structures. Other lateral roots of diameters down to the convenient 0.5 cm dividing line can be treated similarly, but with much labor; or coring devices of sufficient strength can provide a broader sampling if soils are not stony and the trees are not extremely massive (cf. Harris *et al.*, pp. 116–122). Roots smaller than 0.5 cm call for the core approach plus meticulous separations, and problems of interpretation like those discussed by Coupland (pp. 44–49). I also stress the active role and turnover of a very fine fraction (e.g., below 0.1 cm, cf. Olson, 1968), and urge attention to the very special role of myccorhizae in the rhizosphere of the forest ecosystem.

Continual or episodic death of each root size class is probably underestimated. Income must make up for root mortality and for debris and exudates cast off by still-living roots, in addition to providing for net increase of root mass over the years. Especially since the IBP root and rhizosphere symposium (Ghilarov *et al.*, 1968), the nature of this problem has been appreciated. Working hypotheses that a fairly high underground production allows for this replenishment have been built into some total production estimates by Rodin *et al.* (pp. 13-26). How these hypotheses will be refined numerically is one of the larger issues requiring attention during the synthesis of IBP research, and the extensions of newer methods (like isotope tagging) in later years.

HETEROTROPHIC PROCESSES

In many of the ecosystems utilized by man, harvesting removes products for ultimate decay or burning at some distance from where they were produced. A few IBP projects have had the opportunity for elaborating studies of animal food chains which respire some of the produced organic materials before decomposers oxidize the remainder, e.g., beech forests of Germany and Denmark and subalpine (*Tsuga diversifolia*) woodland of Shiga Heights, Japan. The animals not only release some fraction of what they consume and assimilate, but hasten the change of additional plant material from "live" to "dead," thereby initiating decomposer activity earlier and at a higher rate than might have occurred without aggressive herbivores, predators and omnivores.

Herbivores

Rafes (1970), Franklin (1970) and McCullough (1970) together review the wide interest in primary consumers in forests. The conventional wisdom is that very few percent of net primary production is channeled through animal food chains in "normal" years. Yet observation of our *Liriodendron tulipifera* forest in Oak Ridge, Tennessee, over a decade suggests that geometrid caterpillars, weevils and aphids may each take turns in different years drawing off more of the flow of organic carbon or energy than their share in years of average population. When all consumption of all canopy horizons is summed, with or without rough estimation of the underground feeding (Ausmus *et al.*, in press) on roots, early estimates of consumption (e.g., Reichle and Crossley, 1967; Reichle *et al.*, 1973b; Van Hook and Dodson, 1974) may well be increased.

Studies of oak forests near Grange-over-Sands, England, and east of Cracow, Poland, have provided relatively complete energy flow budgets. The latter example (Medwecka-Kornas *et al.*, 1973) was marked by conspicuous defoliation by lepidopteran larvae, typically followed by a second flush of oak (*Quercus robur*) foliage that is presumably produced with labile or previously stored photosynthate. In 1969, consumption studies allocated 0.032×10^6 and 0.108×10^6 gram calories per m^2 of ground area for *Tortrix viridana* and other caterpillars respectively; the sum of 0.14×10^6 calories per m^2 accounts for most of the 0.156×10^6 cal/m^2 ($30.6 g/m^2$) estimate of area removed from leaves from May to November. The two-year average of estimated consumption was 0.41×10^6 $cal/m^2/yr$ ($80.8 g/m^2/yr$), so 1968 (a year of nearly complete oak defoliation by *Tortrix*) had about 130 $g/m^2/yr$ removed.

Compared with such isolated examples for deciduous forest, we expect the conifer forests (both temperate and boreal) to show even greater contrasts in consumption between peak years of a forest pest cycle and the many intervening years. Airplane views of Siberia and of North America show the vast scale (and sometimes sharp boundaries) of devastation; essentially starting new cycles of ecosystem development by succession instead of minor perturbations on previously existing stands. Forestry research in many countries, of course, is analyzing details on the biology of both pests and stands, but synthesis integrating a balanced understanding of stands, pests and their controlling agents is high on the list of priorities for post-IBP years.

Predators and Parasites

The examples just given of defoliators and of other pests illustrate the role of secondary and tertiary consumers in having a "leverage" on the quantity and quality of ecosystems' primary production—out of proportion to the flow of matter and energy through these consumers. One of the reasons why research must consider areas broader than individual stands is the wide range of some of the larger predators. Those parasites and diseases which pass through a population cycle only in local refugia and then spread out also require some attention to regions large enough to illustrate all significant stages of the population cycle, and of the landscape patterns which control it.

Obviously not all stands, or even all ecosystem types in a large or heterogeneous region can be investigated in equal detail—especially in such detail as IBP case studies have sought. Yet the local detail seems necessary to reveal the normal role of predators and parasites as they relate to herbivores and primary producers. We need not only to summarize and model on different scales (cf. Olson, 1971), but also to couple subsystems which must (by their nature) be investigated quite locally with problems which call for sampling and probability statements (at least) on one of the large regional scales.

Scavengers and Omnivores

Among secondary consumers, not only predators and parasites but also scavengers channel some fraction of the materials and energy from the herbivorous trophic level. By hastening the return and redispersal of nutrients over the ground, these organisms, too, play a role of somewhat broader significance than would be indicated by the fraction of primary production which is assimilated or the much larger fraction consumed by them. Neither vertebrates nor invertebrates can be dismissed as generally unimportant; but measures of their absolute and relative importance can be derived only from the more detailed ecosystem analyses.

Many animals (including some social insects, small mammals and certain birds that are "granivorous" when seeds are available) derive only parts of their diet from other animals. We need better estimates of what part—and when the consumption is switched—during the seasonal cycle and perhaps the life cycle.

Among omnivores we can highlight man. His role as a consumer within ecosystems has long attracted interest of environmental anthropologists, and began to get broader public attention in the late 1960's. In the U.S., testimony to the Congress related to IBP itself was among the many channels by which the environmental concern reached points where major policy decisions on the human environment were to be made.

Decomposition

The coupling of subsystems to large—even global—systems is especially challenging for soil processes. Microbes, mesofauna and many media are small enough to be subject to experimental manipulation indoors as well as in the field. In forest stands, even more than in some other systems, however, the rhizosphere is readily distorted by the techniques of study (Ghilarov *et al.*, 1968). On landscapes, production from the uplands may become balanced in part by respiration from leaves which have blown downhill or humus which has washed downstream.

Yet as we integrate over larger areas, microbial and other heterotrophic respiration (and fire in many regions as well) must come to approximate more closely the autotrophic production. For a whole large ecosystem or regional complex, we are tempted to view the implications of balance (or of imbalance as the case may be) which was reviewed some years ago with more particular regard to litter and soil (Olson, 1963). IBP's own contributions to the study of heterotrophic processes are left for others to review.

PRODUCTION OF FORESTS AND OTHER BIOME AREAS

A geographic index of world ecosystems (Olson, 1970b) was complemented by a generalized map of living organic carbon (inside back cover of Reichle, 1970). That map represents continental patterns approximating present climatic conditions (i.e., following the major postglacial migrations onto formerly glaciated areas) but before the major clearing of forests by man. Although labelled "prior to the Iron Age" the map perhaps approximates patterns and magnitudes corresponding with the early Neolithic. Late Neolithic and Bronze Age societies may have already reduced the areas and especially the average mass per unit area of forests on the lands they occupied.

Woodland Areas

There is considerable discrepancy between the areas estimated by foresters to be either woodlands or forest, and the larger areas estimated by Bazilevich, Rodin and Rozov (1970) from careful study of the USSR Physical-Geographical Atlas of the World (Senderova, 1964). These Soviet sources indicate about 17 million km^2 for boreal, subalpine and various "semiboreal" forest zones (including "hemiboreal") and over 18 million km^2 for temperate wooded zones. Included in the latter "temperate" area is about 1 million km^2 of "moist site" woodlands and open communities (as on floodplains) in semiarid to arid climates. The tropics include another 17 million km^2 of forest and woodland and 14 million km^2 of woody savanna and scrubland. Thus a total of 66×10^6 km^2 could well have been considerably wooded 5,000 to 10,000 years ago (early Neolithic time). However, this atlas-measuring approach does not yet hint at how much of each area was locally in nonforest cover, nor how much of the 15×10^6 km^2 of cultivated land would have to be subtracted from

TABLE 2 Location, Extent, Biomass, and Productivity of Some Major Biomes (Preagricultural)

Major Ecosystem Complexes Typical Plant/Soil Habitats	Map Zone[a]	Area[b] (10^6 km^2)	Dry Live Plant Mass[c] (metric tons/km^2)	Total Biomass (10^9 metric tons)	Net Primary Productivity (metric tons/km^2 × yr)	Total Production (10^6 metric tons)
Tundra and related ecosystems						
Polar tundra complex						
Polar barren, mountain; spotty tundra[d]	1	1.146	20	0.023	4	4.6
Tundra: glei soil, permafrost[d]	2	3.768	2,800	10.52	150	565
Mountain tundra (subpolar)	3	2.946	700	2.06	70	206
SUBTOTALS		7.860		12.60		776
Bog and mire (treeless)						
Polar wetland: permafrost	2–4	0.188	2,500	0.47	220	41
Boreal and other: deep peat	5–10a	1.112	3,500	3.89	350	389
SUBTOTALS		1.300		4.36		430
Other tundra-like						
Mountain "tundra": black soils	3–9a	1.575	3,500	5.51	1,000	1,575
Wooded bog and other scrub: peat	4–10a	1.446	7,100	10.33	482	697
Wooded tundra: glei permafrost soil	4	1.219	10,000	12.19	400	488
Herb-wood complex: volcanic soil	4a	0.135	10,000	1.35	1,000	135
SUBTOTALS		4.375		29.38		2,895
TOTALS (Tundra, etc.)		13.535		46.34		4,091
Boreal and semiboreal ecosystems						
Taiga conifers, small broadleaved woods						
Northern taiga parkland: glei podsols	5	0.880	12,500	11.00	500	440
North, middle taiga woodland: permafrost	5–6	2.463	20,000	49.26	600	1,478
Middle taiga woodland: podsolic soils	6	3.571	26,000	92.85	700	2,500
Southern taiga forest: sod-podsolic soils	7	3.182	30,000	95.46	750	2,386
SUBTOTALS		10.096		248.57		6,804
Semiboreal forests and woodland						
Subalpine woods: permafrost, etc.	5a–6a	3.040	16,000	48.64	500	1,520
Other montane conifers: podsols	7a–9a	2.696	17,000	45.83	600	1,618
Mixed forest: yellow podsols	7–8a	0.393	35,000	13.75	1,000	393
Mixed forest: yellow carbonate	5–9a	0.301	35,000	10.53	1,000	301
Small broadleaved: gray podsolic	9	0.227	20,000	4.54	800	182
Mixed forest: gray mountain soils	8–9a	0.252	30,000	7.56	750	189
SUBTOTALS		6.909		130.85		4,203
TOTALS (Boreal, etc.)		17.005		379.42		11,007

Temperate forest or woodland						
Cool conifers: montane, valley soils[e]	8a	3.269	32,000	104.45	1,200	3,269
Giant and coastal conifers: podsolic[e]	34	0.500	70,000	35.00	2,506	1,253
Cool; mostly broadleaved forest						
Northern: gray and podsolic soils	10a	0.711	37,000	26.31	800	569
Central: brown, gray-brown soils	10	2.485	40,000	99.40	1,300	3,230
Central: redzina soils	10	0.027	37,000	0.99	1,200	32
Wetland: muck and glei soils	10	0.346	28,800	9.96	1,360	469
Other complexes	10	0.193	9,000	1.78	1,200	238
SUBTOTALS		7.531		277.89		9,060
Warm; evergreen and deciduous						
Montane: yellow, red soil	11,31–33	2.537	41,000	104.02	1,800	4,567
Lowland: red, yellow soil	11, 31	1.977	45,000	88.96	2,000	3,954
Lowland: redzina, terra rosa	11	0.172	38,000	6.54	1,600	275
Wetland: flood and swamp soils	11	0.126	40,000	5.04	2,200	277
Wetland: flood and delta soils	30	0.269	20,000	5.38	13,000	3,497
Other special complexes	11	0.682	25,000	17.05	4,000	2,728
SUBTOTALS		5.763		226.99		15,298
Warm or montane, semiarid woodlands						
Montane woods: brown soils	12, 22	2.236	12,000	26.83	1,300	2,907
Other dry-season woods	12, 35	1.593	17,000	27.08	1,600	2,549
SUBTOTALS		3.829		53.91		5,456
Warm; lowland: moist, rich sites						
Semiarid regions	30	0.629	25,000	15.72	4,000	2,728
Arid regions	30	0.439	20,000	8.78	9,000	3,951
SUBTOTALS		1.068		24.50		6,679
TOTALS (Temperate, etc.)		18.191		583.29		36,493

[a] Map zones of Bazilevich and Rodin (1967), as modified by Olson (1970b) or Students of Earth's Future (1971), are schematic.
[b] No effort has been made to readjust area estimates of Bazilevich, Rodin and Rozov (1970), although some regrouping and relabelling has been done (their "subboreal" ≈ cool temperate; their "subtropical" ≈ warm temperate).
[c] Grams per m^2 and metric tons (10^3 kg)/km^2 are numerically equal, and seem appropriate units for statistical comparisons among contrasting ecosystems. (Multiplication by 10 gives kg per hectare, the units sometimes preferred in forestry and agriculture.)
[d] Plant mass and/or productivity averages are here considered as lower than estimates cited in footnote *b* for the ecosystem areas noted. Otherwise data of Bazilevich *et al.* are followed here. It seems likely that a reassessment will lower the preferred estimates of some other lines when more IBP data are integrated.
[e] Very tentative breakdown of Bazilevich's "mountain forest on brown mountain forest soils" subject to revision by the Western Coniferous Biome of the US/IBP.

each kind of original woodland to account for subsequent conversion from forest to agriculture and grazing.

Starting from various United Nations sources, estimates of forested areas range from 40 to 44 $\times 10^6$ km^2. These inventories apparently exclude some extensive open woodland ecosystems that are valued (and are therefore reported) primarily for grazing, i.e., as rangeland. This omission probably outweighs the incongruous naming of shrubland as forest, in countries where wood is scarce or where even small stems are especially important for fuel and light construction.

Between the extremes of 40 and 66 $\times 10^6$ km^2, my estimate (Olson, 1970a) of 48 $\times 10^6$ km^2 for current forest and woodland has since been adjusted slightly (by 2 $\times 10^6$ km^2 to include partially wooded "moist sites" in Olson *et al.*, 1970, or by 1 $\times 10^6$ km^2 in the 1973 reprinting of Olson, 1970a). Several other recent estimates reviewed by Whittaker and Likens (1973) are generally comparable for the temperate and boreal zones. I believe that they imply, however, a greater area for tropical rainforest than is warranted by general observation and by Ovington's recent review (see footnote, p. 33).

To improve much on the present estimates of forest areas will probably call for more sophisticated forestry and ecological survey statistics, especially in tropical regions. Eventually all our estimates of vegetative cover will need to be evaluated by remote sensing imagery and multispectral scanning, partly by satellites but partly from aerial photography and extra ground surveys in areas where the interpretation of landscape pattern is critical.

Boreal Woodlands Compared With Polar Zones

Bazilevich, Rodin and Rozov (1970) estimated less than 500 g/m^2/yr of production and less than 7,000 g/m^2 of biomass for tundras and bogs (mostly in zones 1–3).* Higher production occurs on mountain "tundra" meadows on black soils and higher biomass on certain wooded tundra (Table 2). Table 2 also compares these tundra-like ecosystems (mostly Polar zone) with forests and related communities of the Boreal zone. In zones 5 to 7 (northern, middle and southern taiga), estimated average productivity increased southward from 500 to 750 g/m^2/yr but live biomass (including estimated roots) increased even more from 12,500 to 30,000 g/m^2. Similar ranges exist for values for subalpine and montane conifer woodlands, but 1,000 g/m^2/yr productivity and 35,000 g/m^2 of dry matter were estimated for "mixed forest" (zone 8) on the richer soils.

Biomass, productivity and location of forest ecosystems are broken down further by typical plant/soil habitats for each major ecosystem complex in Table 2. By multiplying area times mass per unit area and summing over zones, we obtain an estimate for taiga of over 6.8 $\times 10^9$ metric tons per year as the production rate (net primary productivity) and a live organic mass of 2.5 $\times 10^{11}$ metric tons. For the varied systems lumped as "semiboreal" (including extensive birch, aspen and some other hardwoods in zones 7–9 especially), the total production rate was estimated as over 4.20 $\times 10^9$ metric tons per year by an inventory of about 1.3 $\times 10^{11}$ metric tons. Summing boreal and semiboreal values gives estimates of 1.1 $\times 10^{10}$ metric tons exchanging on a total of 3.79 $\times 10^{11}$ metric tons for the combined original area of 17 million km^2 (much of Canada, Fennoscandia and the USSR).

Comparing these estimates with the corresponding subtotals for all the high latitude tundras and related ecosystems (including bogs) at the top of Table 2, we find the boreal and semiboreal ecosystems occupying only 25 percent more area. However, net primary production is here estimated as 270 percent and dry mass as 820 percent as high as the tundra and related ecosystems taken together.

Temperate Woodlands

Compared with the "semiboreal," boreal and polar zones, the temperate forest has consistently higher productivity and biomass both on a unit area and total basis (Table 2). Preliminary pre-agricultural estimates of Bazilevich *et al.* (1970) for both the cool temperate conifer forest (especially in the northern Cordillera ranges and valleys of western North America) and for the mostly deciduous forest (of eastern North America and western Europe) were each nearly the same in area (~3.8 million km^2) and in productivity (4.5 $\times 10^9$ metric tons/year) and in mass (~1.3 $\times 10^{11}$ metric tons dry matter).

The appropriate estimates for area of "giant and coastal conifers" (mostly from northwestern North America) will be revised by work now underway in the Coniferous Biome program of the US-IBP. The mass and productivity per unit area for this category will be higher than previously supposed (cf. Fujimori, 1971).

Most temperate forest is broadleaved deciduous in prevailing aspect, but different coniferous groups can be important in the broad transitions toward both the Boreal and the Subtropical zones. For example, the "Northern Forest" of eastern North America ("Northern Hardwoods" of foresters) includes shade tolerant *Tsuga canadensis* (eastern hemlock) in many habitats that are moist and/or free of fire; this species extends south to Alabama in the Appalachian Highlands (Olson, 1971). In the Great Lakes St. Lawrence forests and Acadian portions of the "Northern forests" which straddle much of the eastern Canadian-USA boundary, various pines were important since early postglacial time, and probably expanded by wildfire and by man's influence over many areas that otherwise might have

* The map zones referred to are those of Bazilevich and Rodin (1967) and modifications by Olson (1970b).

developed through ecological succession toward a mosaic of deciduous and mixed forests.

It is still difficult to judge whether the biomass and productivity estimates of both the Cool and Warm Temperate zones in Table 2 are appropriate averages over wide areas or not. More massive and productive stands than those for the average stand do exist, and these may have been given a disproportionate representation among the diverse forests (Whittaker and Woodwell, 1971; Olson, 1971) which influenced Bazilevich *et al.* (1970) in their calculations for broad geographic averages. Warm, mostly humid, temperate forests with varying mosaics of deciduous and/or evergreen broadleaved forest and some conifers (e.g., southern pines) were taken as having higher production rates than the combined cool temperate forests (per unit area and total). Although these forests have higher biomass per unit area, their smaller area (5.8 million km^2) gives a slightly lower total biomass of 2.27×10^{11} metric tons.

Semiarid climatic types include open woodlands (e.g., of pine, juniper and oak) and also the dense but somewhat stunted forests of regions with winter rain and summer drought (Mediterranean-type). In arid and semiarid climates, as in the warm humid ones with moist sites such as floodplains having abundant nutrient reserves, extraordinarily high production rates (4,000 to 13,000 $g/m^2/yr$) are characteristic of quick-growing trees, thickets and luxuriant herbs. Biomass per unit area in these ecosystems is assumed to be lower than for other forest types because the upper canopy is typically less dense.

Together the estimated 18 million km^2 originally wooded in the temperate zone were inferred to have an annual net primary productivity total of about $\sim 3.6 \times 10^{10}$ metric tons/km^2/yr, on an inventory of 5.82×10^{11} metric tons/km^2. These figures may be high, because relatively well-developed stands were chosen for study and then extrapolated over large areas which included medium and poor growth as well as very productive stands.

For tropical woodlands and many other systems better estimates can be anticipated from other biome programs. Preliminary estimates (Olson, 1974) will be discussed only briefly in terms of carbon dioxide exchange.

Forests in the Global Carbon Balance

In Table 3, preliminary estimates of organic carbon inventory and exchange are based on assumptions of carbon percentages ranging from 43 to 49 percent for major plant parts. While improved conversion factors will modify details, such

TABLE 3 Net Primary Production, Carbon Pool, and Turnover for the Major Terrestrial Ecosystems with Emphasis on Forests (From Olson, 1974)[a]

Ecosystem Complexes	Production[b] (10^9 metric tons of carbon per year)	Live Carbon Pool (10^9 tons)	Turnover Fraction[c] (per year)
Nonwoodland			
Tundra and bog (treeless)	0.88	8.08	0.109
"Tundra" meadow and scrub	1.46	13.20	0.1
Grassland (subalpine to tropical)	10.27	30.89	0.332
Deserts (excluding local moist sites)	3.37	8.58	0.394
TOTAL	15.98	60.75	0.26[d]
Forest, woodland			
Boreal taiga	3.33	121.80	0.0275
Semiboreal forest, woodland	1.93	64.12	0.0301
Cool temperate, montane conifer	2.08	68.38	0.0304
Cool temperate, mostly deciduous	2.09	67.88	0.0308
Warm temperate, mostly broadleaf	4.05	97.76	0.0414
Warm temperate wetland	2.97	10.32	0.2878
Warm, montane, woodland (semiarid)	2.40	24.80	0.0968
Warm temperate wetland (arid to semiarid)	3.14	12.82	0.2449
Tropical rich wetland (arid to semiarid)	0.79	2.66	0.2970
Tropical scrub, woodland savanna	10.52	139.13	0.0756
Tropical montane forest	4.08	99.62	0.0410
Tropical lowland rainforest	11.17	83.86	0.1331
Other tropical forest	10.82	216.26	0.0500
TOTAL	59.37	1,009.41	0.059[d]
TOTAL LAND	75.35	1,070	0.07[d]

[a]Equivalent to the gross primary production less the respiration of all parts of green plants.
[b]Includes carbon tied up in all living plants.
[c]Annual loss of carbon from the live carbon pool to dead organic matter.
[d]Total turnover estimates are not additives of individual values, since total average turnover fractions are weighted by pool sizes for each ecosystem complex.

adjustments are not likely to be as critical as improvements in the estimates of biomass per unit area already considered in Tables 1 and 2, and assignment of numbers to expressions of ecosystem area in Table 2. I have made slight downward adjustments from Bazilevich, Rodin and Rozov (1970) for tundra systems, which have little influence on the total. However, my preliminary readjustment for tropical rain-forest more significantly affects the estimated world carbon balance. Otherwise Tables 2 and 3 intentionally retain these authors' estimates for other ecosystems pending later revision.

Personally I still consider these estimates as upper bounds, and this impression has just been confirmed by the Woodland Workshop. Undoubtedly many stands are as massive and productive as outlined for each zone in Table 2. However, some of these may represent the upper range of a frequency distribution of mass per unit area. Mean or expected values should be smaller—even for the pre-agricultural conditions of postglacial times.

Man's own activities then diminished both the extent of forests and mass or carbon per unit area. We have only preliminary estimates of what man's input on the terrestrial ecosystem carbon budget has been (Olson, 1974), but even these should help lead to a broader perspective on technology's role in changing global geochemistry.

Annual turnover, in terms of the net primary production and equivalent loss of plant organs, was apparently only a few percent of the biomass or carbon inventory for all but some tropical forests. Of course, respiration of green and nongreen organs would add very short-lived components to the nearly continuous spectrum of residence times which, by definition, cover such a wide range for all the systems which we call forests or woodlands.

REFERENCES

Andersson, F. 1972. Systems analysis in northern coniferous forests–IBP workshop. Ecological Research Committee Bulletin No. 14, Swedish Natural Science Research Council. 194 p.

Ausmus, B. S., J. M. Ferris, D. E. Reichle, and E. C. Williams. In press. The role of primary consumers in forest root processes. US-IBP Interbiome Symposium: the belowground ecosystem: a synthesis of plant-associated processes.

Bazilevich, N. I., and L. E. Rodin. 1967. Maps of productivity and the biological cycle in the earth's principal terrestrial vegetation types. Izvestiya Geographicheskogo Obshchestva, Leningrad 99(3):190–194.

Bazilevich, N. I., L. E. Rodin, and N. N. Rozov. 1970. Geographical aspects of biological productivity. Papers of the Fifth Congress of the Biological Society, Leningrad, USSR. 28 p. (Translated in Soviet Geography: Review in Translations, May 1971, p. 219–317)

Duvigneaud, P. (ed.) 1971. Productivity of forest ecosystems. Proceedings, Brussels Symposium, October 1969. UNESCO, Paris. 707 p.

Franklin, R. T. 1970. Insect influences on the forest canopy, p. 86–99. *In* D. E. Reichle (ed.) Analysis of temperate forest ecosystems. Ecological studies 1. Springer-Verlag, Berlin–Heidelberg–New York.

Fujimori, T. 1971. Primary productivity of a young *Tsuga heterophylla* stand and some speculations about biomass of forest communities on the Oregon coast. Pacific Northwest Forest and Range Experiment Station, Forest Service, USDA, Portland, Oregon. 11 p.

Ghilarov, M. S., V. A. Kovda, L. N. Novichkova-Ivanova, L. E. Rodin, and V. M. Sveshnikova (ed.) 1968. Methods of productivity studies in root systems and rhizosphere organisms. International symposium, USSR, August–September 1968. Nauka, Leningrad, USSR. 240 p.

Hozumi, K., K. Shinozaki, and Y. Tadakai. 1968. Studies on the frequency distribution of the weight of individual trees in a forest stand. 1. A new approach toward the analysis of the distribution function and the -3/2th power distribution. Jap. J. Ecol. 18(1):10–20.

Hozumi, K., K. Yoda, and T. Kira. 1969. Production ecology of tropical rain forests in southwestern Cambodia. 2. Photosynthetic production in an evergreen seasonal forest. Nature and Life in Southeast Asia 4:57–81.

McCullough, D. R. 1970. Secondary production of birds and mammals, p. 107–130. *In* D. E. Reichle (ed.) Analysis of temperate forest ecosystems. Ecological studies 1. Springer-Verlag, Berlin–Heidelberg–New York.

Medwecka-Kornas, A., A. Lomnicki, and E. Bandola-Ciolczyk. 1973. Energy flow in the deciduous woodland ecosystem, Ispina project, Poland, p. 144–150. *In* D. E. Reichle, R. V. O'Neill, and J. S. Olson [comp.] Modeling forest ecosystems. Report of International Woodlands Workshop, IBP PT Section, August 1972. EDFB-IBP-73-7 Oak Ridge National Lab., Tenn.

Olson, J. S. 1963. Energy storage and the balance of producers and decomposers in ecological systems. Ecology 44(2):322–332.

Olson, J. S. 1964. Gross and net production of terrestrial vegetation. J. Ecol. (Suppl.) 52:99–118.

Olson, J. S. 1968. Distribution and radiocesium transfers of roots in a tagged mesophytic Appalachian forest in Tennessee, p. 133–138. *In* M. S. Ghilarov, V. A. Kovda, L. N. Novichkova-Ivanova, L. E. Rodin, and V. M. Sveshnikova (ed.) Methods of productivity studies in root systems and rhizosphere organisms. International symposium, USSR, August–September 1968. Nauka, Leningrad, USSR.

Olson, J. S. 1970a. Carbon cycle and temperate woodlands, p. 226–241. *In* D. E. Reichle (ed.) Analysis of temperate forest ecosystems. Ecological studies 1. Springer-Verlag, Berlin–Heidelberg–New York.

Olson, J. S. 1970b. Geographic index of world ecosystems, p. 297–304. *In* D. E. Reichle (ed.) Analysis of temperate forest ecosystems. Ecological studies 1. Springer-Verlag, Berlin–Heidelberg–New York.

Olson, J. S. 1971. Primary productivity: temperate forests, especially American deciduous types, p. 235–258. *In* P. Duvigneaud (ed.) Productivity of forest ecosystems. Proceedings, Brussels Symposium, October 1969. UNESCO, Paris.

Olson, J. S. 1974. Terrestrial ecosystem. Encycl. Brit. 18:144–149.

Olson, J. S., J. B. Hilmon, C. D. Keeling, L. Machta, R. Revelle, W. W. Spofford, and F. Smith. 1970. Appendix: Carbon cycle in the biosphere, p. 160–166. *In* Report of Study of Critical Environmental Problems (SCEP). Man's impact on the global environment. Assessment and recommendations for action. MIT Press, Cambridge, Mass.

Ovington, J. D. 1962. Quantitative ecology and the woodland ecosystem concept. Adv. Ecol. Res. 1:103–192.

Ovington, J. D. 1965. Organic production, turnover and mineral cycling in woodlands. Biol. Rev. 40:295–336.

Rafes, P. M. 1970. Estimation of the effects of phytophagous insects on forest production, p. 100–106. *In* D. E. Reichle (ed.) Analysis of temperate forest ecosystems. Ecological studies 1. Springer-Verlag, Berlin–Heidelberg–New York.

Reichle, D. E. (ed.) 1970. Analysis of temperate forest ecosystems. Ecological studies 1. Springer-Verlag, Berlin–Heidelberg–New York. 304 p.

Reichle, D. E., and D. A. Crossley, Jr. 1967. Investigation on heterotrophic productivity in forest insect communities, p. 563–587. *In* K. Petrusewicz (ed.) Secondary productivity of terrestrial ecosystems (principles and methods). Vol. II. Proceedings, Working Meeting, Jablonna, 1966. Polish Acad. Sci.

Reichle, D. E., B. E. Dinger, N. T. Edwards, W. F. Harris, and P. Sollins. 1973a. Carbon flow and storage in a forest ecosystem, p. 345–365. *In* G. M. Woodwell and E. V. Pecan (ed.) Carbon in the biosphere. Proceedings of Symposium, Brookhaven National Lab., Upton, New York, May 1972. U.S. Atomic Energy Comm. CONF-720510.

Reichle, D. E., R. A. Goldstein, R. I. Van Hook, Jr., and G. J. Dodson. 1973b. Analysis of insect consumption in a forest canopy. Ecology 54(5):1076–1083.

Reichle, D. E., R. V. O'Neill, and J. S. Olson [comp.] 1973c. Modeling forest ecosystems. Report of International Woodlands Workshop, IBP PT Section, August 1972. EDFB-IBP-73-7, Oak Ridge National Lab., Tenn. 339 p.

Reiners, W. A., L. H. Allen, Jr., R. Bacastow, D. H. Ehaalt, C. S. Ekdahl, Jr., G. Likens, D. A. Livingstone, J. S. Olson, and G. M. Woodwell. 1973. Appendix: Summary of world carbon cycle and recommendations for critical research, p. 368–382. *In* G. M. Woodwell and E. V. Pecan (ed.) Carbon and the biosphere. Proceedings of Symposium, Brookhaven National Lab., Upton, New York, May 1972. U.S. Atomic Energy Comm. CONF-720510.

Richardson, C. J., B. E. Dinger, and W. F. Harris. 1973. The use of stomatal resistance, photopigments, nitrogen, water potential, and radiation to estimate net photosynthesis in *Liriodendron tulipifera* L.—a physiological index. EDFB-IBP-72-13, Oak Ridge National Lab., Tenn. 130 p.

Senderova, G. M. (ed.) 1964. Physical-Geographical Atlas of the World. USSR Akad. Sci. and Main Admin. of Geodesy and Cartography.

Students of Earth's Future. 1971. Earth's ecology action zones. SEF, Oak Ridge, Tenn. 1 p.

Van Hook, R. I., and G. J. Dodson. 1974. Food energy budget for the yellow-poplar weevil *Odontopus calceatus* (Say). Ecology 55(1):205–207.

Whittaker, R. H., and G. E. Likens. 1973. Carbon in the biota, p. 281–302. *In* G. M. Woodwell and E. V. Pecan (ed.) Carbon in the biosphere. Proceedings of Symposium, Brookhaven National Lab., Upton, New York, May 1972. U.S. Atomic Energy Comm. CONF-720510.

Whittaker, R. H., and G. M. Woodwell. 1971. Measurement of net primary production of forests, p. 159–175. *In* P. Duvigneaud (ed.) Productivity of forest ecosystems. Proceedings, Brussels Symposium, October 1969. UNESCO, Paris.

Woodwell, G. M., and E. V. Pecan (ed.) 1973. Carbon and the biosphere. Proceedings of Symposium, Brookhaven National Lab., Upton, New York, May 1972. U.S. Atomic Energy Comm. 400 p. CONF-720510.

Young, H. E. (ed.) 1968. Symposium on primary productivity and mineral cycling in natural ecosystems. Proceedings, AAAS 13th Annual Meeting, December 1967. Univ. Maine Press, Orono. 245 p.

Young, H. E. 1971. Forest biomass studies. Life Sci. and Agri. Exp. Sta., Univ. Maine Press, Orono. 250 p.

PRODUCTIVITY OF GRASSLAND ECOSYSTEMS

R. T. COUPLAND

ABSTRACT

The contributions made during IBP to the techniques of studying structure and function of grassland ecosystems are discussed and some of the findings within the various trophic levels are indicated. The present state of international synthesis within the grassland biome is summarized.

INTRODUCTION

My objective in this paper will be to assess the present state of our knowledge concerning structure and function of herbaceous ecosystems ranging from dry subhumid to arid climates, particularly with respect to the contributions that IBP has made to their understanding. A wide variety of ecosystems is included within the national projects coordinated through the Grassland Working Group under Section PT. These are both tilled and natural, and range from wet meadows to semidesert; they are mostly dominated by perennial grasses. The Arid Lands Working Group also includes, within its framework, some studies of semidesert and semiarid grasslands and extends through the arid shrubby types of ecosystems. Since these working groups have been more successful in accumulating results on biological productivity in natural and seminatural systems my present evaluation of IBP impact will avoid the man-imposed or tilled ecosystems.

Primary production has been a basic parameter measured in all studies, while other ecosystem processes, such as secondary production and decomposition, have been more limited. Consequently, this paper emphasizes plant production but includes some discussion of the roles of consumers and decomposers in grassland. According to Bazilevich *et al.* (1970), "We now know that the living matter on earth is made up almost entirely of autotrophic, photosynthesizing plant organisms, which account for more than 99 percent of the total amount." I wonder how much this conclusion reflects insufficient knowledge of the abundance of residents in other trophic levels. The total ecosystem studies of some grasslands have revealed living biomass of decomposers and consumers in excess of 50 percent of that of the primary producers (Clark and Paul, 1970).

In some IBP projects synthesis is well advanced while in others it is just beginning. Regular progress reports have been prepared for some but not even all of these are widely distributed and readily available. Usually only summary information is available and for only one or two has final analysis and synthesis reached the publication stage. Attempts to compare data in international workshops began in 1970, but were ineffective until a concerted effort was made in the grassland-tundra workshop held in Fort Collins (August 14–26, 1972). At this workshop 39 scientists from 18 countries joined with 23 Americans in an intensive session (International Biological Program, 1972). Interproject comparisons of results were made using a data bank containing information from 25 sites in 16 countries. Modeling sessions were devoted to development of simulation models.

This present report is not a final assessment of achievements in studies of grasslands and arid lands under IBP. It is a progress report and cannot do justice to the vast array of information acquired during the IBP. The working groups are arranging for international analysis of data and its synthesis in published form; final evaluation of the impact of IBP must await these volumes. Meanwhile, in this preliminary account, I must draw disproportionately on information concerning those studies that are better known to me. Although many projects have converted their results to energy values, others have not. Accordingly, my discussion is based on biomass values; biomass is defined as organic material, whether living or dead (and weights are of oven-dry material including ash).

PRIMARY PRODUCTIVITY

In all grassland studies, biomass (harvest) methods have been used to estimate dry matter production, usually on an annual basis. In a few, measurements of CO_2 exchange under canopies have been included. Perhaps in only one or two instances have CO_2 gradients (in and above the canopy) been a basis of estimation. Biomass techniques quantify the rate of production by sorting harvested materials into categories at each of several times in the growing season. Changes in the amounts in the various categories are interpreted in terms of production. In some instances aboveground (and sometimes underground) parts are divided taxonomically into groups with distinctive seasonal growth patterns, while in others the principal division is into living and dead material.

More success has been experienced with these methods aboveground, since difficulties persist in finding a satisfactory basis for distinguishing living from dead structures in the soil. It is unfortunate that sampling techniques necessitate individual estimates of biomass of aboveground and belowground parts of the same organisms. The effect of translocation on these estimates is not usually recognized.

The detailed analyses of plant growth, that have been undertaken as a means of estimating the rates of energy flow in and out of producers, have given values much greater than those derived from traditional measurements which are designed to estimate only harvestable yields. This was only partially predictable. While the detailed analysis of such workers as Wiegert and Evans (1964) for aboveground and of Dahlman and Kucera (1965) for belowground had suggested the degree of magnitude by which harvestable yield underestimates net biological production, intensification of study under IBP has uncovered a number of factors not taken into account in even these previous studies. New concepts have been developed, especially in respect to events that take place during the intervals between shoot harvests and that bear on the changes of plant materials from one category to another. Recognition of the importance of these processes will have a major effect in designing future studies.

Shoot Production

The degree to which production estimates are increased by species separation of biomass into taxonomic categories presumably is related to differences in timing of their maximum standing crops. It is also expected that the longer and more environmentally variable the period (or periods) of growth, the greater is the effect of differential species activity on annual shoot production. Estimates of shoot production vary appreciably even if individual species contributions are considered, e.g., Wiegert and Evans (1964) had an increase of 26 percent by this method in Michigan, while this yielded only a 10 percent increase in Saskatchewan (Coupland, 1971).

Even with short intervals between harvests, the extent of losses from the green standing crop (and changes from dead standing crop to litter) often has been grossly underestimated or ignored in "ungrazed" sites and in control areas in "grazing" impact studies. The very apparent effect of domesticated herbivores on grassland has caused us to assume that in protected areas the effect of herbivory is minimal and need not be measured. However, in at least one study site, the biomass of invertebrates feeding on shoots in the "ungrazed" areas is at least equal (3 to 4 g/m^2) to cattle grazing on the adjacent range (Mook, 1969; Coupland, 1972a).

Another inadequately evaluated factor in relation to estimates of aboveground net primary production is the mode of growth of grass shoots. Emphasis on annual crops often has led us to assume that the growth pattern of perennial grass plants is associated with one crop of shoots per growing season, at least in areas where environmental conditions suitable for growth are of short duration. This presumably has justified the use of "hay" yields to compare productive capacities of different swards, even as pastures.

Misconceptions on length of growing season have resulted in some estimates of shoot production which do not allow for early and late growth. For example, pre-IBP clipping studies in Saskatchewan which were terminated in September (Lodge and Campbell, 1971) indicate that 95 percent of growth takes place by the time maximum standing crop of green shoots is reached. We have since found with studies extended into November that 30 to 40 percent of shoot production takes place after the period (Coupland, 1973). However, evidence from photosynthetic measurements in the same site suggest that in spring and fall gains in aboveground biomass are at the expense of underground reserves (Redmann, 1973).

The biomass data available at the 1972 IBP grassland tundra workshop (International Biological Program, 1972) give a measure of the range in productivity of the variety of herbaceous ecosystems represented. Maximum standing crops of green shoots range from 100 to 3,000 g/m^2 with a trend of increased production with higher mean annual temperatures between $-2°C$ and $26°C$. But this is not a direct measure of productivity. At the workshop we applied uniform techniques of estimating annual net aboveground primary production by accumulating all gains in standing crop (dead, as well as green). This procedure had a differential effect in the various environments and resulted in a greater increase of production values over standing crops in temperate climates (Table 1). It would appear that the values for shoot production will be higher relative to underground than those previously estimated for the above types of ecosystems by Bazilevich *et al.* (1970).

Underground Parts

Assessment of net primary productivity by biomass methods necessitates an estimate of the net movement of photosynthates into underground plant organs. The method involves washing of soil and plant material through screens, and requires recognition of both living and dead parts. The retrieved plant material inevitably contains a portion of dead roots, rhizomes and shoot bases which generally has not been separable from live materials. This problem has been addressed at two symposia on root systems during the IBP, in the USSR in 1968 (Academy of Sciences of the USSR, 1968) and East Germany in 1971, but it still remains the most limiting factor in obtaining reliable estimates of net primary productivity.

Dahlman and Kucera's (1965) method has probably been the one most commonly applied in herbaceous vegetation during IBP. It presumes that the difference between maximum and minimum underground plant biomass gives a reasonable (but minimal) estimate of plant material developed underground during the growing season. Measurements made by this means agree with those made by laborious sorting of apparently live and dead parts at the same site (Kucera *et al.*, 1967). This latter method requires a large number of very time-consuming samplings and many times the results do not justify the effort.

The method of Dahlman and Kucera does not account for losses due to degredation of plant materials during the measurement period. For example, in cool temperate grasslands it seems reasonable that the period of photosynthate translocation underground coincides with the most active period for soil microorganisms and detritus feeders; thus, the net movement underground can be highly underestimated, if losses to consumers are ignored. This possibility was suggested by results in a study of the tilled version of our Canadian grassland system (Coupland, 1972b). To correct for the dead organic materials present at the time of seeding of the annual wheat crop, we measured the mass of underground organic materials immediately after seeding and near maturity (assuming that an increase could be used as an estimate of underground biomass additions during the 80-day period). The results showed lower values at maturity than at seeding suggesting greater organic matter losses than gains.

TABLE 1 Annual Shoot (Aboveground) Production Values for the Major Types of Herbaceous Ecosystems. Data Taken from the Report of the Grassland–Tundra Workshop (International Biological Program, 1972)

Type of Ecosystem	Annual Production (g/m^2)
Tropical and subtropical grasslands and savannas	600–4,600
Temperate steppe	600–1,300
Temperate meadow	700–3,400
Subarctic meadow	300–500

Biomass methods fail to account satisfactorily for losses due to underground plant respiration. During the IBP, there has been considerable interest in using "soil respiration" to estimate production, particularly in ecosystems that are sufficiently stable, i.e., where inputs approximate outputs. In the Canadian study, CO_2 flow rates from the soil system suggest that biomass methods may be underestimating the amount of carbohydrate translocated underground by as much as 50 percent (MacDonald, 1973). Attempts are being made to partition underground respiration between roots and decomposers. Several studies have used methyl bromide to suppress soil invertebrates so as to estimate the activity of microorganisms.* Coleman (1970) has separated microbe and root respiration in the laboratory by comparing the activity of separated fractions (roots, litter and soil) with that of intact soil cores. Results of these studies enable only a very generalized apportionment of underground organic matter pools and respiratory activity.

Tracers have been used to follow the fluxes of carbon after fixation in photosynthesis. Dahlman and Kucera (1968) devised a means of labeling a small plot of grassland so that quantitative sampling of shoots and underground organs is possible over a period of several years. By this means net rates of carbon turnover have been estimated. In the Canadian project this approach has been modified to provide short-term translocation rates of ^{14}C from shoot to root to soil atmosphere. Interestingly, this approach permits monitoring of the ^{14}C content of soil atmosphere at various depths, and such measurements suggest a much more rapid rate of carbon release from the plant than expected—10 to 15 percent in the first three weeks after fixation (Warembourg and Paul, 1973).

Intersite comparisons of underground biomass at the 1972 IBP grassland–tundra workshop indicate an increase in the ratio of underground parts to green standing shoots along the latitudinal-temperature gradient from the subarctic to tropics; the ratio ranges from 2 to 15 or more. This root-shoot ratio declined over a similar range along a gradient of increasing precipitation (from 100 to 2,300 mm per annum).

* Personal communication with M. Numata of the Japanese IBP grassland study.

SECONDARY PRODUCTIVITY

Consumption by small herbivores has been assumed to be negligible in many previous studies of grassland productivity. Where biomass methods are used to estimate primary productivity, it is customary to exclude large herbivores from control ("ungrazed") areas which then are compared with "grazed" areas supporting domesticated herbivores. However, while the activities of large herbivores have been intensively studied, the biomass consumed by small herbivores (grasshoppers, rodents, birds) may be a significant factor. Differences between "grazed" and "ungrazed" areas in small herbivore populations may confound measurements of the effect of the managed herbivores. For example, a preliminary assessment in the Canadian study indicates that invertebrates may ingest and drop to litter as much as 80 percent of the amount consumed by cattle (Coupland, 1972). Activities of small subterranean herbivores (especially invertebrates) have been given even less attention, although their biomass may be five to six times that of the cattle supported by the system.

Estimates of biomass and energy flow in consumer populations are not as refined as those for producers. One major difficulty is the large number of species in some groups (e.g., invertebrates). Another is the frequent presence of omnivory, which complicates distinctions between primary and secondary consumption. Population studies have, of necessity, been of groups (e.g., orders of insects), to which energetic values have been applied on the basis of laboratory energy budgets of important species. No accurate estimate has been made of the reliability of this approach in overall assessment of the activities of the consumer components of the ecosystem. Progress appears greater for aboveground than for underground invertebrates. The subterranean environment is capable of supporting a host of very small invertebrates whose interrelationships are only now beginning to be revealed as a result of intensive investigations in IBP, e.g., nematodes in the Canadian Matador site comprise 60 percent of the underground invertebrate biomass (which totals 4 to 6 g/m^2)* while soil zoologists expected arthropods to predominate (Zacharuk and Burrage, 1968).

DECOMPOSITION

The role of organic matter in the ecosystem is of particular concern in grasslands because of the generally high content of soil humus, particularly in temperate regions. The quantity of dead organic matter in a grassland system at equilibrium under steady state conditions depends upon complex interactions between rates of primary production, respiration, consumption and decomposition. We have much to learn concerning the relative activities of consumers and decomposers in degrading organic material through the various stages to soil humus, as well as the functional relationships of organic matter within the system. Perhaps the IBP stimulated efforts in total ecosystem modeling will reveal to what extent organic content is important to the sustenance of a system and its influence as a storage location for plant nutrients.

The balance between the rates of production and degradation of plant materials above the soil surface seems to be more critical in grasslands in subhumid climates (e.g., the forest steppe of southern Russia [Kursk] and the portion of the True Prairie adjacent to the eastern deciduous forests of the USA) than in dry subhumid and semiarid ones (such as the mixed prairie region of North America and Fescue-Stipa steppes of the Ukraine and northern Kazakstan). Under protection from large herbivores, rapid accumulation of surface litter occurs in subhumid climates to the extent that repeated fire is necessary to maintain the existing dominant plant species. In the drier regions, however, decomposer organisms are able to deal rapidly with even wide annual variations in shoot production; consequently, less than one year of shoot production is accumulated as coarse litter on the surface.

Under IBP we have learned much about the population dynamics and biomass of different groups of soil microbes in various grassland systems. Their abundance is staggering. In temperate grasslands with large herbivores excluded the biomass of soil microbes can be as much as 100 times that of the carrying capacity of domesticated animals (Clark and Paul, 1970; Coupland, 1972a). Measuring microorganism metabolism in the field is very difficult. It is not immediately obvious how the available energy supply supports microorganisms at levels of metabolic activity. While rates of CO_2 flux from the soil surface are related to organic processes within the soil, there are considerable difficulties in partitioning this flux between root respiration, consumer metabolism and microbial decomposition. There are indications that much microbial activity takes place within the rhizosphere, where plant substances are apparently exuded in quantities greater than generally supposed. Techniques are being developed in IBP projects, e.g., (Doxtader, 1970) in rapid chemical determination that may provide some answers regarding decomposer substrates. There are indications that much of the microbial activity occurs in spurts associated with certain metabolic events optimizing the cycles of nutrient elements in the ecosystem (Clark and Paul, 1970).

NUTRIENT SUPPLY

It seems clear that the two major limiting factors in grassland production are climate and nutrient supply. The more favorable the climate is for growth of grasses, the more critical is the supply of nutrients. The natural grasslands

* Personal communication from J. R. Willard.

of the world occur in climates where low amounts and erratic distribution of precipitation limit the invasion of trees from more humid and subhumid regions. Under these conditions, one assumes that the amount of soil moisture is a more important factor in growth than is soil nutrients. It follows that the drier the area, the less likely it is that the supply of soil nutrients will be limiting. These are general concepts with which we are all familiar; international comparisons between projects will do much to support or modify these concepts. There are some grassland workers who are more concerned with nutrient supply than with moisture supply, probably as a result of nutrient deficiencies in the systems with which they are working.

Nitrogen and phosphorous have received most attention, probably because it is inevitable that fertilizing practices for arable land would be used to test responses of natural ecosystems. Additional nitrogen is usually effective in stimulating grass growth, and phosphorous indirectly increases the nitrogen supply by stimulating legumes. The extent to which nitrogen can be beneficially applied (in terms of production) depends on the amount of soil moisture, the results being more erratic in arid and semiarid than in dry, subhumid regions. An important output of IBP ecosystem analyses is expected to be an evaluation of the degree to which the nitrogen economy can be altered without sacrificing, unduly, the stability of the system. It seems that increased nitrogen supply favors plant dominants characteristic of stages of succession preceding climax and favors the entrance of annual species into the natural perennial vegetation cover. In areas where litter accumulates on the soil surface, there is concern that the amount of nutrients taken temporarily out of circulation will limit the productivity of the system. This implies that it might be desirable to speed up the rate of cycling of these nutrients. Analyses using models should provide a better comparative measure of alternative ways to achieve this, i.e., by increasing the rate of use by consumers or by stimulating decomposition through fertilization.

Under IBP, nutrient budgets have received considerable attention. We are beginning to understand better the relationships between available and nonavailable forms of various nutrients, the magnitudes of nutrient "sinks", and the nutrient release rates from these sinks. In many instances, it seems that the rate of exchange from unavailable to available forms affects the vigor of plant growth. Perhaps the major contributions in this area have been consideration of nutrient cycles under a wide variety of grassland conditions. Studies of nitrogen fixation have apparently provided much new information on inputs of nitrogen into grassland ecosystems and have stimulated the search for other sources of origin, fixation and internal conversion. For example, the organisms responsible for asymbiotic nitrogen fixation have been identified and the intensity of their activity in natural grasslands has been found to add 1 to 3 kilograms of nitrogen per hectare per year (Paul *et al.*, 1971). Rainfall supplies approximately twice the amount provided by asymbiotic microorganisms. These inputs are rather insignificant in comparison with the nitrogen fixed symbiotically where legumes are abundant in grasslands. Grassland ecosystems with a low natural input of nitrogen, but with a high component of stored nitrogen in organic matter, can be increased in productivity by processes that increase the rate of breakdown of organic matter. We must be careful not to exploit organic reserves in these ecosystems without taking into consideration the long-term consequences of these actions on site fertility. Are we sure that artificial applications of nitrogen later will compensate fully for the loss of organic matter?

CONCLUSION

IBP studies in herbaceous ecosystems from subhumid to arid climates are revealing complex trophic structure and a diversity of factors that determine capacity for biological productivity. Many new principles will evolve during the analysis and synthesis of the vast amounts of data from projects, and comparison of parameters across regional and geographical scales. Data sharing has been accepted by investigators and makes possible the future development of simulation models of wide application to resource management. A very sound basis has been developed for planning and executing of grassland studies under UNESCO's Man and the Biosphere program that will provide a background for management of grassland at a sustainable level of production.

REFERENCES

Academy of Sciences of the USSR. 1968. Methods of studies of productivity of root systems and rhizosphere organisms. Proceedings of a symposium held at Moscow, Leningrad and Dushambe, August 28–September 12, 1968.

Bazilevich, N. I., L. E. Rodin, and N. N. Rozov. 1970. Geographical aspects of biological productivity. Papers of the Fifth Congress of the Biological Society, Leningrad, USSR. 28 p. (Translated in Soviet Geography: Review in Translations, May 1971, p. 219–317).

Clark, F. E., and E. A. Paul. 1970. The microflora of grassland. Advances in Agronomy 22:375–435.

Coleman, D. C. 1970. A compartmental analysis of total soil respiration, p. 126–128. *In* R. T. Coupland and G. M. Van Dyne (ed.) Grassland Ecosystems: Review of Research. Range Sci. Dep. Sci. Ser. No. 7. Colorado State University. 208 p.

Coupland, R. T. 1971. Biomass measurements in native grassland, p. 19–33. *In* Fourth Annual Report of the Matador Project. University of Saskatchewan, Saskatoon, Canada.

Coupland, R. T. 1972a. Operational phase (1967–1972): A summary of progress. Technical Report No. 1, Matador Project, University of Saskatchewan, Saskatoon, Canada.

Coupland, R. T. 1972b. Biomass measurements in wheatland, p. 25–26. *In* Fifth Annual Report of the Matador Project. University of Saskatchewan, Saskatoon, Canada.

Coupland, R. T. 1973. Plant biomass production, p. 93–97. *In* Measurement and Modelling of Photosynthesis in Relation to Productivity. Proceedings of the CCIBP/PP-Ps Workshop at the University of Guelph, December 8 to 10, 1972.

Dahlman, R. C., and C. L. Kucera. 1965. Root productivity and turnover in native prairie. Ecology 46:84–89.

Dahlman, R. C., and C. L. Kucera. 1968. Tagging native grassland vegetation with carbon-14. Ecology 49:1199–1203.

Doxtader, K. G. 1970. Biomass determination of soil microorganisms, p. 107–108. *In* R. T. Coupland and G. M. Van Dyne (ed.) Grassland Ecosystems: Reviews of Research. Range Sci. Dept. Sci. Ser. No. 7. Colorado State University. 208 p.

International Biological Program, PT Section, Grassland and Tundra Working Groups. 1972. Report of the modelling and synthesis workshop held at the Natural Resource Ecology Laboratory, Colorado State University, USA, August 14 to 26, 1972.

Kucera, C. L., R. C. Dahlman, and M. R. Koelling. 1967. Total net productivity and turnover on an energy basis for tallgrass prairie. Ecology 48:536–541.

Lodge, R. W., and J. B. Campbell. 1971. Management of the western range. Canada Dept. of Agriculture, Publication 1425.

MacDonald, K. B. 1973. Modelling soil respiration, p. 205–211. *In* Measurement and Modelling of Photosynthesis in Relation to Productivity. Proceedings of the CCIBP/PP-Ps Workshop at the University of Guelph, December 8 to 10, 1972.

Mook, L. J. 1969. Surface invertebrates, p. 45–52. *In* Second Annual Report of the Matador Project. University of Saskatchewan, Saskatoon, Canada.

Paul, E. A., R. J. K. Myers, and W. A. Rice. 1971. Nitrogen fixation in grassland and associated cultivated ecosystems, p. 495–507. *In* Plant and Soil, Special Volume on Biological Nitrogen Fixation in Natural and Agricultural Habitats.

Redmann, R. E. 1973. Carbon dioxide assimilation model, p. 187–193. *In* Measurement and Modelling of Photosynthesis in Relation to Productivity. Proceedings of the CCIBP/PP-Ps Workshop at the University of Guelph, December 8 to 10, 1972.

Warmebourg, R. F., and E. A. Paul. 1973. The use of $C^{14}O_2$ canopy techniques for measuring carbon transfer through the plant-soil system. Plant and Soil 38(2):331–345.

Wiegert, R. C., and F. C. Evans. 1964. Primary production and the disappearance of dead vegetation on an old field in southeastern Michigan. Ecology 45:49–63.

Zacharuk, R. Y., and R. H. Burrage. 1968. Subsurface invertebrates, p. 38–41. *In* First Annual Report of the Matador Project, University of Saskatchewan, Saskatoon, Canada.

THE IMPORTANCE OF DIFFERENT ENERGY SOURCES IN FRESHWATER ECOSYSTEMS

K. W. CUMMINS

INTRODUCTION

Despite a shift resulting from the activities of industrialized man, photosynthesis and respiration (Machta, 1971) generally balance in the biosphere. Some ecosystems, or compartments of ecosystems, depending upon the somewhat arbitrary conceptualization of system dimensions, are known to be autotrophic, i.e., producing reduced carbon compounds in excess of the amount that is respired. It follows that other compartments or ecosystems are heterotrophic, with oxygen consumption exceeding photosynthetic oxygen production, i.e., soil communities, aphotic aquatic subsystems and woodland streams.

Autotrophic communities are characterized by a ratio of photosynthesis to respiration greater than 1 ($P:R > 1$) (Odum, 1956) with the excess production exported to heterotrophic compartments or systems. For heterotrophic communities P : R is less than 1 ($P:R < 1$). Clearly systems in which $P:R = 1$ must be combinations of autotrophic and heterotrophic compartments. Fisher (1971) (Fisher and Likens, 1972) suggested a modification of the P : R expression such that:

$$I + P = R + E + \Delta S$$

where I = input; E = export; ΔS = annual change in organic-matter standing crop. Of course, if $I = E$ and $\Delta S = O$ over the annual cycle, as may often be the case in streams, the P : R system is valid as previously used.

GENERALIZATIONS ABOUT AQUATIC ECOSYSTEM ENERGETICS

Recently it has become clear that striking analogies are to be found between terrestrial soil communities, lentic benthic communities (hydrosoils) and running water systems in general. Excluding large rivers, particularly when extensive impoundment has occurred along the drainage, running waters resemble terrestrial soil communities, since the above-sediment portion supports little primary production constituting essentially a transport system. Analogies between lentic (standing water) and lotic (running water) ecosystems are clear only when certain system compartments are considered. Because lentic environments have previously received the majority of attention by freshwater scientists, running waters have been emphasized in the present discussion.

Lentic Systems

Energy inputs, i.e., light, and photosynthetically initiated organic matter from the terrestrial surroundings, and functionally compartmentalized biomass, are shown in highly simplified form in Figure 1. Both the terrestrial supply system, and the littoral and planktonic compartments of standing waters produce organic matter in excess, that is, $P:R > 1$. This excess of reduced carbon compounds from the landscape, from phytoplankton and from vascular

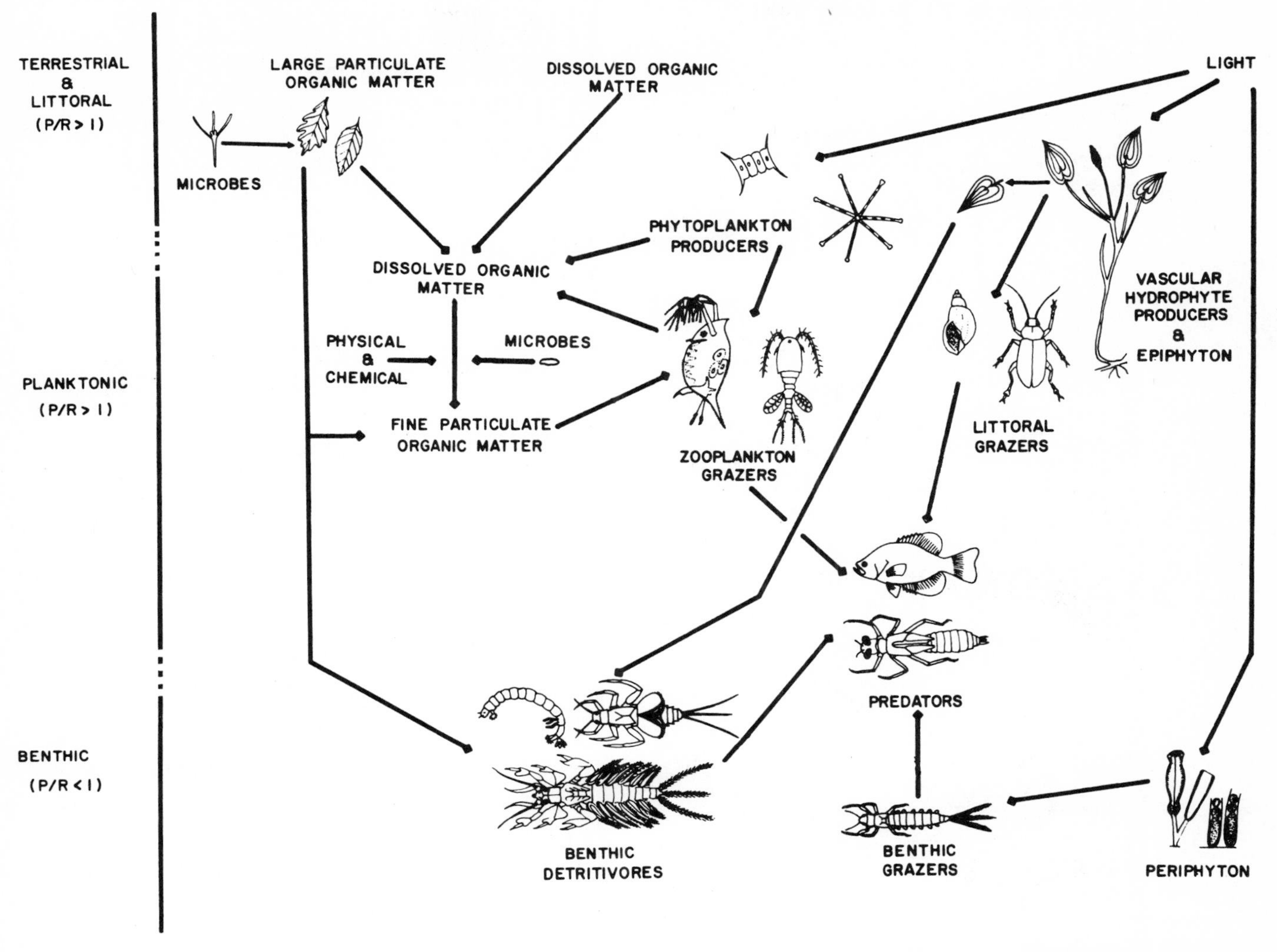

FIGURE 1 A simplified representation of lentic ecosystem functional compartmentalization. Emphasis is on the processing of energy inputs—terrestrial organic matter and photosynthesis. Photosynthesis to respiration ratios are shown at the left.

hydrophytes (plus epiphytes) "feeds" the benthic community—a compartment characterized by P : R < 1.

As in aboveground terrestrial communities, the majority of the biomass resulting from photosynthesis (planktonic algae and vascular hydrophytes and associated epiphyton) is not processed by grazing animals. Upwards of 80 to 90 percent of lentic primary production enters the benthic system without passing through grazers (for a current treatment and review of detrital processes in lakes see Saunders, 1972a). Many, if not all, lentic grazers are actually detritivore-herbivores with food habits restricted by particle size and food texture rather than the presence or absence of functional chlorophyll (Cummins, 1973). Planktonic grazers are at least as dependent upon particulate detritus (and the associated microflora) as upon living algae.

Lotic Systems

The general pattern by which organic matter is made available to, and processed by, the communities of stream ecosystems of temperate-zone woodlands is now fairly clear (Cummins, 1972a; Fisher and Likens, 1972) although many critical details (particularly rates) remain to be delineated. As shown in highly simplified form in Figure 2, the energy supply for streams, similar to lakes, can be partitioned into two general components: the input of particulate and dissolved organic matter of terrestrial origin, and in-stream photosynthetic carbon fixation. In running waters the former is generally quantitatively much greater.

The pool of large particulate organic matter (approximately > 1 mm) is maintained primarily by the terrestrial

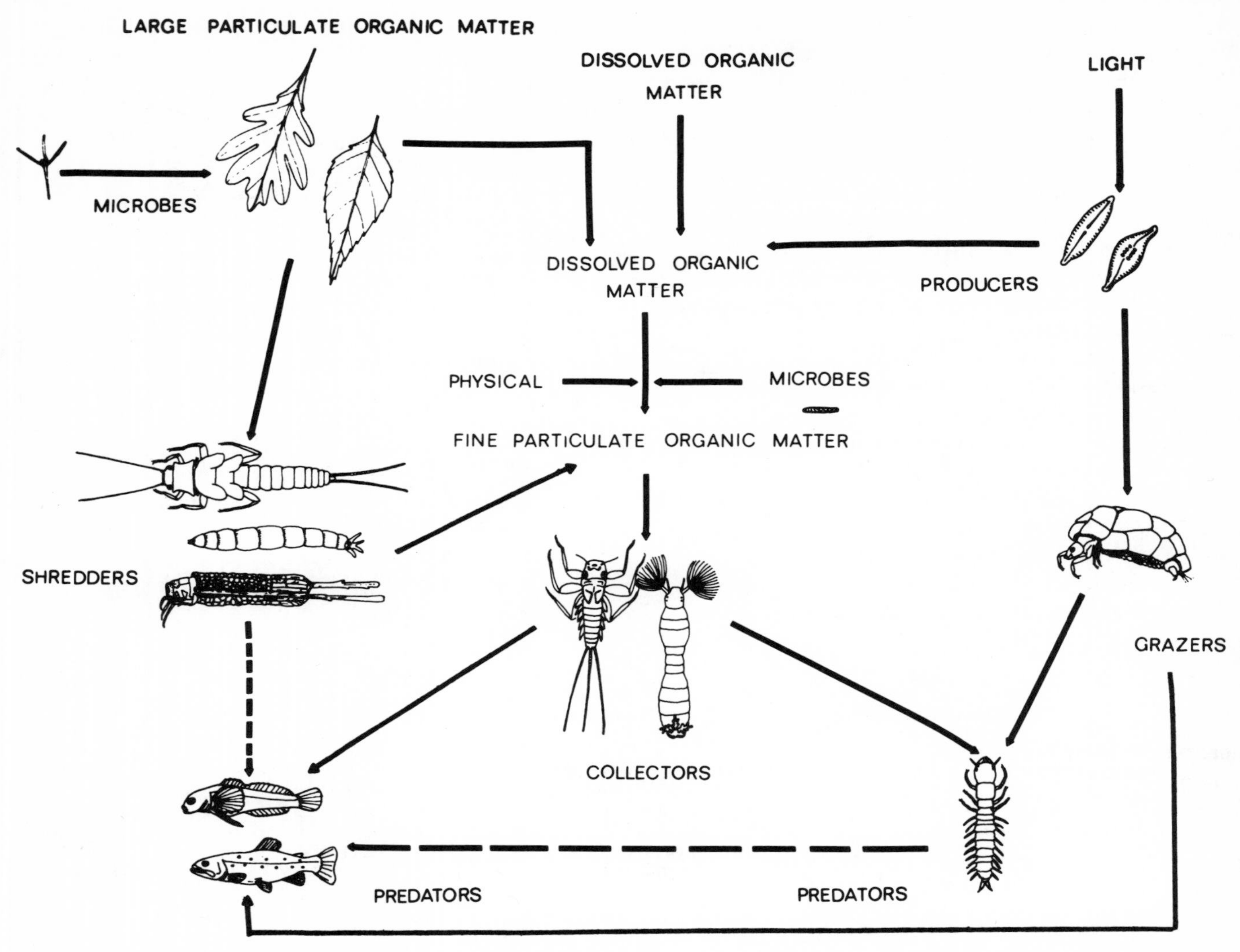

FIGURE 2 A simplified representation of the functional compartmentalization of a lotic ecosystem. Emphasis is on the processing of energy inputs—terrestrial organic matter and phytosynthesis. After Cummins, 1973.

input of vascular plant tissue, e.g., leaves, bud scales, flowers, fruits, bark, twigs, branches and large material such as logs which are subject to very slow processing and, therefore, have long residence times. An important feature of streams is their function as concentrators of this material, since they act as "sticky traps" (H. B. N. Hynes, University of Waterloo, personal communication). As in the lentic systems, if vascular hydrophytes are present (e.g., watercress) they tend to enter the processing cycle primarily as if they were of terrestrial origin rather than through the grazing activities of stream herbivores. The pool of fine particulate matter (approximately < 1 mm) is derived from in-stream processing of large particles (e.g., fragments broken loose through animal activities or physical abrasion by sediments in transport, and animal feces), from algal cells sloughed from surfaces, from the microorganisms always associated with fine organic particles, and from the physical-chemical flocculation of dissolved organics. Other fine particles of terrestrial origin, such as feces of insects feeding in the forest canopy, are also part of this pool and may be quantitatively very important during the summer. The dissolved organic pool is replenished from the leaching of terrestrial particulate organics, either before entry (runoff) or after the particulates are in the stream, and from algal (plus microbial and animal) excretions. Leaching of leaf litter is rapid (essentially complete 24 hours after the litter is wetted) and quantitatively significant (5 to 40 percent of the dry weight of the litter) (Cummins *et al.*, 1972; Petersen and Cummins, in press).

Leached vascular-plant tissue represents an organic substrate composed of "resistant" carbon compounds (about half cellulose) and little nitrogen. Although propagules of terrestrial fungi and bacteria are present on the litter before it enters the water, rapid colonization and growth by aquatic forms (e.g., aquatic hyphomycetes and gram-negative bacterial rods) dominate in the stream. The metabolism of these microorganisms is characterized by cellulose degradation and nitrogen uptake from the water. The presence of the associated microflora converts the litter to a suitable food source for animal metabolism (especially through increase in protein). Labile organics in the leachate are undoubtedly rapidly taken up by microorganisms associated with the sediments and organic particles, as well as those in transport (Cummins *et al.*, 1972).

Soon after, and probably in response to, microbial infestation of the large particulate organic matter such as leaf litter, large particle detritivores ("shredders") move into accumulations of such material and begin feeding. A variety of experiments have shown that stream detritivores selectively feed upon the detritus maximally colonized with microorganisms, particularly fungi (Triska, 1970; Kostalos, 1971; Mackay, 1972). The general effect of shredder feeding is to reduce the average particle size, thereby providing additional substrate surface for microbial (especially bacterial) colonization and metabolism, and food for fine particle detritivores ("collectors") (Cummins, 1973; Cummins *et al.*, 1973).

Primary production, usually minimal in woodland streams, is dominated by diatoms which are shade-adapted and may function as facultative heterotrophs. Correspondingly, the grazer species feeding on the periphyton are characteristically few, and often dependent on the intake of significant amounts of fine particle detritus which accumulates in the interstices of the periphyton cover (e.g., Cummins, 1973).

Populations of collectors and grazers are reduced by stream predators (both invertebrate and vertebrate) throughout the growing portion of their generations, while the shredders are subject to significant predator mortality in the early part of the growth period.

The bodies of animals, dying otherwise than by predation, animal exuviae and feces quickly enter the fine particle pool. The conversion of reduced carbon compounds to CO_2 is efficient, being at least 80% of particulate and 50% of dissolved materials on an annual basis and, strikingly, a significant portion is accomplished at temperatures below 10°C. This is in sharp contrast to terrestrial and, usually,

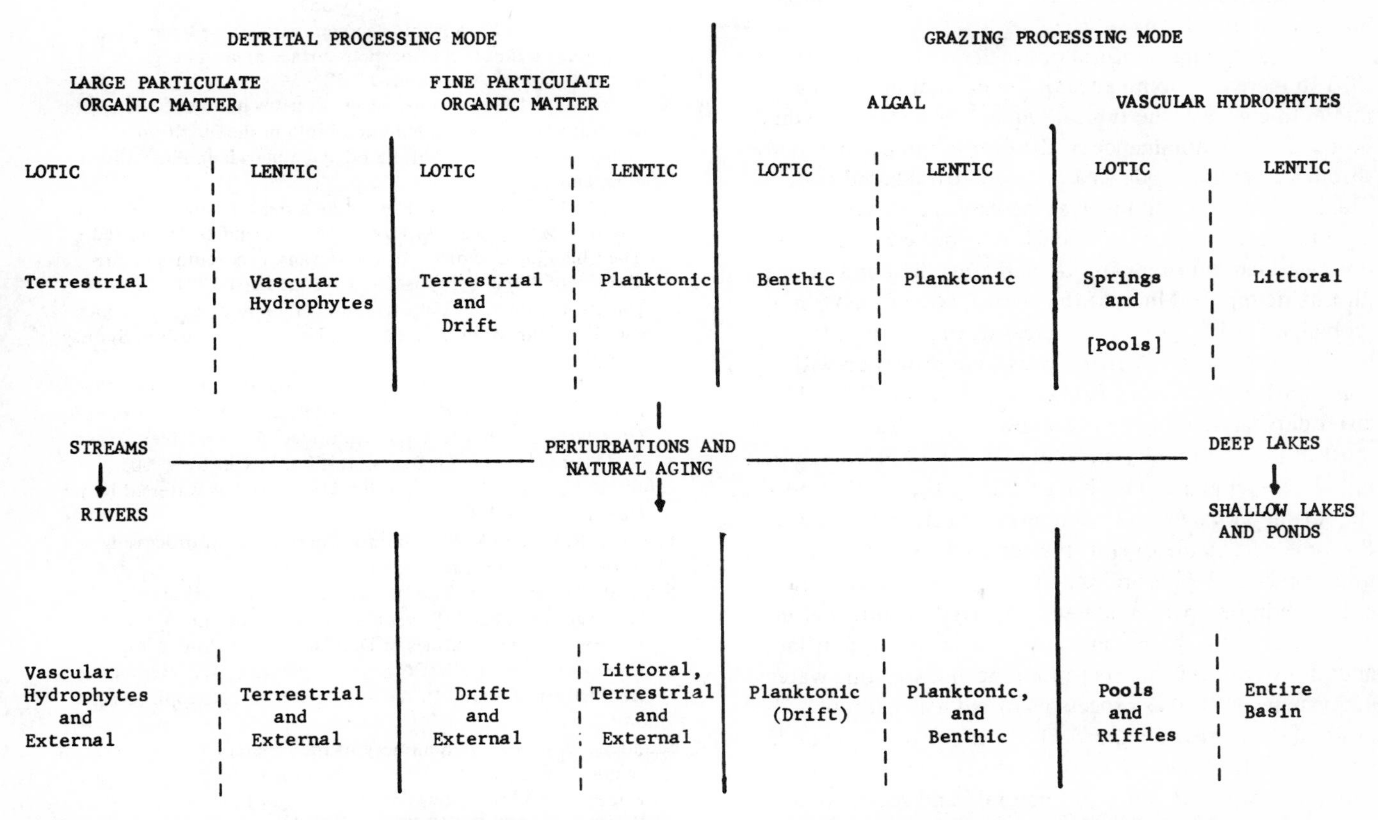

FIGURE 3 A simplified summary of energy inputs to freshwater systems. The columns are arranged according to consumer energy processing modes (and particle size) and lotic or lentic systems. The horizontal organization of the figure is directed at the primary source for the system or the location within the system of the energy supply. A transition is indicated from the upper half of the figure, "young" and/or insignificantly perturbated systems, to the lower, "mature" and/or perturbated systems.

standing water systems of the temperate zone (Cummins *et al.*, 1972).

COMPARISONS OF SYSTEM ENERGETICS

Lotic and lentic systems have been compared, according to the processing of energy, inputs, in Figure 3. Detrital and grazing processing functions are compared, with lotic systems dominated by the former and lentic systems dominated by the latter. The primary origin or localization of the energy source of a lotic or lentic system operating predominantly in either a detrital or grazing processing mode is given along the horizontal axis of Figure 3. Changes in the dominance of inputs and processing mode relative to size of stream or river and depth of lake or pond are compared from the upper to lower portion of the figure. As indicated, these differences may result from natural "aging" (that is "young" head water streams and deep and/or clear lakes to "old" lower drainage rivers and shallow lakes and ponds) or accelerated "aging", i.e., man-engendered perturbations. "External" energy sources (lower half of Figure 3) refer to inputs of reduced carbon compounds resulting from human-related accumulations (urban, agricultural and industrial) as opposed to normal inputs from the drainage.

Through "aging", natural or accelerated, lentic systems shift to increased dominance of the detrital processing mode. In contrast, the typical sequence for streams would be a shift from dominance of the detrital processing mode through a grazing mode to a second detrital-based system. The intermediate step involves increases in filamentous algal forms and, often, in vascular hydrophytes. This stage is accelerated in proportion to increasing light and nitrate-phosphate inputs. Much of the animal, and probably microbial, diversity is lost and increasing organic inputs, leading to the second detrital phase, are processed with comparative inefficiency (Cummins, 1972a, 1972b). The latter detrital-based system operates in the absence of most of the community structure important in the natural initial detrital phase. This is particularly true of that portion of the biota which in nonperturbed streams utilizes the large particulate organic matter, such as leaf litter, since human-related inputs are usually characterized by effluents high in fine particulate and dissolved organic matter.

It seems clear that quantitative assessment of detrital and photosynthetic inputs to running and standing water ecosystems, viewed as processors of reduced carbon molecules, is a fruitful approach—both for the basic understanding of freshwater ecosystem structure and function, and as a sensitive monitor of "natural" and accelerated system changes. It should be particularly fruitful to contrast detritus processing modes in lotic and lentic systems (Figure 3), especially community respiration associated with detritus particles. The relationships between respiration, particle size and the qualitative nature (e.g., cellulose, lignin, chitin, etc., content) of detritus, when viewed in reference to community photosynthesis, should provide new and useful insights in P : R comparisons (Hargrave, 1972; Petersen and Cummins, in press).

REFERENCES

Cummins, K. W. 1972a. Predicting variations in energy flow through a semi-controlled lotic ecosystem. Mich. State Univ. Inst. Water Res. Tech. Rep. 19:1–21.

Cummins, K. W. 1972b. What is a river?—zoological description, p. 33–52. *In* R. T. Oglesby, C. A. Carlson, and J. A. McCann (ed.) River ecology and man. Academic Press, New York. 465 p.

Cummins, K. W. 1973. Trophic relations of aquatic insects. Ann. Rev. Ent. 18:183–206.

Cummins, K. W., J. J. Klug, R. G. Wetzel, R. C. Petersen, K. F. Suberkropp, B. A. Manny, J. C. Wuycheck, and F. O. Howard. 1972. Organic enrichment with leaf leachate in experimental lotic ecosystems. BioSci. 22(12):719–722.

Cummins, K. W., R. C. Petersen, F. O. Howard, J. C. Wuycheck, and V. I. Holt. 1973. The utilization of leaf litter by stream detritivores. Ecol. 54:336–345.

Fisher, S. G. 1971. Annual energy budget of a small forest stream ecosystem, Bear Brook, West Thornton, New Hampshire. Ph.D. dissertation, Dartmouth College, 97 p. (Unpubl.)

Fisher, S. G., and G. E. Likens. 1972. Stream ecosystem: organic energy budget. BioSci. 22(1):33–35.

Hargrave, B. T. 1972. Aerobic decomposition of sediment and detritus as a function of particle surface area and organic content. Limnol. Oceanogr. 17:583–596.

Kostalos, M. S. 1971. A study of the detritus pathway: the role of detritus and the associated microbiota in the nutrition of *Gamarus minus* Say (Amphipoda:Gammaridae). Ph.D. dissertation, Univ. of Pittsburgh. 152 p.

Machta, L. 1971. The role of the oceans and biosphere in carbon dioxide cycle, p. 121–145. *In* D. Dyrssen and D. Jagner (ed.) The Changing Chemistry of the Oceans. Proceedings of the Twentieth Nobel Symposium, 16–20 August 1971, Aspenäsgården, Lerum and Chalmers Univ. of Technology, Göteborg, Sweden. John Wiley & Sons, Inc., New York, London, Sydney. 365 p.

Mackay, R. J. 1972. The life cycle and ecology of *Pycnopsyche gentilis* (McLahlan), *P. luculenta* (Betten), and *P. scabripennis* (Rambur), (Trichoptera:Limnephilidae) in West Creek, Mont St. Hilaire, Quebec. Ph.D. dissertation, McGill Univ. 103 p.

Odum, H. T. 1956. Primary production in flowing waters. Limnol. Oceanogr. 1:102–117.

Petersen, R. C., and K. W. Cummins. In press. Leaf processing in a woodland stream. Freshwater Biol. 4.

Saunders, G. W. 1972a. The transformation of artificial detritus in lake water, p. 262–288. *In* U. Melchiorri-Santolini and J. W. Hopton (ed.) Proceedings of Detritus and Its Role in Aquatic Ecosystems. IBP-UNESCO Symp. Pallanza, 1972. Memorie Dell' Istituto Italiano Di Idrobiologia. Vol. 29 Suppl. 1972. 540 p.

Saunders, G. W. 1972b. Summary of the general conclusions of the symposium, p. 533–540. *In* U. Melchiorri-Santolini and J. W. Hopton (ed.) Proceedings of Detritus and Its Role in Aquatic Ecosystems. IBP-UNESCO Symp. Pallanza, 1972. Memorie Dell' Istituto Italiano Di Idrobiologia. Vol. 29 Suppl. 1972. 540 p.

Triska, F. J. 1970. Seasonal distribution of aquatic hyphomycetes in relation to the disappearance of leaf litter from a woodland stream. Ph.D. dissertation, Univ. of Pittsburgh. 162 p.

TERRESTRIAL DECOMPOSITION

DENNIS PARKINSON

ABSTRACT

The importance of the decomposition of organic matter in the maintenance of soil fertility has been long known. However, since the concept of organic matter decomposition as merely the chemical oxidation of organic compounds gave way to a more dynamic biological interpretation, accurate, coordinated data on the interactions of soil microflora and fauna in this complex process have, for various reasons, been long delayed. This present contribution does not aim to review comprehensively the subject of decomposition of organic matter in terrestrial environments, rather it aims to point out the problems encountered in such studies.

INTRODUCTION

Over several decades agricultural and forestry studies have yielded much data on organic matter decomposition under specific cultural conditions and from which schema, such as that given by Paul (1970), have been built up. A variety of organic materials serve as substrates for microbial growth and energy production (Paul, 1970, has reviewed the field of plant components and soil organic matter). The metabolic versatility, with respect to ability to degrade a wide range of organic compounds, of the bacteria and fungi led to the almost tacit assumption that soil contained microorganisms capable of adapting to the addition of any compound; however, Alexander (1965) in reviewing this concept points out that various compounds have chemical properties which render them difficult to degrade, thus the concept of microbial infallibility must be modified. While much attention, in this regard, has centered on the degradation of herbicides, etc., it should be remembered that the humus complex is a natural example of this type of substance.

Despite the work on earthworms, and on animal activities in mull and mor forest soils, little attention was given to the role of soil fauna in organic matter degradation in applied ecological investigations. However, the inception of integrated ecological studies under the IBP/PT has allowed the development of studies on organic matter decomposition and of detailed studies on the soil organisms involved (and the interactions between these organisms). These IBP/PT studies are demonstrating that the decomposer cycle, with accompanying nutrient cycling is one of the great driving forces of ecosystems—of greater importance than many ecologists hitherto appear to have realized.

MAIN FACTORS AFFECTING DECOMPOSITION

As a result of many of the IBP/PT studies, various factors have been defined as important in affecting either the potential or the observed rates of decomposition: chemical qualities of the organic matter, temperature, moisture content, available decomposer organisms, being the factors which have received most consideration.

Chemical Quality

Plant and animal debris provide complex mixtures of organic materials which serve as substrates for biological activity and therefore decomposition. For many studies, detailed chemical analyses of such materials are lacking; therefore detailed studies on the rates of disappearance of individual compounds have not been followed. One exception to this generalization is the case of cellulose decomposition, detailed studies on which are being followed in all

IBP biomes. Factors such as high polyphenol and tannin content, high content of soluble carbohydrates, high lignin content, etc., are known to affect the time course and the by-products of decomposition. Local concentrations of inorganic nutrients and trace elements alter the activities of decomposer organisms (e.g., the importance of nitrogen content in affecting rates of cellulose degradation).

Temperature

This is a well-known controlling factor for biological processes and, therefore, acts as a control of the rate of organic matter decomposition. For organic matter in the litter layer (at the soil surface) in open situations, marked diurnal and annual fluctuations in temperature may occur; these fluctuations can occur independent of air and soil temperatures (because of radiation fluxes, etc.). For organic matter incorporated in the mineral soil (e.g., in the rooting zone) such short-term temperature fluctuations are not seen; however, freezing–thawing cycles in soil systems are known to be important in the release of nutrients from soil organic matter.

Moisture

Moisture content has been shown to be of great importance in controlling biological activity in litter and soil. The effects of extreme moisture conditions on the rates and products of decomposition are well known (thus waterlogging causes changes in pH, redox potential and oxygen concentration—all of which affect the pattern of decomposition). Desiccation and remoistening of soil affects nutrient release from soil organic matter.

In IBP/PT projects a good deal of detailed data on the controlling effects of temperature and moisture are being accumulated from field studies. However, it is necessary that these studies should be accompanied by experiments on decomposition rates under controlled environment conditions in order that the relative roles of temperature and moisture availability as controlling factors in decomposition can be more precisely defined. This type of study is of particular importance in studies of decomposer activity in more rigorous climatic zones, e.g., tundra, where human activity and colonization is increasing, with a consequent addition of organic materials (natural and "bizarre") to the ecosystem. Obviously a detailed study on the dynamics of organic matter decomposition is necessary.

Available Organisms

Prior to the IBP/PT projects there were very few comprehensive studies on soil organisms in natural communities (notable exception being the New Zealand Soil Bureau work on tussock grassland soils). Now, for most terrestrial biomes, detailed and integrated soil biological data are being obtained. Thus, information on species present (and their frequency of occurrence), biomass and (where possible) the rates of growth and metabolic activity of these species is being obtained.

One of the features of these studies has been the emergence of the fungi as, presumptively, the major group of decomposer organisms. However, this comment must be tempered with the knowledge that methodology for the study of soil fungi and, in particular with respect to the estimation of amounts of live hyphae, is, to say the least, imperfect.

For both soil bacteria and fungi there is still a lack of knowledge on rates of turnover of populations, and on the physiological activities of individual groups. Inferences on the latter problem can be made from pure culture studies on individual species, however, such inferences must be made with care (Harley, 1971).

The view that the principal roles of the soil fauna is to predispose (through fragmentation) the organic matter to microbial attack, and to act as agents of spread of soil microorganisms, is well known and is being reconfirmed in soil studies in IBP.

METHODOLOGY

In the various IBP/PT studies, a variety of methods of varying sophistication, have been applied to assess rates of organic matter decomposition. In most biomes there does not appear to have been any attempts at standardizing such methods—the tundra biome being an exception. However, in most sites field studies on rate of dry weight loss studies have been carried out using litter samples (leaves, stems or roots) as have similar studies using pure substrates (with cellulose the prime example of this type of substance). In these studies, the substrates being studied are held in nylon mesh bags and placed on the litter surface and also at different depths in the litter and soil. In a restricted number of national projects field studies of the type mentioned above are being accompanied by similar studies in litter and soil samples held, in the laboratory, under a range of constant (and defined) environment conditions (e.g., temperature, nutrient status, moisture). Reference has been made earlier to the need for more studies of this type.

Soil and litter respiration in both field and laboratory situations has been used as a basis for decomposer cycle studies (Macfadyen, 1971) and for study of the effects of environmental conditions on rates of decomposition (particularly of specific substrates). Laboratory experiments, using a variety of micro- or macrorespirometers, on soil and litter respiration and on the effects of soil amendments (using pure or "natural" substrates), can give a measure of potential decomposer activity and can indicate the effects of such environmental variables as temperature, moisture, etc., on the decomposition rate. However, such methods have been frequently criticized because of the gross disturbance of the system and the artificial conditions under which the experiments are usually carried out. Field measurement of soil and litter respiration has been discussed by Macfadyen (1971) and Parkinson *et al.* (1971), where problems are encountered because of the imprecision of calculation of the amount of respiring material being measured and because of the role of living roots in the recorded respiration. Hopefully, as a result of IBP/PT studies, more data will be available to allow better assessment of these field studies.

More precise data on the degradation of pure substrates may be obtained using radiorespirometry, i.e., $^{14}CO_2$ evolution from ^{14}C-labelled substrates. Mayaudon (1971) has reviewed these methods. Edwards *et al.* (1970) discuss the values of labelling (with radioisotopes) organic debris in order to measure loss of elements during decomposition and there are numerous examples of use of isotopes in studying nutrient cycling in natural ecosystems.

SOME PROBLEMS IN TERRESTRIAL DECOMPOSITION STUDIES

One of the major group of problems facing the soil biologist studying organic matter decomposition is that of the heterogeneity of the environment, the heterogeneity of the decomposable substrates entering the system, and the heterogeneity (in nature and distribution) of the decomposer organisms.

To allow for most efficient study, decomposition sites such as pure stands of conifers with no associated plant species, are much preferable to the complex angiosperm population structure of mixed deciduous woodland (where a bewildering array of materials are presented to the decomposer organisms). However, even in the former example where the litter input appears very uniform in quality and quantity, it must be appreciated that this litter represents a complex mixture of substrates for microbial colonization and exploration, and a complex environment for soil fauna–macroflora interactions. Also differences in physiological-biochemical condition between individuals of the same litter producing species will cause differences in the chemical quality in the organic matter entering the decomposer system. Another factor contributing to "substrate heterogeneity" is the time of the year at which the material becomes available for decomposition—changes in chemical and physical quality of plant parts at different times in the growing season are being followed in various IBP/PT projects, and these factors affect patterns of initial biological colonization and subsequent decomposition rate.

Effects of topography in ecology are well known, and their effects also are important in studies on decomposition rates. In areas of varied topography, e.g., some tundra situations, it is very difficult to provide meaningful generalizations on rates of decomposition of organic matter. In such situations marked temperature differences can be recorded in the litter layer on the north- and south-facing slopes of small hummocks, and (at least in summer) cellulose degradation is more rapid on the warmer south-facing slopes. Thus, microtopography may have marked effects on decomposition rates.

Against this background of substrate and environmental heterogeneity is the heterogeneity in distribution and activity of the decomposer organisms (cf. Burges, 1960). Soil biologists have developed the view of soil, with its litter layer, as comprising a mosaic of microhabitats for biological activity. These microhabitats support varying numbers and species of organisms. From the microbiological viewpoint the small fragments of organic matter (varying in their chemical and physical state) may demonstrate very different decomposition rates.

Another factor which can affect the quality of dead organic matter, and therefore the decomposition pattern and rate, is the time after death that the organic matter falls to the litter surface. This raises another problem in the study of organic matter degradation—the so-called "standing dead." In many plant species there is prolonged attachment of dead parts to the parent plant and it may be years before, under the influence of physical factors (wind, precipitation, herbivore activity, etc.), these parts fall to the litter surface of the ground. These standing dead tissues are colonized by decomposer organisms, this colonization being from spores on the surface or from cells already active on or in the living tissues prior to death (e.g., the potential role of weak facultative parasites in the initial phase of leaf decomposition and the possibility of phyllosphere microorganisms being active in the initial phase of decomposition). However, standing dead tissues are frequently exposed to more extreme climatic fluctuations than occur in the litter layer and materials are leached from them—this leaching has been shown to be an important nutrient input in some studies. The standing dead is also prone to drying out, and in some situations the standing dead appears to comprise dry sclerophyllous tissue of high tannin content.

Many of the foregoing comments refer specifically to decomposition of organic matter at or above the soil surface. However, root production and death are major factors in adding decomposable organic materials to the soil. During root growth, the sloughing off of dead cells and the production of root exudates all well known factors increasing the amount of soil organic matter, but these input factors are difficult to quantify and studies on their rates of decomposition have seldom been attempted. The rhizosphere and attendant root region phenomena (e.g., root surface microfloras, mycorrhizas, etc.) have been intensively studied, but the relation of these phenomena to decomposition rates of the materials mentioned above are not known. However, the presence of active and numerous rhizosphere microfloras are, for most plants, generally accepted.

Following death of roots there is little information on their rates of decomposition—"litter bag" studies, in which dry weight loss of dead roots buried in the rooting zone is being followed, are attempting to provide data on this question. Studies have been made on the qualitative nature of microorganisms colonizing moribund and dead roots of various higher plants. If the root region is the zone of high microbial activity and production that many microbiologists have suggested it to be, then the absence of detailed studies on this zone may be a defect in IBP decomposer cycle studies.

In terrestrial situations there is considerable consumption of plant materials by vertebrate herbivores, and a good portion of this consumption is returned to the litter surface as dung. This partially decomposed material deposited frequently in local patches, because of its peculiar physicochemical characteristics supports a specific decomposer flora and fauna of high biomass. This material may markedly affect the rate and pattern of decomposition of organic matter in the soil in its vicinity (because of the diffusion of nutrients into the soil from the dung). There appear to be few studies where the factor of herbivore dung decomposition is being studied. In many areas invertebrate consumers far exceed the mammals in consumption of the products of primary production or in the consumption of litter. (In the litter the role of fauna in fragmenting organic matter has been mentioned.) Again much of this consumed material is returned to the litter as feces. Again much more work is needed on the microbial attack of such substrates.

The degradation of nitrogen-rich excreted materials has been studied in agricultural systems, but has received little attention in IBP/PT projects.

The examples just given are further cases of the heterogeneity of materials entering the decomposer cycle and providing substrates for microbial exploitations. Any detailed field observations in any biome would provide a much longer list of substrates worthy of study in relation to their rates of decomposition (e.g., no specific mention has been made in this contribution of wood decomposition).

NEEDED EXPERIMENTS

A good deal of discussion has centered on the relative contributions of the different components of the soil biota to organic matter decomposition and to energy flow. Phillipson (1966) quoted the general consideration that microorganisms account for as much as 90 percent of the energy flow through an ecosystem. Earlier in this contribution reference has been made to the view that the majority of organic matter decomposition is effected by the microflora.

The work of Edwards and Heath (1963), in which the decomposition rates in soil of oak leaf discs held in litter bags of different mesh size (ranging from 0.003 mm to 7.0 mm), is of interest in this regard. This technique was one of progressive exclusion of components of the decomposer biota with decreasing mesh size (i.e., in 7.0 mm mesh bags all microorganisms and invertebrates had free access to the leaf discs; in the 0.003 mm mesh bags only the microorganisms had free access). The data obtained for discs in the bags of 0.003 mm mesh indicated no visible decomposition of the leaf discs over a 9-month period. Thus, it is accepted that the role of the soil fauna is of considerably greater importance than the bare energy flow figures suggest. This importance may result from their role in litter fragmentation, mixing and perhaps causing chemical changes—all these phenomena enhancing the ability of the microorganisms to actively decompose the litter. The matter of chemical change and even enzymic degradation of components of litter by soil fauna are matters requiring more study (e.g., Luxton, 1972).

The addition of inhibitory substances has been used, in field experiments, to assess the role of soil arthropods in litter decomposition (Witkamp and Crossley, 1966). Similar techniques have been suggested, even attempted, to allow distinction between fungal, bacterial and fauna activities in soil and litter. This approach has strong initial attractions, however, Parkinson *et al.* (1971) have summarized the dangers attendant on selective inhibitor experiments with respect to soil biological activity, i.e., it is extremely difficult to ensure that the inhibitor used is brought into contact with all the microhabitats supporting the sensitive organisms; it may be that the inhibitor used may itself act as a substrate for certain groups of soil organisms; the dead remains of inhibitor-sensitive organisms will themselves become substrates for decomposition; and the removal of one group of soil organisms may release other groups of organisms for competitive influences.

Another possible approach is that of using laboratory models where sterilized litter or soil is reinoculated with a known group or groups of soil organisms and the decomposition rate of added substrates is followed (by weight loss or respiration measurements). One of the major technical problems in this type of work is the appropriate sterilization technique—Parkinson *et al.* (1971) have discussed this. The realistic interpretation of data obtained from such experiments will be difficult, however the use of such simple model systems may provide valuable data on various problems regarding the activity of individual groups of decomposer organisms.

As has been stated earlier soil and litter support large and diverse microbial populations, therefore the microbes themselves provide considerable amounts of organic matter available for decomposition. As yet no good data are available on rates of microbial productivity or on death rates, and although direct observations have demonstrated lysis of hyphae in soil there is no data on rates of hyphal or bacterial cell decomposition. Here is another area needing experimental examination.

SUMMARY

The foregoing comments indicate that, for proper studies on decomposition processes, much data on other components of the ecosystem are required—these data are principally those on primary production and from meteorological studies. Eventually synthesis of data on decomposition and soil processes will be attempted at the site, biome and interbiome levels. Heal (1971, 1972) has discussed the possibilities for synthesis of data on decomposition in the tundra biome, and various modelling ventures are now in progress (others are far better qualified than the author to comment on these aspects).

Despite the apparent multiplicity of problems in terrestrial decomposition the IBP/PT studies, which have already emphasized the vital role of the decomposer cycle, will also provide valuable (important) data of a base line type in this major subsystem of the ecosystem. These data are vital to applied studies on human impact on ecosystems, and will allow the more precise definition of questions in future programs.

REFERENCES

Alexander, M. 1965. Biodegradation: problems of molecular recalcitrance and microbial fallability. Advance. in Appl. Microbiol. 7:35–80.

Burges, A. 1960. Time and size as factors in ecology. J. Ecol. 48:273–285.

Edwards, C. A., and G. W. Heath. 1963. The role of soil animals in breakdown of leaf litter, p. 76–84. *In* J. Doeksen and J. van der Drift (ed.) Soil Organisms. North Holland Publishing Co., Amsterdam.

Edwards, C. A., D. E. Reichle, and D. A. Crossley. 1970. The role of soil invertebrates in turnover of organic matter and nutrients, p. 147–172. *In* D. E. Reichle (ed.) Analysis of Temperate Forest Ecosystems. Springer-Verlag, New York. 304 p.

Harley, J. L. 1971. Fungi in ecosystems. J. Ecol. 59:653–668.

Heal, O. W. 1971. Decomposition, p. 262–278. *In* O. W. Heal (ed.) Tundra Biome Working Meeting on Analysis of Ecosystems. Kevo, Finland, September, 1970. 297 p.

Heal, O. W. 1972. Decomposition studies in tundra, p. 93–97. *In* F. E. Wielgolaski and T. Rosswall (eds.) Tundra Biome Proceedings IV. International Meeting on Biological Productivity of Tundra. Leningrad, USSR, October 1971. 320 p.

Luxton, M. 1972. Studies on the Oribatid mites of a Danish wood soil. I. Nutritional biology. Pedobiologia 12:434–463.

Macfadyen, A. 1971. The soil and its total metabolism, p. 1–13. *In* J. Phillipson (ed.) Methods of study in quantitative soil ecology. IBP Handbook No. 18. Blackwell Scientific Publications. Oxford and Edinburgh. 297 p.

Mayaudon, J. 1971. Use of radiorespirometry in soil microbiology and biochemistry, p. 202–256. *In* A. D. McLaren and J. Skujins (ed.) Soil Biochemistry, Vol. 2. Marcel Dekker, Inc., New York.

Parkinson, D., T. R. G. Gray, and S. T. Williams. 1971. Methods for studying the ecology of soil microorganisms. IBP Handbook No. 19. Balckwell Scientific Publications, Oxford and Edinburgh. 116 p.

Paul, E. A. 1970. Plant components and soil organic matter. Recent Advances in Phytochem. 3:59–104.

Phillipson, J. 1966. Ecological energetics. St. Martin's Press. New York. 57 p.

Witkamp, M., and D. A. Crossley. 1966. The role of arthropods and microflora in the breakdown of white oak litter. Pedobiologia 6:293–303.

MEASUREMENT OF PRIMARY PRODUCTIVITY BY GAS EXCHANGE STUDIES IN THE IBP

RICHARD B. WALKER

INTRODUCTION

In its broad sense, gas exchange includes the uptake and evolution of CO_2 in photosynthesis and respiration, concomitant evolution and uptake of O_2 in these processes, and the loss of water vapor in transpiration. However, the emphasis in this review on primary productivity focuses attention principally on the CO_2 exchange, with discussion being limited to terrestrial plants.

By definition, CO_2 *uptake* by green cells is the basis of plant productivity, always modified by evolution of CO_2 in *respiration*. There has been interest in CO_2 fluxes for over a century. In the 1920's and 1930's the development of equipment for more or less continuous measurement made possible closer examination of the relationship between CO_2 uptake and the external and plant factors affecting it. The introduction of infrared CO_2 analyzers about 1950 gave further impetus to CO_2 exchange studies in general and the factors affecting rates in particular.

Because of this substantial interest in CO_2 exchange, a considerable body of information existed before the advent of IBP research in the mid-1960's. Most of the studies were autecological involving a variety of agricultural, forest, and herbaceous species. Generally, the technique of enclosing foliage in an assimilation chamber or cuvette was used, with the change in CO_2 content of the gas stream being measured.

NEED FOR FURTHER GAS EXCHANGE STUDIES UNDER THE IBP

The emphasis from the start in the IBP studies was upon vegetation and ecosystems. It was immediately evident that the existing autecological information was inadequate for proper evaluation of gas exchange of these more complicated systems.

Marked improvements in technical aspects became available or in wide use by the middle or late 1960's: viz air-conditioned cuvettes, aerodynamic CO_2-gradient methods, improved measurements of environmental variables and of foliar temperatures. The need for better assessment of those conditions and responses of the plant material affecting CO_2 exchange has become more fully appreciated. Thus, adequate stirring of air to minimize boundary-layer effects, monitoring of stomatal aperture, and surveillance of water status by psychrometry or Scholander techniques have become standard practices. The IBP has not only benefitted from such technical advancements, but has actively promoted their development and acquisition in many cases. In particular, the IBP has brought home the necessity of improved absolute accuracy in measurements, so that values obtained at one site may be effectively compared with those obtained at other sites usually with different equipment and methods. The many advancements of the recent years in equipment and techniques are covered in the authoritative treatise edited by Šesták, Čatský and Jarvis (1971).

Recognizing the need for extension of previous autecological studies to the vegetation and ecosystem level, and the necessity of incorporating the improvements in technology, many countries included gas exchange studies in their IBP plans and efforts. (See Dinger and Harris, 1973, for a review of U.S. activities.)

EXTENT OF THE IBP GAS EXCHANGE STUDIES

Seventeen countries are sponsoring gas exchange studies at over 60 sites (Table 1). These sites are concentrated in Europe and North America, with locations in addition in

TABLE 1 Studies Being Conducted in the Framework of the International Biological Program Using Gas Exchange Techniques

Country	Location	Biome	Investigator(s)	Institution	Nature of Work[a]
Austria	Tyrol Alps:				
	Patscherkofel	Alpine	W. Larcher, A. Cernusca *et al.*	Univ. of Innsbruck	PS, TR, R of dwarf scrub heath
	Nebelkogel	Alpine	W. Moser, W. Brzoska	Univ. of Innsbruck	PS, R of high alpine cushion plants
	Patscherkofel	Forest	W. Tranquillini, Havranek	Austrian Forest Res. Service	PS, TR, R of conifers at the treeline
	Neusiedlersee	Aquatic	K. Burian, R. Biebl *et al.*	Univ. of Vienna	PS of *Phragmites* and *Utricularia*
Australia	Canberra	Forest	O. T. Denmead, J. D. Ovington	Australian Natl. Univ.	Comparison of biomass with aerodynamic CO_2 exchange
Canada	Saskatchewan Matador site	Grassland	R. E. Redmann, E. A. Ripley, and R. T. Coupland	Univ. of Saskatchewan	Aerodynamic and cuvette–PS of native grasses
	Kingston	Forest	D. T. Canvin	Queen's Univ.	PS vs biomass in *Populus* in growth room
	Montreal	Agric.	K. M. King		PS vs biomass in maize
	Devon Island	Tundra	J. M. Mayo, L. C. Bliss	Univ. of Alberta	PS by cuvette of *Dryas et al.*
Czechoslovakia	Bratislava	Agric.-Forest	M. Duda, J. Kolek, and Š. Eged	Bot. Inst. of the SAV	PS by aerodynamic method in field crops and deciduous woods
	Brno	Grassland	J. Gloser	Bot. Inst. of the CAS	PS, R, TR of *Stipa* and *Bromus*
	Kroměříž	Agric.	L. Nátr	Cereal Crop Res. Inst.	PS in relation to mineral nutrition in cereals
	Praha	Agric.	B. Slavík, J. Čatský, J. Václavik, and Z. Šesták	Inst. of Expt. Bot. of the CAS	PS in relation to leaf-aging and H_2O stress in crop plants
Finland	Helsinki	Forest	P. Hari, O. Luukkanen	Univ. of Helsinki	PS in field and lab.–*Picea, Betula, Alnus*
	Oulanka Res. Sta.	Forest	P. Havas	Univ. of Oulu	PS, TR of *Vaccinium;* R of forest floor plants
	Kevo Res. Sta.	Forest	P. Kallio	Univ. of Turku	PS of lichens
France	Montpellier	Agric.; Forest	F. E. Eckardt *et al.*	C.N.R.S. Montpellier	PS, TR of field crops and orchard trees
	Versailles	Agric.	Ph. Chartier, A. Perrier	I.N.R.A. Versailles	Aerodynamic PS, TR of maize
Germany, East	Gatersleben	Agric.	M. Tschäpe	Zentralinst. für Genetik	PS, TR of various spp. in growth room
	Graupa über Pirna	Forest	G. Neuwirth	Inst. für Forstwiss. Eberswalde	PS, TR of *Pinus* stands
Germany, West	Solling site	Forest and Grassland	O. L. Lange, E. -D. Schulze, and W. Ruetz	Univ. of Würzburg	PS, TR of beech, *Picea*, grasses and *Plantago*
	Munich (Ebersberg)	Forest	W. Koch, A. Baumgartner, W. Kerner	Univ. of Munich	Comparison of cuvette and aerodynamic CO_2 exchange in *Picea*
Ireland	Glenamoy	"Tundra" (Blanket bog)	G. Doyle, J. J. Moore	Dept. of Botany, Univ. College, Dublin	PS, TR, R of dominant spp. in undisturbed and in reclaimed bog
Israel	Negev Desert	Desert	O. L. Lange, W. Koch, E. -D. Schulze, and M. Evenari	Univ. Würzburg and Hebrew Univ.	PS, TR of desert shrubs
Japan[b]	Various	Agric.		Various	Cuvette and aerodynamic study of mulberry and various field and veg. spp.
	Various	Forest		Various	PS, TR, and resp. of *Populus* clones, *Magnolia*, and several understory spp.
	Natl. Grassland Res. Inst.	Grassland	T. Okubo *et al.*		PS, resp. of pasture spp.
Netherlands	Wageningen	Agric.	W. Louwerse, J. L. P. van Oorschot	Inst. for Biol. and Chem. Res. on Field Crops	PS, TR, R of crop plants and grasses
	Wageningen	Agric.	J. Groen, E. C. Wassink	Lab. for Plant Physiol.	Comp. PS of *Calendula* and *Impatiens* in growth rm.
Norway	Hardangervidda	Tundra	F. E. Wielgolaski *et al.*	Univ. of Oslo	PS: monocots, shrubs, lichens and mosses
	Hardangervidda	Forest	F. E. Wielgolaski *et al.*	Univ. of Oslo	PS of understory *Vaccinium* and *Empetrum*
Sweden	Umeå	Forest	S. Linder, P. Halldal	Univ. of Umeå	PS action spectrum of *Pinus* from nurseries

(Continued overleaf)

TABLE 1 (continued)

Country	Location	Biome	Investigator(s)	Institution	Nature of Work
United Kingdom	Aberdeen	Forest	P. G. Jarvis *et al.*	Univ. of Aberdeen	Aerodynamic PS in *Picea* stand; PS, TR in lab.
	Aberystwyth	Grassland	C. F. Eagles, D. Wilson	Welsh Plant Breeding Sta.	PS, R, TR of cultivated grasses
	Cambridge	Agric.	F. G. Lupton	Cambridge Univ.	PS, R of cereals
	Nottingham	Agric.	J. L. Monteith, P. V. Biscoe	Univ. of Nottingham	Aerodynamic PS and TR of barley fields
	Hurley, Maidenhead, Berks	Grassland	E. L. Leafe, W. Stiles *et al.*	Grassland Res. Institute	Cuvette PS, TR in grass swards
	Little Hampton, Sussex	Agric.	J. Warren Wilson, D. Charles Edwards	Glasshouse Crops Res. Inst.	PS, R of glasshouse crops
United States	NE Colo. Pawnee site	Grassland	A. J. Dye, P. G. Risser	Colo. State Univ.	PS, TR of native grasses using cuvettes
	Pt. Barrow, Alaska	Tundra	L. L. Tieszen *et al.*	Augustana College	PS of monocots and *Salix* using cuvettes and $C\text{-}140_2$
	Niwat Ridge, Colorado	Tundra (alpine)	M. M. Caldwell, R. Moore	Utah State Univ.	PS, TR of monocots
	Tucson, Ariz.	Desert	D. T. Patten *et al.*	Arizona State Univ.	PS, TR of sajuaro cactus
	Curlew Valley, Utah	Desert	M. M. Caldwell, R. Moore	Utah State Univ.	PS, TR of *Artemisia* spp., *Atriplex*, and *Eurotia*
	Oak Ridge, Tenn.	Forest	B. Dinger	Oak Ridge Nat'l. Lab.	PS, R of *Liriodendron*, PS *Quercus*, *Acer*, *Carya*, TR *Carya*, *Liriodendron*, *Pinus*
	Durham, N.C.	Forest	B. R. Strain, R. Kinerson, K. Knoerr, K. Higginbotham	Duke Univ.	Comparison of aerodynamic and cuvette PS, TR in *Pinus taeda*
	Seattle, Washington	Forest	R. B. Walker, L. J. Frischen	Univ. of Washington	Comparison of aerodynamic and cuvette PS, TR in *Pseudotsuga menziesii*
	California	Mediterranean scrub	H. A. Mooney, A. T. Harrison, E. L. Dunn	Stanford Univ.	PS, R, TR of various shrubs–also comparisons with So. America
USSR	Taimyr Peninsula	Tundra	V. M. Shvetzova, V. L. Voznesensky, O. V. Zalensky	Komarov's Bot. Inst.	PS over the year of some 10 spp. of herbaceous plts and shrubs
	Kyzylkum Desert	Desert	I. L. Zakharyants, S. Fazylova	Uzbek Akad of Sci.	PS and limiting light intens. for many spp.
			O. A. Semikhatova, L. N. Alekseeva	Komarov's Bot. Inst.	Comp. physiology of resp. of several spp.
	Tajikistan	Desert	Yu. S. Nasyrov, G. P. Lebedeva, V. K. Kichitov, K. P. Rakhmania	Tadjik Academy of Sci.	PS, TR of a large number of spp.
	Karakum	Desert	V. L. Voznesensky, V. M. Sveshnikova, N. I. Bobrovskaya	Komarov's Bot. Inst.	PS, TR of some 20 to 30 spp.
	East Pamir	High mt. desert	N. N. Izmailova, N. P. Litvinova	Komarov's Bot. Inst.	TR of several spp.

[a] PS = photosynthesis (net), TR = transpiration, R = respiration.
[b] The Japanese group of some 46 investigators working at some 12 universities and institutes is involved in a wide range of projects under the chairmanship of M. Monsi, Univ. of Tokyo. Full listings are published in annual reports of JIBP/PP PS Level III Gp.

Australia, Israel, Japan, and asiatic U.S.S.R. A wide variety of vegetation is likewise included that may be said to represent in a broad sense the agricultural, desert, forest, grassland, and tundra (both alpine and arctic) biomes. Further, some 200 scientists are engaged in the work at these sites over the world.

Already a substantial amount of work has been finished, but much more is still under way, since the assembly or acquisition of equipment, the development of field sites, and collection of data over seasons or even years consume much time. Further, the analysis of extensive data may require long additional periods. Fortunately, however, the existence of internal IBP reports and summaries makes preliminary data widely available, and the various IBP symposia and synthesis meetings have made oral presentation and discussion of studies possible even in their earlier stages.

Certainly gas exchange studies have greatly increased in number, and usually in intensity and extent as well, under the stimulus and support of the IBP. Funding has been substantial; in fact many of the programs could never have been planned or brought to fruition without this financial support. International planning has made for sharper goals, and widespread comparisons of methodology. These have enhanced the value of comparisons of resultant data.

JUSTIFICATIONS FOR THESE EXTENSIVE GAS EXCHANGE STUDIES

Logically one can ask what values can be attained from gas exchange studies which justify the large investment in money and scientific effort outlined above. In short, these can be stated as:

1. Only through careful studies of net photosynthesis and both dark and light respiration can gross productivity be determined. Differences between species and varieties in gross as well as net productivity is of substantial practical interest in agriculture and forestry.

2. Information on the influence of external factors (light intensity, air temperature, vapor pressure deficit) and plant factors (especially leaf temperature and stomatal and other leaf resistances) on net photosynthesis and respiration is vital to the development of models of terrestrial plant productivity. If these are to have good predictive values, they must take into account the influence of wide fluctuations in these factors between different days, between the seasons of the year, and between one year and another. Models are just now being intensively worked on and are often as yet based on inadequate data on these environmental and plant factors.

3. A major strength of the IBP lies in the integrated studies of all components of ecosystems. Here gas exchange studies made of terrestrial plants—especially root respiration—are of particular use to those studying metabolism of the soil, influences of parasites, consumer organisms, etc.

CONCLUSIONS

Extensive studies using gas exchange techniques have been conducted in many countries under the auspices of the IBP. Some of these studies are completed, but most are either still under way or the data are still being analyzed. However, preliminary reports and summaries certainly indicate that major progress has been made, especially in assessing the influences of both environmental and plant factors on assimilation and respiration, so that good models of terrestrial plant productivity may be envisioned. The completion of all of the ongoing studies, and their analysis and publication will greatly enhance theoretical and applied knowledge in this field, so that prediction and modeling will be on a firm basis.

AKNOWLEDGEMENTS

Preparation of this paper was supported in part by the University of Washington and in part by the National Science Foundation, Grant No. GB-20963, and is Contribution No. 160 of the Coniferous Forest Biome, U.S. Analysis of Ecosystems, International Biological Program. Warm appreciation is expressed to all those who furnished information about IBP programs in the various countries.

REFERENCES

Dinger, B. E., and W. F. Harris. 1973. Terrestrial primary production. Proceedings of Interbiome Workshop on Gaseous Exchange Methodology. Oak Ridge National Laboratory, April 13–14, 1972. 184 p.

Šesták, Z., J. Čatský, and P. G. Jarvis (ed.) 1971. Plant photosynthetic production, manual of methods. The Hague. Dr. W. Junk N. V. Publishers. 818 p.

THE ROLE OF HERBIVORE CONSUMERS IN VARIOUS ECOSYSTEMS

K. PETRUSEWICZ and W. L. GRODZIŃSKI

INTRODUCTION

Before proceeding to the main theme of this paper, it should be pointed out that: (1) this paper is a review based on already published data assumed to be correct; we do not take into consideration the sampling procedures and techniques used for calculation of the data collected; (2) the data presented here cannot be regarded as in anyway exhaustive; and (3) the role and significance of herbivores are considered in the context of the ecosystem.

The theoretical importance of herbivorous animals in any ecosystem depends on a diminution of plant biomass by grazing, and production of highly organized matter—the biomass of the herbivore (Figure 1). Generally, the nontrophic effects, i.e., pollination, seed dispersal, soil aeration, etc., of herbivores are not taken into account. Although these processes are necessary for some plants, their role in the total ecosystem is of secondary significance.

Herbivore activity can affect many ecological processes, such as an increase or decrease in the primary productivity, a decrease in the level of plant biomass, an increase in food for predators, including man (Figure 1). An increase in diversity and complexity of ecosystem organization can be the result of herbivore pressure. Thus, by increasing the energy (information) transfer within the system, increased ecosystem stability results. The intensification of productivity is usually at the expense of plant *standing crop* (biomass); a part of primary production covers the cost of ecosystem stabilization.

In discussing these functions of the phytophagous animals (Figure 1) the following aspects should be considered:

1. The greatest influences by far exerted by phytophages in an ecosystem are connected with the consumption of plants.

2. The ecological roles of phytophages are generally not alternatives. They are not of the "either-or" type, but they are of "this-and-this" or the "this-and/or-that" type of relationship.

3. When considering plant biomass diminution, consumption (C)—the amount of organic matter consumed in a unit of time from a unit of space—is not the best measure; often the better measure is the organic matter removed from the standing crop (biomass) of the plants (Figure 2) (Petrusewicz and Macfadyen, 1970). Unfortunately, there are very few data on this aspect of ecology. Therefore, consumption (C) must be taken as our basis rather than the matter removed (MR).

HERBIVORE CONSUMPTION: HERBIVORE IMPACT ON PRIMARY PRODUCTION

Obviously, if a herbivore is to play a significant role, it must remove enough plant material from the plant to effect the plant community and add sufficient biomass to its own body weight to initiate the grazing food chain, and hence to increase the diversity and complexity of the system as a whole.

It is difficult to predict what percentage of biomass removal would exert an important influence on the functioning of the ecosystem. Under some conditions even heavy grazing may be of minor importance (e.g., the Colorado beetle consuming 20 percent of potato plant leaves decreases the

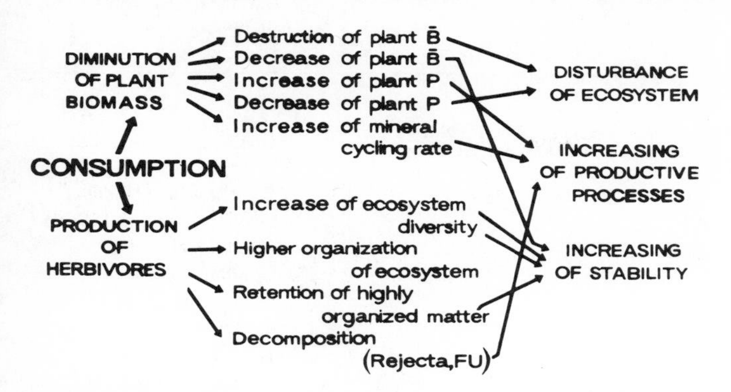

FIGURE 1 The nature of herbivorous influences on ecosystems.

yield by only 1 to 2 percent). In other situations, very limited grazing may have a very great influence; e.g., Varley's (1967) investigation documentation of the wood increment losses of oak trees many times greater than consumption by caterpillars.

It is also difficult to imagine an ecosystem without primary producers, and similarly without decomposers. Without producers there would be no energy stored for life. Without decomposers, the surface would be covered with a layer of dead organic matter in a short time and most important nutrient elements would be removed from biological circulation. In contrast, at least theoretically, a biosphere functioning without consumers can be imagined. It would be a very poor biosphere and probably life within it would be very uninteresting. Nevertheless, such an ecosystem can be imagined. Many ecologists consider that herbivore consumption in ecosystems is negligible. For example, a frequently used method of estimating primary production is to measure the maximum standing crop and to ignore the grazed plant biomass.

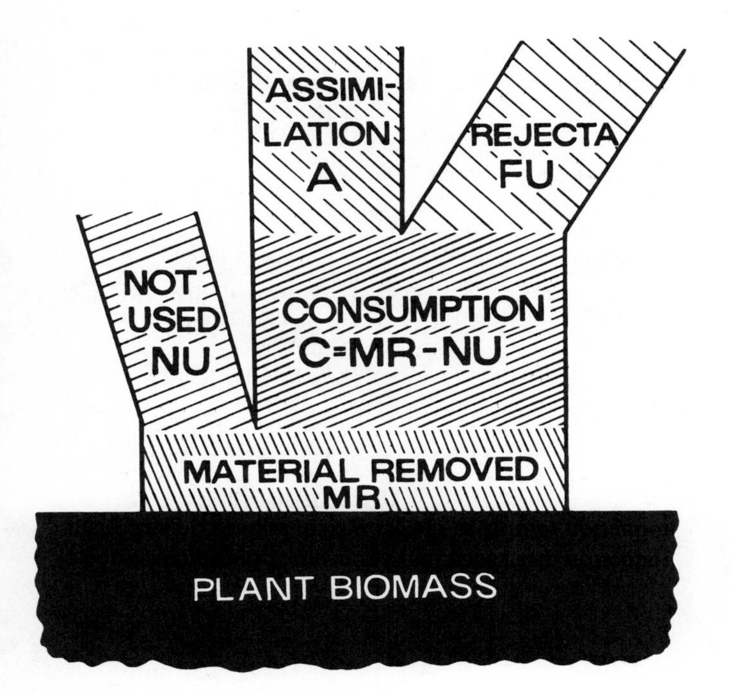

FIGURE 2 Material removed versus consumption—a comparison of trophic exchange and terms used to evaluate herbivore impact on plants.

Consequently, a first effort was made to compile some available data on herbivore consumption. Data collected on small rodent consumption in different ecosystems are presented in Table 1. In comparing the consumption figures in these ecosystems it is obvious that: (1) a single phytophage group, the rodents, can consume more than 10 percent of the available food, and (2) in spite of the variations, the simpler and poorer the ecosystem, the higher the proportion that is removed. It should be stressed that the ecosystems summarized here are not ones commonly exposed to the action of outbreak species.

The concept of "food available" has been mentioned. The food available to some herbivores has been defined as that "food which is easy to find, is being chosen and being eaten by these animals" (Grodziński, 1968). The food available to consumers was described in the well-known studies on productivity of a beech forest in southern Poland, initiated by Dr. A. Medwecka-Kornaś (Medwecka-Kornaś and Lomnicki, 1967). We present, after Droźdź (1967), a relationship between total aboveground primary production of the beech forest and the food available to rodents (Table 2). In this forest, the food available amounts to 4.5 percent of total primary production. It is clear that this proportion varies, but generally it may be said that in the herbaceous ecosystems the proportion of food available is much higher than in forest ecosystems. In meadows and fields, almost the entire aboveground production can be considered as potential food for rodents and other herbivores.

To illustrate the overall influence of herbivorous animals, we analyzed several of the better-known ecosystems in which the action of a larger number of phytophagous animals has been studied (Table 3). Unfortunately, we were not able to find any one ecosystem in which the consumption by all phytophages had been studied. Therefore, we indicate to which groups of phytophages upon which the data are based.

The following points emerge:

1. The value of phytophage consumption is considerable; it amounts to 8 to 20 percent (2 to 5×10^6 kcal/ha-yr) in terrestrial ecosystems.
2. The comparison decidely supports the well-known division of Odum into "detrital" (oak–hornbeam forest) and "grazed" (meadow and cultivated field) ecosystems.
3. Even in the terrestrial ecosystems, considered as belonging to the typical grazing food chain ecosystems, only a minor part of the plant biomass (13 to 20 percent) passes

TABLE 1 Rodent Consumption in Various Ecosystems

Ecosystem Type		10^3 kcal/ha-yr Food Available (F_a)	Consumption (C)	C/Fa × 100 (%)	References
Temperate forest					
(4 pine)		1,024–16,190	21–105	0.6–1.9	Ryszkowski (1969)
(2 deciduous)		1,950–2,050	40–129	2.0–4.6	Grodziński *et al.* (1969) Grodziński (1971)
Grassland, forest plantation (4 ecosystems)		2,395–19,700	131–699	1.5–5.6	Grodziński *et al.* (1966) Myllymäki (1969) Golley (1960) Hansson (1971)
Northern taiga (Alaska)		1,320	179	13.5	Grodziński (1971)
Microtus in farmland	N = 80/ha	40,700	365	0.9	Trojan (1969)
	N = 1,000/ha		4,600	11.3	

TABLE 2 Primary Production and Food Available to Rodents in a Beech Forest (After Drożdż, 1967)

	10^3 kcal/ha-yr Plant Production (P_p)	Food Available (F_a)	F_a/P_p × 100 (%)
Herbs	1,080	920	85
Tree seeds	225 (28–456)	202 (22–360)	90
Tree leaves	13,425	670	5
Trunks and branches (twigs, bark)	29,140	110	0.4
Fungi and invertebrates		53	
TOTAL	43,870	1.955	4.5

through this chain. The grazing food chain is considerably more important in the grasslands than in the forest ecosystem (13 to 20 percent vs. 7.7 percent). Nevertheless, in both forest and meadow ecosystems the detritus food chain is more important than the grazing food chain. Only in an oligotrophic lake, in pelagic (plankton) ecosystems, is the major part of the biomass (90 percent) grazed. Such an ecosystem can be considered as an "eugrazing" type.

A further illustration of the magnitude of herbivore impact on vegetation may be made by comparing available data on forest and grassland herbivore consumption (Figure 3). In this comparison, forest clearings up to 4 years old are included in the grassland ecosystems, because the pioneering stage of secondary succession of forest is similar to the grass-

TABLE 3 Primary Production Consumed in Some Terrestrial and Freshwater Ecosystems in Poland

	Food Available (F_a)	Known Consumers	Consumption	C/F_a × 100 (%)	References
Oak-hornbeam forest (Niepolomice) (10^3 kcal/ha-yr)	21,940	Mammals, birds *Tortrix*, other Lepid. larvae	1,700	7.7	Medwecka-Kornaś, Lomnicki, and Bandola-Ciolczyk (1973)
Uncultivated meadow (Dziekanów) (10^3 kcal/ha-yr)	20,750	Orthoptera, Homoptera, Diptera, Lepidoptera, rodents	2,827	13.1	Breymeyer (1971)
Rye field (Turew) (10^3 kcal/ha-yr)	41,700	*Microtus,* Diptera, Colorado beatle	4,500	10.9	Trojan (1967)
Potatoes (Turew) (10^3 kcal/ha-yr)	26,600	*Microtus,* Diptera, Colorado beatle	4,870	19.8	Trojan (1967)
Meso-oligotrophic (Pilakno Lake) (cal/24 h)	0.77	Plankton	0.69	90	Gliwicz and Hillbricht-Ilkowska (in press)
Mesotrophic (Taltowisko Lake) (cal/24 h)	0.87	Plankton	0.38	44	Gliwicz and Hillbricht-Ilkowska (in press)
Eutrophic (Mikolajki Lake) (cal/24 h)	0.38	Plankton	0.22	10	Gliwicz and Hillbricht-Ilkowska (in press)

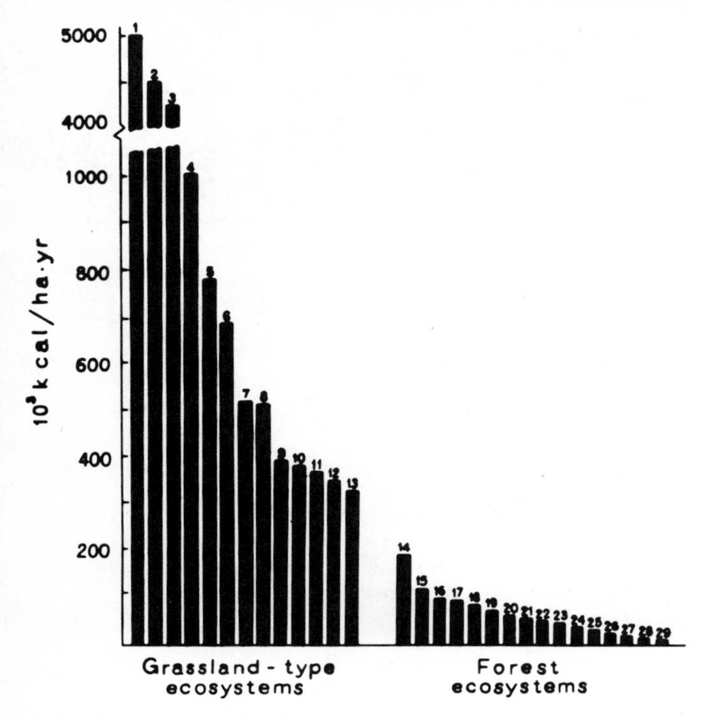

FIGURE 3 Examples of herbivore consumption in grassland and forest ecosystems (10^3 kcal/ha-yr). 1. *Microtus* (Myllymäki, 1969); 2. Colorado beatles (Trojan, 1967); 3. Plant hoppers (Wiegert and Evans, 1967); 4. Grasshoppers (Wiegert and Evans, 1967); 5. Orthoptera (Wiegert and Evans, 1967); 6. Cicadella (Andrzejewska, 1967); 7. Rodents (Hansson, 1971); 8. *Myrmica* (Pętal, 1967); 9. Orthoptera (Gyllenberg, 1969); 10. *Microtus* (Gębczńska, 1970); 11. *Myrmica* (Pętal *et al.*, 1971); 12. *Pogonomyrmex* (Wiegert and Evans, 1967); 13. *Microtus* (Trojan, 1969); 14. Rodents (Grodziński, 1971); 15. Rodents (Ryszkowski, 1969); 16. Rodents (Grodziński *et al.*, 1969); 17, Rodents (Ryszkowski, 1969); 18. Rodents (Grodziński *et al.*, 1969); 19. Lepidoptera (Winter, 1971); 20. Ground squirrel (Wiegert and Evans, 1967); 21. Curculionidae (Funke, 1971); 22. Orthoptera (Wiegert and Evans, 1967); 23. Rodents (Ryszkowski, 1969); 24. Lepidoptera (Funke, 1971); 25. Sparrows (Wiegert and Evans, 1967); 26. Rodents (Ryszkowski, 1969); 27. Spittlebugs (Wiegert and Evans, 1967); 28. Curculionidae (Funke, 1971); 29. Peromyscus (Wiegert and Evans, 1967).

land type. Savannas and semideserts have been also included in the grassland ecosystem category. This above comparison (Figure 3) leads to two conclusions:

1. The value of plant biomass consumed is sometimes considerable and attains the level of several million kcal/ha-yr.
2. Based on the absolute values of consumption, evidence is that in grassland-type ecosystems consumption (grazing food chain) is more important, and that in the forest-type ecosystem the detritus food chain is more important.

Andrzejewska and Wóicik (1971) showed that after the total elimination of phytophagous insects on 4 m^2 meadow plots there was a 40 percent increase in the primary production compared to control plots. Since the experiment which involved using pesticides lasted only one growing season, it is difficult to predict the relationships after several years. However, if the 40 percent increase in primary production is attributed solely to consumption an abnormally high value is derived. The examples discussed provide illustrations of the complex interactions between phytophages and producers in an ecosystem. Their direct impact on the primary production is shown by a consumption rate of the order 4 to 40 percent of primary production; more complex relationships are obscure.

HERBIVORE CONSUMERS AND CYCLING OF MATERIALS

In the beginning of this paper it was indicated that phytophages may also play a role in accelerating the cycling of material by, for instance, comminution (crumbling) of the plant biomass, thus making it easier for the mineralization process to be completed by reducer organisms (bacteria and fungi). These processes are dependent, to a great extent, upon the assimilation efficiency of phytophagous animals.

The assimilation efficiency of herbivores has been variously expressed. For comparison of some data from the literature, see Table 4. The range of values for food energy rejected as feces and urine is very broad. Of the food energy consumed, rodents assimilate about 90 percent, the elephant 40 percent, and some lepidopteran larvae only 20 percent (Table 4).

Obviously, there is little need to emphasize that the importance of phytophages compared with decomposers will depend on their standing crop, i.e., mean numbers and mass of individuals feeding during a given period of a year [number of individual days ($\overline{N}T$)]. Consideration must be given

TABLE 4 Rejecta (Feces and Urine) Returned into the Cycle of Material by Different Herbivore Consumers

Consumer Species	FU/C × 100 (%)	References
Field mouse (*Apodemus agrarius*)	11	Droźdź (1968)
European hare (*Lepus europaeus*)	22	Myrcha (1968)
Wild boar (*Sus scrofa*)	24	Gere (1957)
Orthoptera (*Chortippus dorsatus*)	30	Chlodny (1969)
Lepidoptera (*Pieris brassicae*)	31	Nakamura (1965)
Elephant (*Loxodonta africana*)	60	Petrides and Swank (1966)
Lepidoptera (*Croesus septentrionalis*)	79	Janda (1960)

not only the numbers of individuals consuming, but also their size. So we come to the very important value, quantity of biomass-days ($\bar{B}T$), i.e., the amount of consumer biomass during a given period of a year. It may happen that rodents, which reject only about 11 percent of the food consumed, may be able to excrete a greater quantity of feces and urine (FU) than *Croesus* larvae, which reject 80 percent of the food consumed during a short time period only. The data for consumption and the comparison of assimilation efficiencies (Table 4) are expressed in the most comparable units, i.e., in terms of energy—calories rejected over calories consumed.

Here we should point out the problem sometimes neglected by ecologists: of the two components of rejecta (FU), feces (F) contain the major share of energy in comparison with urine (U). Thus, often urine (U) is ignored in the study of energy flow. Quite different conclusions prevail if one considers nutrient cycling. Considerably more nitrogen is contained in urine than in the feces (Table 5). The nitrogen in urine may be assimilated directly by plants without the action of decomposers.

To illustrate the contribution of herbivore consumers in the decomposition of organic matter, data were compiled on the annual inflow of dead organic matter of plant origin in four types of ecosystems (Table 6 and Figure 4). This comparison was possible using the following assumptions. In meadows and forest 20 percent of the total annual root

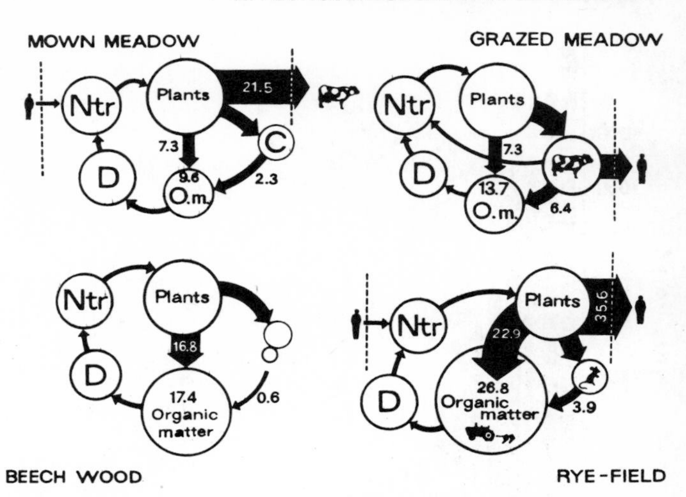

FIGURE 4 Organic matter recycled in the ecosystem (10^6 X kcal/ha-yr). O.m., organic matter in soil; D, decomposer; Ntr, plant nutrients—plant organic matter returned to soil; left arrow, dead plant matter; right arrow, plant matter comminuted by herbivores.

production dies yearly. Five percent of the annual wood production returns to the ecosystem as dead organic plant mass. In the grazed meadows 30 percent of the plants consumed return to the ecosystem as rejecta (FU) from grazing animals in the form of crumbled organic matter (if the cattle are on the pasture for 24 hours per day). We were not able to find any suitable single collection of empirical data for the same ecosystem; calculations were based on empirical data taken from various sources. In analyzing these comparisons (Figure 4) the following features become evident:

1. The beech-wood forest and grazed-meadow ecosystems are typical examples of natural balanced detrital and grazed food chain ecosystems, respectively. In the detrital food chain ecosystem only a minimal amount (less than 10^6 kcal/ha-yr) returns to the soil in the comminuted form. Most of the decomposition processes are due to the action of detritophagous organisms. But even in a "classical"

TABLE 5 Nitrogen Efflux (Expressed as Percentages of Consumption) by a Vole and Pig, in a Steady State of Nitrogen Balance (without Retention for Growth and Reproduction)

	Vole	Pig
Consumption	100	100
Feces	26	30
Urine	74	70

TABLE 6 Primary Production and Organic Matter Returned to Ecosystem (10^6 kcal/ha-yr); Assumptions Used are Indicated by Parentheses

	Production			Annual Return to Soil		
Ecosystem Type	Foliage	Wood	Roots	Conditions	Subtotal	Total
Mowed meadow	25.8	–	23.7	10% unmowed (20% of roots)	7.3	
				90% as FU and bodies from 10% grazed	2.3	9.6
Grazed meadow	25.8	–	23.7	10% unmowed (20% of roots)	7.3	
				30% as FU from 80% grazed	6.4	13.7
Beechwood forest	13.6	43.0	8.6	95% of green parts (5% of wood) (20% of roots)	16.8	
				5% of grazed leaves	0.6	17.4
Rye field	43.9	–	18.5	10% unmowed green parts, all roots	22.9	
				90% as FU and bodies from 10% grazed	3.9	26.8

grazed ecosystem, as in a grazed meadow, more than 50 percent of plant biomass passes through the detritophage food chain (7.3×10^3 vs. 6.4×10^3 kcal/ha-yr).

2. The ecosystems chosen differ distinctly in the amount of organic matter returning to the ecosystem.

3. The mown meadow and rye-field ecosystems, i.e., man-made or man-changed ecosystems, are different when compared with natural or seminatural ones.

4. Significantly smaller amounts of organic matter contribute to mineral cycling in the mown-meadow ecosystem, the balance of plant mineral nutrients here being negative. The nutrients necessary for life of the ecosystem must be imported via the hydrological route (flooding of meadows, etc.) or by the help of man in the form of fertilizers.

5. The flux of organic matter in the two naturally balanced ecosystems (forests and pastures grazed by big herbivorous animals) has a mean value of 1.4 to 1.8×10^4 kcal/ha-yr.

6. Much higher fluxes of organic matter occur in field crops, because about 1.5 to 2 times more organic matter returns to the cultivated fields (the frequent addition of dung not being taken into account) as compared with the pasture and forest ecosystems. Even the amount of the annual flux of noncrumbled (not comminuted) organic matter in the ecosystem is higher than the total organic matter flux in the forest and pasture ecosystems. This is the result of the annual killing and ploughing under of the whole root system which, in other ecosystems, occurs as a continuing dying-off process. It is worth drawing attention to the very high level of microorganism activity which must be involved.

Cultivated fields and mown meadows occupy, at least in Central Europe, more than 50 percent of the total area. In these strongly altered ecosystems intensive ecological processes are taking place, and it seems to us that the work of the ecologists, at least concerning the cultivated fields, should be concentrated (besides on pest control) on the search for ways of intensifying the decomposition processes.

CONCLUSION

It can be clearly demonstrated that, even in normal ecosystems without mass outbreaks of pests, the impact of herbivores can be significant. Herbivores have been stated to be responsible for very different ecological processes, such as increase or decrease of productivity, stability, etc. An attempt will now be made to show the relation between the ecological and/or physiological properties of herbivores and some ecological processes (Figure 5).

The most important consumer population property is abundance, i.e., mean yearly biomass and turnover. To be significant as a consumer, the herbivore population must have a high mean biomass. To be efficient as producer, the herbivore must have high mean biomass and/or high turnover rate (Figure 5). Excessive consumption (grazing) can cause a smaller or greater disturbance in the functioning of an ecosystem, or even destroy it. Moderate grazing often may increase primary production and productivity processes through more rapid mineral cycling.

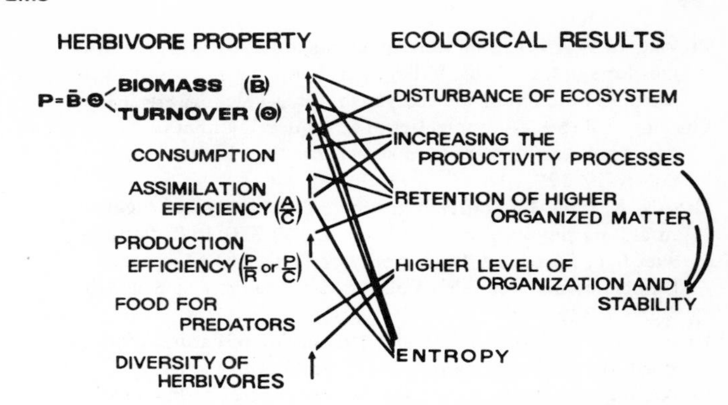

FIGURE 5 Physiological and ecological properties of herbivorous animals in relation to their impact on ecosystems.

Efficiency of assimilation can increase secondary productivity and, hence, increase the retention of total information in the ecosystem; it can, but it need not. Even the very efficient assimilator with only a short life span (i.e., high turnover) can increase productivity but not the retention of highly organized biomass. Inefficient assimilators increase the cycling of materials in the system.

Efficient producers with low turnover contribute to the retention of highly organized organic matter, and in this way increase the complexity of system organization. Efficient producer populations with a high turnover can offer to predators a considerable amount of food and hence start a food chain but at a high price in terms of energy. Such a consumer will accelerate productivity and increase the entropy. The entropy should then be higher if such a herbivore has a high metabolic rate.

By increasing biomass retention and/or intensification of productivity processes within an ecosystem, herbivores contribute to the diversity and higher organization of the system. Unfortunately, in the present state of ecological knowledge, all these processes can only be described in general, semiquantitative terms.

REFERENCES

Andrzejewska, L. 1967. Estimation of the effects of feeding of the sucking insects, *Cicadella viridis,* on plants, p. 791–806. *In* K. Petrusewicz (ed.) Secondary Productivity of Terrestrial Ecosystems (Principles and Methods). Vol. II. Polish Academy of Sciences, Warsaw. 879 p.

Andrzejweska, L., and Z. Wójcik. 1971. Productivity investigation of two types of meadows in the Vistula Valley. VII. Estimation of the effect of phytophagous insects on the vascular plant biomass of the meadow. Ekol. Pol. 19:173–182.

Breymeyer, A. 1971. Productivity investigation of two types of meadows in the Vistula Valley. XII. Some regularities in structure and function of the ecosystem. Ekol. Pol. 19:249–261.

Chlodny, J. 1969. The energetics of larval development of two species of grasshoppers from genus *Chortippus* Fieb. Ekol. Pol. A. 17:391–407.

Droźdź, A. 1967. Food preference food digestibility and the natural food supply of small rodents, p. 323–330. *In* K. Petrusewicz (ed.) Secondary Productivity of Terrestrial Ecosystems (Principles and Methods). Vol. I. Polish Academy of Sciences, Warsaw. 379 p.

Droźdź, A. 1968. Digestibility and assimilation of natural foods in small rodents. Acta Theriol. 13:367–389.

Funke, W. 1971. Food and energy turnover of leaf eating insects and their influence on primary production, p. 81–93. *In* H. Ellenburg (ed.) Integrated Experimental Ecology, Methods and Results of Ecosystem Research in the German Solling Project. Springer-Verlag, Berlin–Heidelberg–New York. 214 p.

Gere, G. 1957. Productive biologic grouping of organisms and their role in ecological communities. Ann. Univ. Sci. Budapest Sect. Biol. 1:61–69.

Gębczyńska, Z. 1970. Bioenergetics of a root vole population. Acta Theriol. 15:33–66.

Gliwicz, Z. M., and A. Hillbricht-Ilkowska. (In press) Utilization of nannoplankton production by filter feeders community. Verhandl. int. Verein. Limnol. 16.

Golley, F. B. 1960. Energy dynamics of a food chain of an old-field community. Ecol. Monogr. 30:187–206.

Grodziński, W. 1968. Energy flow through a vertebrate population, p. 239–252. *In* W. Grodziński and R. Z. Klekowski (eds.) Methods of Ecological Bioenergetics. Warszawa–Kraków.

Grodziński, W. 1971. Energy flow through populations of small mammals in the Alaskan taiga forest. Acta Theriol. 16:231–175.

Grodziński, W., A. Górecki, K. Janas, and P. Migula. 1966. Effect of rodents on the primary productivity of alpine meadows in Bieszczady Mountains. Acta Theriol. 11:419–431.

Grodziński, W., B. Bobek, A. Droźdź, and A. Górecki. 1969. Energy flow through small rodent populations in a beech forest, p. 291–298. *In* K. Petrusewicz and L. Ryszkowski (eds.) Energy Flow Through Small Mammal Populations. Polish Scientific Publisher, Warsaw. 298 p.

Gyllenberg, G. 1969. The energy flow through a *Chortippus parallelus.* Zett. *Orthoptera.* Population in a meadow in Tvärminae, Finland. Ann. Zool. Fennici 123:1–74.

Hansson, L. 1971. Estimates of the productivity of small mammals in a South Swedish spruce plantation. Ann. Zool. Fenn. 8:118–126.

Janda, V. 1960. Gesamtstoffwechsel der Larven einiger Blattwespenarten. Hym. Tenthrenoidea, p. 190–194. *In* The Ontogeny of Insects. London.

Medwecka-Kornas, A., and A. Lomnicki. 1967. Discussion of the results of ecological investigations in the Ojców National Park. Studia Naturea A. 1:199–213.

Medwecka-Kornas, A., A. Lomnicki, and E. Bandola-Ciolczyk. 1973. Energy flow in the deciduous woodland ecosystem, Ispina Project, Poland, p. 144–150. *In* L. Kern (ed.) Modeling Forest Ecosystems. Report of International Woodlands Workshop, International Biological Program/PT Section, August 14–26, 1972. Oak Ridge National Laboratory, Oak Ridge, Tenn. 339 p.

Myllymäki, A. 1969. Productivity of a free-living population of the field vole, *Microtus agrestis* L., p. 255–265. *In* K. Petrusewicz and L. Ryszkowski (eds.) Energy Flow Through Small Mammal Populations. Polish Scientific Publisher, Warsaw. 298 p.

Myrcha, A. 1968. Winter food intake in european hare, *Lepus europaeus* Pallas, 1778 in experimental conditions. Acta Theriol. 13:453–459.

Nakamura, M. 1965. Bioeconomic of some larval populations of pleurostict scarabeidae on the flood plain of the River Tamagawa. Jap. J. Ecol. 15:1–18.

Pętal, J. 1967. Productivity and the consumption of food in the *Myrmica laevinoides* Nyl. population, p. 841–857. *In* K. Petrusewicz (ed.) Secondary Productivity of Terrestrial Ecosystems (Principles and Methods). Vol. II. Polish Academy of Sciences, Warsaw. 879 p.

Pętal, J., L. Andrzejewska, A. Breymeyer, and E. Olechowicz. 1971. Productivity investigation of two types of meadows in the Vistula Valley. X. The role of ants as predators in a habitat. Ekol. Pol. 19:213–222.

Petrides, G. A., and W. G. Swank. 1966. Estimating the productivity and energy relations of an African elephant population, p. 831–842. *In* Proceedings Ninth International Grasslands Congress, Sao Paulo.

Petrusewicz, K. 1967. Concepts in studies on the secondary productivity of terrestrial ecosystems, p. 17–49. *In* K. Petrusewicz (ed.) Secondary Productivity of Terrestrial Ecosystems (Principles and Methods). Vol. I. Polish Academy of Sciences, Warsaw. 379 p.

Petrusewicz, K., and A. Macfadyen. 1970. Productivity of Terrestrial Animals. Principles and Methods. IBP Handbook No. 13. Backwell Scientific Publications. Oxford and Edinburgh. 190 p.

Ryszkowski, L. 1969. Estimates of consumption of rodent populations in different pine forest ecosystems, p. 281–289. *In* K. Petrusewicz and L. Ryszkowski (eds.) Energy Flow Through Small Mammal Populations. Polish Scientific Publisher, Warsaw. 298 p.

Trojan, P. 1967. Investigation on production of cultivated fields, p. 545–563. *In* K. Petrusewicz (ed.) Secondary Productivity of Terrestrial Ecosystems (Principles and Methods). Vol. II. Polish Academy of Sciences, Warsaw. 879 p.

Trojan, P. 1969. Energy flow through a population of *Microtus arvalis,* Pall., in an agrocenosis during a period of mass occurrence, p. 267–279. *In* K. Petrusewicz and L. Ryszkowski (eds.) Energy Flow Through Small Mammal Populations. Polish Scientific Publisher, Warsaw. 298 p.

Varley, G. C. 1967. The effects of grazing by animals on plant productivity, p. 773–777. *In* K. Petrusewicz (ed.) Secondary Productivity of Terrestrial Ecosystems (Principles and Methods). Vol. II. Polish Academy of Sciences, Warsaw. 879 p.

Wiegert, R. G., and F. S. Evans. 1967. Investigations of secondary productivity in grasslands, p. 499–518. *In* K. Petrusewicz (ed.) Secondary Productivity of Terrestrial Ecosystems (Principles and Methods). Vol. II. Polish Academy of Sciences, Warsaw. 879 p.

Winter, K. 1971. Studies in the productivity of Lepidoptera populations, p. 94–99. *In* H. Ellenberg (ed.) Integrated Experimental Ecology Methods and Results of Ecosystems Research in the German Solling Project. Springer-Verlag. Berlin–Heidelberg–New York. 214 p.

SECONDARY PRODUCTIVITY IN THE SEA

D. J. CRISP

INTRODUCTION

The complexity of biological systems makes it necessary to impose limitations on the way in which we look at them if progress towards general principles is to be achieved. The measurement of energy flow through ecosystems has been the dominant theme in much of the International Biological Programme. By making energy content the common denominator for the comparison of different ecosystems, it is hoped that the trophic relationships of various habitats from a great variety of latitudes and regions can be better understood. Inevitably, this abstraction leaves out of account the faunistic composition, community structure, feeding behaviour and food preferences of the organisms concerned. The juicy steak and the old leather boot become equals in the eyes of the calorimeter.

A major problem is the endless variety of units of measure employed in productivity literature (Table 1), and the failure of authors to specify them. Even the great work of Zenkevitch (1963) does not state that the units of biomass used are wet weights—in themselves a crude measure with which to compare the living matter of animals and plants from different groups and environments. I have therefore endeavoured to keep to the following units throughout: biomass in kilocalories, area in square meters and time in years. Where precise comparative calibrations are not readily available I have used the conversion factors in Table 2. They are likely to be quite as precise as much of the raw data.

The idea of ignoring specific diversity and following the broad pattern of changes in nutrient elements and in the population densities of primary producers and herbivores was applied successfully by marine biologists to open-sea ecosystems as far back as the middle and late twenties. The underlying principles established in the classical works of, for example, Atkins (1926), Cooper (1933), Harvey (1928) and Marshall and Orr (1927, 1930) have since been refined and extended to account for seasonal, latitudinal and topographical variations in productivity over the world's oceans. These early successes have to a degree preempted the part that fundamental studies of oceanic productivity might have played in IBP. I doubt whether they should be ascribed to any superiority of marine over terrestrial ecologists; more likely, they can be explained by the circumstances of the environment itself.

CONDITIONS IN MARINE AND TERRESTRIAL ENVIRONMENTS

The open sea is much more difficult to sample and explore than a typical terrestrial environment. It is an inhospitable place and it extends in three dimensions. But even more important is the fact that it is continuously on the move. Large-scale turbulence renders marking and inspection of a small body of water on successive occasions very difficult (but cf. Cushing and Tungate, 1963). In terrestrial and in many freshwater environments specified groups or populations of organisms living in small, manageable and well-defined areas can usually be examined at regular intervals

TABLE 1 Primary Productivity and P/B Values for Open Water (Planktonic), Benthic, and Terrestrial Ecosystems

Habitat	Data as Given		Data in kcal			
	Biomass (m^{-2})	Production Net (m^{-2} yr^{-1} Unless Stated)	Biomass ($kcal/m^2$)	Production ($kcal/m^2/yr$)	P/B	Author
Open water						
Marine						
Fladen ground N. Sea 1953 Sta 1–4	3.5 g C[a]	58 g C (April–October)	35	1,000	29	Steele (1956)
English Channel	4 g dry org. wt[a]	270 g dry org. wt	20	1,350	67.5	Harvey (1950)
Long Island Sound	8 g C[a]	470 g C	80	4,700	58.8	Riley (1956)
Sargasso Sea (Oligotrophic)	0.87 g C[b]	134 g C	8.7	1,340	154	Menzel and Ryther (1961)
Peru current (Eutrophic)	14 g C[c]	10 g C day^{-1}[c]	140	36,500	261	Menzel *et al.* (1971)
Freshwater						
Oligotrophic Lakes	–	7–25 g C/yr	–	70–250	–	Lund (1970)
Eutrophic Lakes	–	75–250 g C/yr	–	750–2,500	–	Lund (1970)
Sewage treatment ponds, California	140–340 ppm to 25 cm depth = 60 g m^{-2}	4,500 g dry wt	240	18,000	75	Goluake *et al.* (1960)
Benthic, littoral, and shallow water						
Marine						
Fucus beds, Woods Hole, full sunlight	2,000 g dry wt = 800 g C	0.4×10^3 ml O_2 day^{-1} = 7,300 g C yr^{-1}	8,000	73,000	9.1	Kanwisher (1966)
Ascophyllum beds, Nova Scotia	8 kg dry wt (max)	2.0–2.6 kg dry wt	16,000[d]	9,200	0.58	MacFarlane (1952)
Cystoceira abies marina community, Canary Isles	630 g C	10.5 g C day^{-1} = 3,836 g C yr^{-1}	6,300	38,360	6.1	Johnston (1969)
Laminaria community, Nova Scotia	98 g C/m shoreline = 265 g C m^{-2}	1,750 g C	2,650	17,500	6.6	Mann (1972)
Thalassia sp., tropical turtle grass beds	5.66 kg dry wt	4,650 g C	22,600	46,500	2.1	Odum, H. T. (1956) Burkholder *et al.* (1959)
	601 g dry wt	5.8 g C day^{-1} = 2,100 g C yr^{-1}	1,962	21,000	10.8	Qasim and Bhattashiri (1971) Moore *et al.* (1968)

Marine coral reef	703 g dry wt (algal biomass)	8,760 g glucose m^{-2} yr^{-2} × 1.2 = 4,200 g C[e]	2,800	42,000	15	Odum and Odum (1955)
		2,900 g C	—	29,000		Kohn and Halfrich (1957)
Estuarine and brackish water						
Spartina marsh, Georgia	2.1 kg dry wt	3.3 kg m^{-2} dry wt		13,200		Odum, E. P. (1969)
		5,200 kcal m^{-2}		5,200		Smalley (1960)
		32 metric tons ha^{-1} org. dry wt	8,400	16,000	1.9	After Westlake (1963)
	42.5 S.N.U.[d]	43 S.N.U.	4,350	4,300	1.0	McFadyen (1964)
Chara sp., Caspian Sea	29,840 g fresh wt = 6 kg m^2 dry wt	30 metric ton ha^{-1} dry wt = 10 tons ha^{-1} org. wt	10,000	5,000	0.5	Kiriva and Shapova (1939) (Quoted in Westlake, 1963)
Freshwater						
Typha (hybrid), Minnesota	4.36 × 100 kg ha^{-1}	>160 × 100 kg ha^{-1}	17,440	6,400	0.37	Bray *et al.* (1959)
Terrestrial						
Field grass, Minnesota	2,000 g root, 200 g shoot, dry wt	50 × 10^6 kcal ha^{-1}	7,400	5,000	0.67	Golley (1960)
	6,400 kg ha^{-1} dry wt	1,400 kg ha^{-1}	6,400	1,400	0.22	Bray *et al.* (1959)
Sugar cane, Java	4.25 kg m^{-2} dry wt[d]	2,300 Pikol $Bouw^{-1}$ = 34,500 kg C ha^{-1}	17,000	34,500	2.0	Giltay (1898)
Woodland deciduous (Birch)	79,100 kg ha^{-1}, 4,400 g m^{-2} dry wt[e]	122,900 kg ha^{-1} in 24 yr 850 g org. wt yr^{-1} [f]	17,600	4,250	0.24	Ovington and Madgewick (1959)
(Alder)	124,696 kg ha^{-1}, 8,500 g m^{-2} dry wt[f]	6,299 kg ha^{-1} yr^{-1} 1,570 g m^{-2} org. wt[f]	34,000	7,850	0.23	Ovington (1956)
Coniferous (*Pinus silvestris*)	9,700 kg ha^{-1} 4,600 g dry wt[d]	21,700 kg ha^{-1} in 23 yr 1,600 org. wt[f]	18,400	8,000	0.44	Ovington (1957)
(*Picea omorika*)	69 S.N.U.	76 S.N.U.	6,900	7,600	1.1	Ovington and Heitkamp (1960) (see McFadyen, 1964)

[a]From chlorophyll measurement.
[b]Mean of quarterly values quoted in Cushing (1959).
[c]Over a period of 5 days of upwelling.
[d]Mean biomass assumed to be half maximum.
[e]Corrected for diffusion, see Kohn and Halfrich (1957).
[f] Adjusted for root, leaf fall, thinning, etc., and maximum biomass halved to give mean biomass.

TABLE 2 Conversion Factors

Conversion	Multiplier
g wet flesh weight to kcal	× 0.5
g dry weight to kcal	× 4
g dry organic (ash-free) weight to kcal	× 5
g carbon (g C) to kcal	× 10

by quadrat or transect methods. In contrast, the marine biologist must accept samples taken at random from a very large study area, such as the English Channel or Long Island Sound, and must in consequence sacrifice finer detail for the sake of generalities applicable to the larger system. Fortunately, the water movements that make station marking so difficult offer compensations. They make the chemical and physical properties of the water mass far more uniform in space and more conservative in time than those of a comparable area of land surface. Consequently, relatively infrequent observations at few stations can often be fitted together making a seasonal picture representative of the whole area. But this advantage does not seem to extend in the same degree to the organisms themselves. Plankton is notoriously patchy in its distribution and its study requires specialized statistical treatments of the kind that have been developed as part of IBP-PM (Cassie, 1962, 1963).

An even more profound difference between oceanic and terrestrial systems arises from the fact that the trophic structure of the former extends in three dimensions. Apart from a scarcely significant aerial plankton and its avian consumers, the biologically important components of the air are disseminules without trophic activity, for example, winged seeds, pollen, spores, flying vertebrates and insects. But aqueous habitats, because they provide buoyancy and a full range of nutrient elements, can support trophic activity throughout their depth. Hence, whereas all trophic levels in a terrestrial ecosystem coexist more or less in the same plane, those in aquatic ecosystems become separated vertically by the sinking of dead organisms or their residues under the action of gravity. Primary production takes place at the well-illuminated fringes or at the surface of oceans and lakes, and successive trophic levels attain greater prominence with increasing depth.

It is unfortunate that planktonic, pelagic and benthic zones have generally been studied separately by marine ecologists, no doubt because each requires different equipment and expertise. As a result, their trophic connections are still very little understood. Nevertheless, it seems clear that an area of sea should be regarded as a single ecosystem, superficially rather uniform, but with a vertical stratification of trophic levels. Indeed, it would be as artificial to separate the benthos from the plankton as it would be to separate ground plants and litter in a woodland ecosystem.

PRIMARY PRODUCTION IN OPEN WATER ENVIRONMENTS

Before I can discuss secondary production, it is necessary to show how the fundamental differences between terrestrial and aquatic ecosystems outlined above influence the character of the primary producers and thereby all those that depend upon them. Plants in aquatic ecosystems exist for the most part as single cells of very small size widely dispersed in the surface waters. They can enjoy the advantages of this form because the medium is buoyant and because the supply of nutrient salts is not limited to a particular stratum. What are the advantages of unicellular dispersion?

In all biological systems, whether animal, plant or social, the larger the size of the organism the longer is the path of diffusion or communication and the less efficiently it can work. Bureacracies, for example, are much less efficient than families; the larger they are the worse they function. In animals, small size is always associated with a high potential rate of metabolism but in plants smallness has two further advantages. For a given quantity of plant tissue a large area to volume ratio not only allows more light to be intercepted but also provides a greater surface area across which nutrients can be mobilized.

Plants on land are less fortunate. Being confined to the plane of the earth's surface, they need roots to obtain water and nutrients and shoots to obtain light. In terms of competition, as distinct from absolute efficiency, there is a premium on size. Indeed, large terrestrial plants have come to resemble some human institutions in which sheer size gives them the advantage of being able to smother their rivals despite the accompanying loss of efficiency and the quantities of dead wood that become built into the supporting system.

Figure 1, taken from McFadyen's (1964) comparative account of early work on terrestrial and aquatic ecosystems, shows that the conversion rate of the same quantity of solar energy into secondary production is about ten times as great in aquatic as in terrestrial ecosystems. This is the well-established index of "ecological efficiency" which can be used to measure the transfer of energy from one trophic level to the next. However, a more spectacular contrast between aquatic and terrestrial ecosystems is seen in the very low standing stock of aquatic ecosystems (10–30 kcal per sq. meter) and the enormously greater quantity of living matter in terrestrial ecosystems (2,000–5,000 kcal m^{-2}). Although the biomass of the aquatic systems is more than two orders of magnitude lower, their production and relative transfer of energy to herbivores is higher. A more complete set of data for a number of ecosystems is given in Table 1, which includes productivity (P) per unit of biomass (B). Values of P/B are generally a function of size of primary producers and it is clear from the table that the micro-

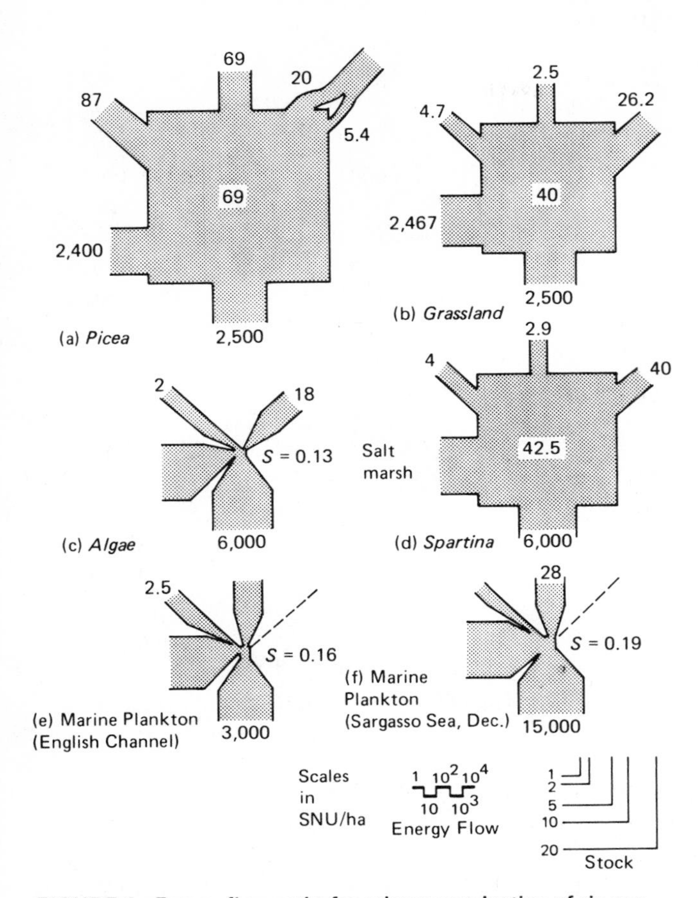

FIGURE 1 Energy flow paths for primary production of six ecosystems, from McFadyen (1964). The boxes represent biomass, the solar input is shown entering at 6 o'clock, energy loss in photosynthesis at 9 o'clock, plant respiration at 10 o'clock, net primary production available to herbivores at 12 noon, and mortality losses available to decomposers at 2 o'clock. Units used are standard nutritional units per hectare which should be multiplied by 100 to convert to kcal m^{-2}.

scopic primary producers of all aquatic systems have very high efficiencies in terms of P/B. Since most of these organisms are unicellular, with divisions taking place every few days, the values of P/B are of the order of 100 yr^{-1}. Not only are rooted plants larger and therefore slower growing than microalgae but they also contain a much larger proportion of material which is metabolically inactive but which is nevertheless included in the biomass. Such substances as cellulose, lignin and cork constitute a high proportion of the biomass of a forest ecosystem.

RELATIONSHIPS BETWEEN PRIMARY AND SECONDARY PRODUCERS

Differences in the physical structure of the terrestrial and aquatic environments have led to primary producers with very different sizes, forms, growth rates and constitution. These differences have in turn given rise to an entirely different relationship between the primary producers and the herbivores that depend upon them (Crisp, 1964). The very large standing stock of plant life in all natural environments on land is immediately obvious so that even an untrained observer could hardly overlook the dominant role played by plants in terrestrial ecosystems. However, the significance might escape him.

Evidently much of the terrestrial plant tissue escapes being eaten by herbivores and is dealt with instead by decomposers. Furthermore, with few exceptions, land-living herbivores are not wholely destructive of the plants on which they feed; they consume part, usually foliage, leaving the rest of the plant to regenerate. Sometimes the damage does not greatly affect potential plant production (e.g., Trojan, 1967) though sometimes it does (e.g., Varley, 1967). The indigestible character of much terrestrial plant tissue may discourage herbivore attack. At all events herbivores do not effectively control the biomass of land vegetation. Consequently the balance between plant and herbivore cannot be a simple matter of the available food supply; some other limitation, such as control of herbivores by carnivores, must exist in plant dominated terrestrial ecosystems.

An entirely different relationship applies to the open water ecosystem. The small size of the primary producers precludes any amicable arrangement whereby the herbivore browses on a permanent field of vegetation capable of continuously making good its losses. Microalgal cells can only be eaten whole. The phytoplankton survives only because the remaining cells become too sparse to support continued herbivore grazing, yet retain a sufficient reproductive potential to recover after the herbivore population has diminished.

The active herbivores of microalgae are the microcrustacea, mainly copepods in the sea and cladocerans in fresh water. From purely mechanical considerations the predators of microalgae must themselves be of small size (Parsons and Le Brasseur, 1970) and will, in consequence, have high metabolic rates and food requirements.

The second group of marine herbivores are the sedentary filter feeders. Since they have evolved fine filtering mechanisms to strain off the microalgae from the water, their size does not have to relate to that of their food particles and so they can attain dimensions considerably greater than those of the microcrustacea. Their food requirement per unit of biomass ought therefore to be lower. But on account of the demands of a pelagic larval strategy (Crisp, 1975) and their often dense accumulations in places where there is a convergence of streams of water from a wide productive area, they are prolific consumers of primary as well as detrital energy, especially in coastal and estuarine regions. Thus, in the sea, both active and sedentary herbivores are capable of consuming very quickly most of the energy that the primary producers make available, and so keep the standing stock of phytoplankton at a relatively low level.

In contrast to the characteristic terrestrial situation in which herbivores are limited and vegetation permanent, in the marine environment herbivores control, and are controlled by, stocks of phytoplankton. Such a simple predator-prey relationship at the base of the food chain is likely to produce severe fluctuations in biomass and production, and may account for the preponderence of opportunistic life styles in the sea (Crisp, 1975).

MICROCRUSTACEAN HERBIVORES

Copepods are the most important herbivores of marine food chains and have duly received most attention. Cushing and Tungate (1963) attempted to measure the grazing rate *Calanus* in a patch of water kept under constant surveillance. The production attributable to the four main diatom species present was estimated from the change in size of frustrules during the period of the survey (Cushing, 1955), and also by integrating solar radiation through depth and time and applying the relationship between division rate and energy demonstrated by Lund (1950). The difference between the observed algal biomass and that expected from the production data represented the total mortality from all causes, and the regression of mortality on herbivore weight was assumed to give the grazing mortality per unit herbivore biomass. When the algal grazing was partitioned between *Calanus* and other herbivores, it appeared that initially the amount eaten was greatly in excess of that required to maintain the observed increase in body weight of *Calanus*, whereas towards the end of the season the amount of algal food available was insufficient for the needs of the copepods. Cushing concluded that during the period of abundant algal food, the copepods were consuming algal cells at a rate greater than their needs (Cushing and Vucetic, 1963). Towards the end of the season, however, they were unable to obtain sufficient food to maintain egg laying. He accused them of "superfluous feeding" (Cushing, 1964).

Marshall and Orr's laboratory studies on *Calanus* failed to give support to this indictment; in contrast they showed that the female *Calanus* was an efficient egg-making machine with a rise in output as the ration of *Skeletonema* cells was increased (Marshall and Orr, 1964).

Cushing's results imply that under natural conditions with abundant food supply copepods have very low assimilation efficiencies. Petipa *et al.* (1970) measured zooplankton feeding rates under natural conditions from the gut content and rate of digestion and also from metabolic requirements. The copepods included under herbivores in their Tables 1 and 2 need a daily ration of up to 200% body weight per day, the figures being very variable but distinctly larger for the rapidly growing nauplius and copepodite stages. These values agree in order of magnitude with the mean of Cushing's determinations. However, the assimilation efficiencies worked out from Petipa's tables (though I am not clear if they are experimentally derived or assumed values) are in the range of 80%, which agrees with the extensive work of Marshall and Orr (1955). Marshall and Orr's values range from 0.6 to 0.95 for the assimilation efficiency of *Calanus finmarchicus* fed on a great variety of diatoms and flagellates. However, their phosphate tracer method may slightly overestimate the ratio. Nevertheless, there is considerable further support from the large number of laboratory experiments which are summarized in Table 3 for the view that copepods, even when passing food rapidly through the gut, can utilize it efficiently.

Growth rate and gross growth efficiencies are similarly high for marine herbivores. Values of P/B published by Mullin and Brookes (1970) for two Pacific copepods show

TABLE 3 Assimilation Efficiencies of Copepods from Laboratory Studies

Species	Food	Method	Mean (A/C)[a]	Author
Calanus finmarchicus	*Skeletonema costatum*	Phosphate tracer	0.69	Marshall and Orr (1955)
	Lauderia (old culture)	Phosphate tracer	0.58	Marshall and Orr (1955)
	Lauderia (young culture)	Phosphate tracer	0.98	Marshall and Orr (1955)
	Chaetoceros	Phosphate tracer	0.93	Marshall and Orr (1955)
	Chlamydomonas pulsatilla		0.95	Marshall and Orr (1955)
	Chromulina pusilla		0.50	Marshall and Orr (1955)
	Peridinium trochoideum		0.76	Marshall and Orr (1955)
	Gymnodinium vitiligo		0.92	Marshall and Orr (1955)
	Skeletonema	Nitrogen balance	0.62	Corner *et al.* (1967)
Calanus helgolandicus	Seston	Difference food–feces	0.82	Corner (1961)
Calanus hyperboreus	*Exuviella* sp.	Difference food–feces	0.72	Conover (1966)
	Exuviella sp.	Inorganic fecal content	0.69	Conover (1966)
		Average	0.70	

[a]A = assimilation; C = consumption.

the expected variation with size and temperature. The smaller nauplius stages weighing in the order of a few micrograms give P/B ratios of about 100 yr^{-1}, while the adults, weighing in the order of 100 micrograms, give values of about 60 at 15 °C and about 30 at 10 °C. Greeze (1970) quotes values of similar order for *Acartia clausii* and *Centropages ponticus* of 0.12 and 0.09 $days^{-1}$ (44 and 33 yr^{-1}) respectively. There is very good agreement generally that gross growth efficiencies, P/C (C = consumption), are also high for copepods. The index P/C is always strongly age-dependent, being much larger in the early, rapidly growing, stages. Since most copepods appear to be reasonably short-lived, the overall growth efficiency must therefore be high. Corner *et al.* (1967) gave a detailed budget for *Calanus finmarchicus* in terms of body nitrogen, separating the treatment of the growing stage from that of the adult stage. The amount of total resource put into growth by the former was 34%, while the reproductive output of the adult female was 14% of the amount of food eaten. Developmental and mature stages were approximately of equal duration, so that the growth efficiency over the whole life span of 10 weeks was, for the female, 24%. Mullin and Brookes (1970) provide exactly similar information for the developmental stages of *Rhinocalanus nasutus* and *Calanus helgolandicus pacificus* reared on two different species of diatoms, with growth efficiencies ranging from 30–45%, not counting the material incorporated in the cast skins.

In most offshore waters, the other marine herbivores and mixed feeders are in a minority compared with the copepods; nevertheless, being organisms of a similar size and longevity their growth efficiencies ought to be similar. Some of the published Russian work suggests that this is indeed so; thus Greeze (1970) and Pavlova (1967) record P/B values for marine cladocerans ranging from 70–110 yr^{-1}, while Petipa *et al.* (1970) give values ranging around 25 yr^{-1} for a variety of herbivores and mixed feeders. In the inshore waters, where there is a large annual influx of very small rapidly growing invertebrate larvae, the production per unit of biomass is bound to be very high. At the other extreme, the relatively large *Euphausia pacifica*, studied by Lasker (1966) and reviewed by Mullin (1969), produces 0.9 mg carbon m^{-2} d^{-1} from an ingestion rate of 5.5 mg carbon m^{-2} day^{-1} at a mean standing crop of 110 mg carbon m^{-2}. The calculated growth efficiency is therefore about 0.16 percent, and the P/B ratio 3.0 yr^{-1}.

Productivity/biomass ratios are strongly dependent both on temperature and food supply. There is a natural tendency for biologists to choose to work in spring and summer when both these factors are optimal so that it would not be surprising if a cursory survey of the literature gave a bias towards higher values. Mullin's (1969) review contains an excellent summary table in which the season and location of the measurements are recorded. Moreover, the text draws attention to the many dubious approximations and assumptions, the ambiguities and downright mistakes that are far from uncommon in this literature. Taken over the year as a whole with allowance for age class distribution and winter starvation, P/B values in temperate waters might well be reduced to 5–20 yr^{-1} according to the size of the species.

COMPARISONS BETWEEN MARINE AND TERRESTRIAL HERBIVORES

Productivity indices of planktonic herbivores may now be compared with those of their terrestrial counterparts. One must here distinguish between cold- and warm-blooded terrestrial herbivores. Grasshoppers serve as a good example of the former, since there have been several studies which agree tolerably well. The assimilation efficiency, as in most insects, is low, the values given for *Melanoplus* (Weigert, 1965) being the highest at 0.37, while slightly lower values are given by Odum *et al.* (1962) and by Smalley (1960) of 0.33 and 0.28 for *Orchelimum*. The utilization of plant food by grasshoppers is therefore only half as efficient as the utilization of phytoplankton by copepods. However, it must be remembered that land plants are different from microalgae in their biochemical composition though the difference might not be apparent in calorimetric measurements. Relatively indigestible material is required for structural strength, for resistance to desiccation and perhaps also to defend the plant against excessive herbivore attack. Such refractory components will result in low assimilation efficiencies unless specially powerful enzymes have been developed. Plant bugs, which restrict their intake to cell sap, perhaps afford a fairer comparison with aquatic herbivores; they have assimilation efficiencies of over 60% (Weigert, 1964). The growth efficiencies of insects similarly compare unfavourably with those of copepods, as one might expect on size criteria alone. Also, the more extreme range of temperature on land implies a greater proportionate reduction in productivity during the cooler months. Weigert and Evans (1967) quote P/B figures for grasshoppers of 3–5 yr^{-1}, an order of magnitude below that of marine herbivores, and their figures for other herbivorous insects are of the same magnitude.

The larger, warm-blooded herbivores present a somewhat different picture. Many have high assimilation efficiencies. From the tables presented by Weigert and Evans (1967) deer mice and ground squirrels have efficiencies over 60%, sparrows greater than 90%, but elephants, notorious for their fecal output, are only 33% efficient (Lamprey, 1964). No doubt choice of food plays an important part in optimizing efficiencies in higher animals, seed-eating birds, for example, being proficient in this respect, while symbionts may greatly assist in providing cellulase and other enzymes capable of breaking down otherwise indigestible materials.

But whatever advantage warm-blooded animals may gain by these refinements, they more than lose by maintaining the body temperature above that of the surroundings. The inevitable result of an over heated biochemical system is high respiratory losses which depress the growth efficiencies well below those of poikilotherms. Generally, the P/A (A = assimilation) values for homeotherms lie between 0.015 and 0.05, only about one-tenth of an average poikilotherm value. The domestic pig, which of all homeotherms must be regarded as having been designed for high productivity, and which has an assimilation efficiency of 76%, can boast of a gross growth efficiency (P/C) of only 10%.

The issue of outstanding importance is the ecological efficiency of marine planktonic herbivores, that is, the extent to which they transfer energy through the food chain. As I have shown, on the basis of their smaller size, higher metabolic and feeding activity, and more efficient ratios of assimilation and conversion, they clearly qualify for a rating well above that of terrestrial herbivores. But whether these merits are in fact put to man's advantage depends upon two vital relationships. First, whether in marine ecosystems a high proportion of primary energy in fact passes through herbivores, and secondly, whether the further transfer to higher trophic levels is efficient in giving yields which are useful to man.

In terrestrial ecosystems, the amount of primary energy passing through herbivores is surprisingly small, even if one includes organisms which feed on dead leaves and wood litter. As Table 4 indicates, herbivores rarely ingest more than a third of the plant tissue available above the ground. Even where the proportion of plants being grazed is favorable, as in naturally grazed grassland, the primary production available to the herbivores is rather small and the herbivores themselves are poor assimilators of ingested energy.

Intuitively, one tends to the view that since litter and detritus of plant origin are conspicuously absent from marine environments, the bulk of primary production passes through the planktonic herbivore chain. But if one attempts to produce a table corresponding to Table 4 for the marine environment, there are no firmly based observations to support it. The reason is clear. For while it is relatively easy on land to account for the loss of plant material and to trace its path through a series of organisms which can be kept under constant observation, to trace the food web from microscopic plants through a sequential pattern of swimming organisms in a turbulent medium is not at present possible. Instead, planktologists have made two alternative or combined approaches: either to rely on laboratory experiments or to erect models based on very crude and often second-hand statistical data. Laboratory experiments on pelagic animals and plants are probably valid only insofar as they can be extrapolated to low densities of organisms in wall-less containers. As examples of such problems, one can cite the difficulty experienced in obtaining copepod filtration rates similar to those in nature (Marshall and Orr, 1955; Corner, 1961) and in measuring natural metabolic (Petipa, 1966) and reproductive (Mullins and Brookes, 1967) activities. Models and theories are, in my view, an even poorer substitute for the relevant facts, though they may have a value as a "Gedenken experiment" provided that all assumed or cannibalized data are critically assessed, clearly exposed, and correctly calculated.

The available literature giving indications of the fraction of phytoplankton consumed by planktonic herbivores has been well reviewed by Mullin (1969). Table 1, column 6, of his paper records values between 0.0002 and 0.58 for this quantity, while the text cites one or two impossible results where the value exceeds unity! Restricting the choice to

TABLE 4 Primary Production and Utilization by Herbivores in Terrestrial Ecosystems

Habitat	Net Primary Production (kcal m^{-2})	Herbivore Ingestion as Fraction of Production Aboveground	Author
S. Carolina grass field	1,075	0.12	Odum *et al.* (1962)
Michigan grass field	1,360	0.01	Weigert and Evans (1967)
Spartina salt marsh	6,585	0.08	Teal (1962)
Savanna			
Tanganyika		0.28	Lamprey (1964)
Uganda	750	0.60	Petrides and Swank (1965)
Managed range maximum exploitation		0.45	Lewis *et al.* (1956)
Forest			
Coniferous	2,150 (litter)	0.19	Kitazawa (1967)
Warm temperate	5,650 (litter)	0.28	Kitazawa (1967)
Temperate (Canada)	1,570 (foliage)	0.05–0.08	Bray (1964)
Temperate deciduous (*Liriodendron*)	1,640 (foliage)	0.056	Reichle and Crossley (1967)
Vaccinio-Myrtilli-Pinetum	3,030 (foliage and litter)	0.096	Kaczmarek (1967)
Pine-oak-alder	5,060 (foliage and litter)	0.17	Kaczmarek (1967)
Potato crop	676 g dry wt. (?) leaves	0.115	Trojan (1967)

those cases where the calculation is based on the whole of the zooplankton, or on the dominant organism present, there remain several in the region of 0.25–0.5, mainly for temperate waters. The intuitive assumption might also be drawn from Menzel and Ryther's (1961) observations that higher values will apply to tropical seas. In this whole field, new methods and more critically determined facts are badly needed. Turning to the question of food chain efficiency beyond the second trophic level, the results are extremely fragmentary and no general conclusion seems at present possible. Steele (1965, 1974) has cogently argued that in order to account for the observed high fishery yields in such areas as the North Sea, it is necessary to assume, not only high ecological efficiency, perhaps approaching the values which Slobodkin (1962) considered maximal for aquatic systems, but also direct food chains, with little branching, between phytoplankton and fish. Support for this view is afforded by the high growth efficiencies of pelagic herbivores and carnivores, as for example in the tropical *Sagitta hispida* which has a growth efficiency of 0.36 on a nitrogen basis throughout its life (Reeve, 1970). However, Reeve's statement that a large part of the marine food chain passes through secondary carnivores, such as chaetognaths, would not be reconcilable with Steele's views. Gulland (1967), in considering potential global fishery yields, also supports Steele's argument that high ecological efficiencies are necessary. Yields are likely to be particularly high where fish can short-circuit the food chain by feeding directly on phytoplankton, e.g., sardine (Lasker, 1970; Ryther, 1969) or can recover energy from detritus, e.g., mullet (Odum, 1970).

DETRITAL FOOD CHAIN

When planktonic organisms die, their remains join the feces, cast skins and various other particulate residues as detritus. Detritus, together with its associated saprophytic organisms, constitutes an important source of energy and ultimately links the plankton food chains with those of the benthos. Another important source of energy frequently overlooked is the leakage of soluble nutrients from aquatic algae, both planktonic and benthic (Fogg, 1971; Ignatiades, 1973; Khailov and Burlakova, 1969; Khailov and Finenko, 1970; Sieburth and Jensen, 1969). The mode of utilization of dissolved material is not understood, but its most probable fate is to become adsorbed on inorganic particles and utilized by bacteria. All these detrital components, therefore, tend to sink to the bottom and provide much of the raw material for benthic feeders. Obviously, with increasing depth, the component of dead phytoplankton increases as the component of living phytoplankton decreases, but below the eutrophic zone both plant and animal detritus diminish in absolute amounts. Furthermore, the quality of detritus becomes modified by the loss of the more biochemically active material as it sinks through the water column (Finenko and Zaika, 1970). Detritus is freely utilized by many herbivores, including copepods, while *Noctiluca*, a saprophytic dinoflagellate, often becomes extremely abundant at times when large algal blooms are decaying, and then itself becomes a major souce of dead organic matter. However, all detritus must originate from primary production at the surface, and cannot be put forward as a source of food for zooplankton supplementing the photosynthetic production of a closed area, as Finenko and Zaika appear to suggest.

Since phytoplankton and detritus are utilized by herbivores and saprophytes as they sink through the water column, it follows that in shallow waters the trophic chain will be abbreviated and more material will become available to animals living on the sea bed. Hence the productivity of the benthos is generally inversely related to depth, while the efficiency of conversion of plant material into zooplankton increases with the depth of the water. Qasim (1970) studied the primary production of a shallow channel in South India and found a large surplus of plant material which was not required by the rather small numbers of herbivores present. The average yearly net primary production was 124 g C m^{-2} yr^{-1} of which only 30 g C m^{-2} yr^{-1} was required by the planktonic herbivores present in the estuary. Qasim concluded that, in such situations, much of the phytoplankton must either die or be utilized directly by benthic invertebrates and that this pathway could profitably be exploited by herbivorous or detritus-consuming animals such as mullet and prawn. It is of course a general principle that shallow coastal waters offer the best conditions for benthic fisheries of all kinds, not only because of the abbreviation of the pelagic food chain but also because such areas are frequently surrounded by productive marshland which exports organic matter to the estuary (Odum, 1959, 1960). It was the recognition of the potential of such areas that led IBP-PM to emphasize their study.

BIOMASS OF THE MACROBENTHOS

Russian workers have long been active in accumulating information on the quantitative distribution of biomass on the sea bed, much of which is summarized in Zenkevitch (1963), and more is being fostered under IBP. Their data is given in terms of grams wet weight m^{-2} of living or preserved macrobenthos, including water, shell and other nonliving matter. Table 5, abstracted from the above source, indicates the tremendous variation in biomass, and hence presumably in benthic production, in different seas of shallow to moderate depth. Leaving aside the obvious feature that the biomass is greater in shallower seas, certain other trends can be readily picked out. It will be noted that the seas with prolonged ice cover have low benthos biomass. Although there is no clearly marked relationship between salinity and benthic biomass, in regions where there are well

TABLE 5 Mean Biomass[a] of the Macrobenthos (From Zenkevitch, 1963)

Sea	Approximate Surface Salinity (%)	Approximate Depth (m)	Mean Biomass (g m^{-2})	Notes
Northern Seas				
Barents S.	32–35	215	100	Pack ice seasonal in N.E.
White S.	25–26	110	20	6 months ice cover
Kara S.				
West	23–35	200–300	50	Short summer ice free period
"Brown grounds"	26–32	100–200	2–5	Short summer ice free period
Shallows	15–26	50	100–300	Short summer ice free period
Baltic Sea				
N. Gulf of Bothnia	3–4	60–140	0.2	5–7 months ice cover
S. Gulf of Bothnia	4–6	100–300	12	
Gulf of Finland	6–7	0–80	57	3–6 months ice cover
Gulf of Riga	6–7	0–60	38	
Baltic, N. of 56°	6–7	100–200	25	Anaerobic deep water
Baltic, S. of 56°	7–8	50–100	60	
Belts and Oresund	8–25	7–30	186	
Southern Seas				
Black Sea	17–19	1,270	35	Anaerobic deep layer
Sea of Azov	10–12	7.2	210–400	Excluding G. Taganrog
Gulf of Taganrog	1–8	4.7	30–55	Receives R. Don
N. Caspian S.	0–12.5	180	120	Receives R. Volga
S. Caspian S.	12.5–13.5	325	20	Anaerobic deep water
Aral Sea	10–14	16	16–23	Impoverished fauna
Atlantic and Mediterranean				
North Atlantic	33–35	1,000 (?)	266	
Adriatic and W. Mediterranean	35	0–2,000	185	
E. Mediterranean, Tyrrhenian Sea	35–37	0–2,000	6	

[a]Biomass is given as gross wet weight. To convert to kcal m^{-2} multiply by 0.5–0.2 according to the water and mineral content.

marked haloclines (the Central and E. Mediterranean) and stagnant bottom water (deeper parts of the Baltic, Black Sea and S. Caspian) the average level is depressed, even though the shallower regions may be rich. Shallow seas receiving large rivers (e.g., Sea of Azov, Northern Caspian) are as rich as shallow ocean basins.

The reduction in benthic biomass with increasing depth is clearly established, not only by the work of Filatova (1960) and Vinogradova (1962), but also by that of Sanders *et al.* (1965). Despite the wide differences in the equipment used and in the numbers of individuals recorded, which makes strict comparison impossible, the investigators agree that the density of animal life in the abyssal plains is less than that on the continental slopes by one or two orders of magnitude, and is particularly sparse where the overlying seas are relatively infertile (e.g., Sargasso Sea, Table 6).

Bottom living species are divisible into (a) the epifauna, including free living demersal fish and invertebrates (shrimps, whelks, etc.) and sessile forms living at the surface; (b) the infauna, comprising animals hidden beneath the surface. Zatsepin (1970) further divides the macrobenthos into filter feeders which utilize suspended food from the water, detritus feeders which consume material loosely accumulated near the surface and deposit feeders that utilize organic remains by ingesting the sediment itself. The abundance of each group can be related to depth and type of deposit. He showed that, over typical areas of the Barents Sea, the total biomass falls from 266 g m^{-2} wet weight near the surface to 40 g m^{-2} at depths exceeding 325 m. With increasing depth the proportion of filter feeders falls while that of deposit feeders and detritus feeders rises. More significant, however, were the changes in feeding type with the character of the deposit. In coarser, gravelly deposits the filter feeders accounted for 70–80% of all animals. Moving into the finer sand and silt deposits, epifaunal filter feeders almost disappeared and the infaunal filter feeders were increasingly displaced by detritus and deposit feeders. Less than 20% of the benthic biomass was drawn from other trophic groups.

PRODUCTION MEASUREMENTS ON MARINE BENTHOS

The trophodynamics of marine benthic organisms is known only from a few studies, some of which are recorded in Table 7. Taking first the values of the assimilation efficiencies, those for browsing herbivores are about the same as or slightly higher than those for insect herbivores such as grasshoppers. The actual value depends much on the food source as Carefoot's (1967b) careful analysis of feeding in *Aplysia* has shown, the softer and more delicate algae being preferred as well as being more easily digested (Table 8). For the same

TABLE 6 Numbers of Benthic Animals per Square Meter from Moderate to Great Depths, Showing Variation with Depth

Sea Area	Depth (m)	Animals (m^{-2})	Observer
E. Mediterranean	100–200	290	Chukhchin (1963)
	200–1,000	21	
	1,000–3,000	<2	
Bering Sea	100–1,000	521	Kuznetsov (1964)
	2,000–3,000	118	
Kurile Island	0–50	102	Kuznetsov (1963)
	50–100	94	
	100–200	111	
	200–500	245	
	500–1,000	284	
	1,000–2,000	26	
N.E. Pacific	<4,000	26	Filatova and Levenstein (1961)
Java Trench	6,000–7,000	25	Belyaev and Vinogradova (1961)
W. Atlantic	100–180	1,790	Wigley and McIntyre (1964)
Off S. New England	350–600	1,170	
W. Atlantic	Shelf	6,000–13,000	Sanders *et al.* (1956)
Bermuda transect	Upper Slope	6,000–23,000	
	Lower Slope	1,500–3,000	
	Abyssal rise	500–1,200	
	Abyssal plain	150–270	
	Ditto Sargasso Sea	31–130	
	Bermuda Slope	140–850	

TABLE 7 Secondary Production by Marine Benthic Invertebrates

Animal	Population Production (yr^{-1})	P/B (yr^{-1})	Assimilation Efficiency	Author
Herbivores				
Littorina irrorata	41 kcal m^{-2}	0.7	–	Odum and Smalley (1959)
Littorina littorea	50 kcal m^{-2}	0.4	0.45	Grahame (1970)
Littorina planaxis	–	–	0.40	North (1954)
Aplysia punctata	–	7.3	0.67–0.74	Carefoot (1967a)
Carnivores				
Navanax inermis	–	–	0.62	Paine (1965)
Dendronotus frondosus	–	18.2	0.86–0.93	Carefoot (1967a)
Archidoris pseudoargus	–	9.0	0.52–0.93	Carefoot (1967a)
Suspension feeders				
Cardium edule	1,150 g (wet wt) m^{-2} 230 kcal m^{-2} (?)	4.0	–	Zenkevitch (1963)
Mytilaster lineatus	900 g (wet wt) m^{-2} 180 kcal m^{-2} (?)	3.22	–	Zenkevitch (1963)
Balanus improvisus	300 g (wet wt) m^{-2} 60 kcal m^{-2} (?)	4.76	–	Zenkevitch (1963)
Modiolus demissus	16.7 kcal m^{-2}	0.3	–	Kuensler (1961)
Tellina tenuis	3.6 kcal m^{-2}	0.7	–	Trevallion *et al.* (1970)
Mixed suspension-detritus feeders				
Pandora gouldiana	6.2 g dry wt m^{-2} 25 kcal m^{-2}	2.0	–	Sanders (1956)
Scrobicularia plana	71 kcal m^{-2}	0.6–0.9	0.45	Hughes (1970)
Syndesmya ovata	377 g wet wt m^{-2} 75 kcal m^{-2} (?)	1.0–2.05	–	Zenkevitch (1963)
Deposit feeders				
Nephthys incisor	9.3 g dry wt m^{-2} 37 kcal m^{-2}	2.16	–	Sanders (1956)
Cistenoides gouldii	1.7 g dry wt m^{-2} 7 kcal m^{-2}	1.94	–	Sanders (1956)
Yoldia limatula	3.2 g dry wt m^{-2} 13 kcal m^{-2}	2.28	–	Sanders (1956)

TABLE 8 Growth of Juvenile *Aplysia punctata* on Different Algal Diets (From Carefoot, 1967b)

Alga	Order of Choice	Assimilation Efficiency (%)	P/B (yr^{-1})[a]
Plocamium coccineum	2	65	7.3
Enteromorpha intestinalis	1	59	7.1
Ulva lactuca	3	75	6.6
Heterosiphonia plumosa	4	71	5.0
Cryptopleura ramosa	5	71	3.9
Delessaria sanguinea	6	45 (?)[b]	3.4
Laminaria digitata	7	53 (?)[b]	0.6
Desmarestia aculeata	8	—	0

[a] P/B calculated from $(W_2 - W_1)/\frac{1}{2}(W_2 + W_1) \times 365/t$, where W_1 = initial mean weight and W_2 = mean weight after t days growth.
[b] (?) indicates results based on insufficient material ingested.

reason the assimilation efficiency of carnivores is high, whereas that of detritus and deposit feeders, whose food is already highly degraded, is likely to be low. Heywood and Edwards (1962) for example found an efficiency of only 4% for the freshwater mud snail *Potamopyrgus jenkinsii*. Furthermore the greater the depth at which benthic animals live, the less readily assimilable is the food material and the lower its energy content. Allen and Sanders (1966) show that such organisms have very large intestines for long retention of material and exceedingly slow growth rates. Their productivity in consequence is likely to be of a much lower order than that of animals to which we are normally accustomed, but for obvious reasons no precise information is available.

The values of P/B for most of the intertidal and shallow water species lie in the region 1–5 yr^{-1}, not really different from those of terrestrial invertebrates of similar size. Browsing intertidal periwinkles appear to have low P/B values, so also have detritus and deposit feeders, while those given for carnivores are relatively high as might be expected. However, account should be taken of the fact that young stages of marine invertebrates are often hard to find and are not usually fully represented in samples, so that the P/B values of field populations may be too low.

Marine benthic herbivores of temperate climates feeding on macroalgae are few in number. Like their terrestrial counterparts, they do not make serious inroads into the mass of algal vegetation that covers the intertidal and shallow sublittoral. The greater part of this vegetation must therefore decay or be destroyed mechanically, entering the food chain by way of detritus-feeding organisms. It is surprising that the relatively soft tissues of the macroalgae are not more heavily browsed; perhaps the biochemically peculiar reserves or the presence of acids and phenolic compounds deter intensive attack. However, those herbivores that live by scraping and sweeping rock surfaces are present in abundance—for example, the various groups of limpets, the littorinas, neritas and trochids. These herbivores are clearly in the ascendant over the plants on which they feed, not allowing them to progress beyond the sporeling stage before they are cleared away. It is a surprising relationship (Southward, 1956) since the evolution of some mechanism of herbivore restraint would allow a much greater supply of plant material to grow on the rock and thereby permit a larger population of herbivores to be supported. The immediate consumption by the herbivores of the minute algal sporelings as soon as they start to grow on the rock face suggests that herbivore productivity must be limited by food supply, and that strong competition for food and space must exist between individuals. This animal–plant relationship therefore resembles the balance between zooplankton and phytoplankton—except that the littoral herbivores are much longer lived and slower growing than microcrustacea.

The most remarkable feature among shallow water invertebrates is the very high biomass and production of the populations of suspension feeders. Secondary production per unit area of such assemblages must be the highest in the natural environment. For example, the *average* density of a population consisting almost exclusively of *Mytilus edulis* on the Murman Coast is given by Zenkevitch (1963) as 5127 g m^{-2}. Assuming a 2-yr turnover (P/B = 0.5) the production would be 2.5 kg m^{-2} yr^{-1} or about 500 kcal m^{-2}. Considerably higher values of biomass have been recorded elsewhere with, presumably, a productivity several times greater. An interesting verification of the above figure is to be found in an older comparison of terrestrial and marine productivity by Johnstone (1908) in which he gives the average annual commercial yield from cultivated mussel beds in Morcambe Bay at 114 g dry wt. yr^{-1}, or approximately 440 kcal m^{-2}. Total production, including losses to natural predators, would presumably be somewhat higher, perhaps in the order of 1000 kcal m^{-2}.

The fact that the production per unit area of populations of marine suspension feeders appears so high compared with that of terrestrial and planktonic herbivores is really because the comparison is an unfair one. Suspension feeders grow in profusion whenever there are strong currents, so that the area which they normally occupy is only a very small part of the area of sea surface which supplies the primary energy on which they feed. Nonetheless they are probably intrinsically efficient convertors of energy. They do little or no work in seeking their food. Unlike higher vertebrates, they do not maintain an internal temperature, with resulting heat loss, but they nevertheless control their metabolic rate and activity efficiently, so that it is acclimated to ambient temperature (Crisp and Ritz, 1967; Widdows and Bayne, 1971).

MEIOBENTHOS AND MICROBENTHOS

Two other groups of benthic organisms deserve mention, but largely because their importance has not been matched by investigation. These are the benthic meiofauna, consisting of protozoa and invertebrates of the size range 0.1 to 5 mm, living interstitially or on the surface of deposits, and the microfauna consisting of bacteria and other microorganisms living within deposits. Both groups may play an important role in the recycling of nutrients that reach the sea bed, in much the same way as bacteria and soil microfauna do on land. Far too little is known of their significance in the tropic chain, but from Wieser and Kanwisher's (1961) observations of the uptake of energy by estuarine mud, a demand of 14–19 mg C m^{-2} h^{-1} (1400 kcal m^{-2} yr^{-1}) seems possible. Even higher values for estuarine shoals of 425 g C m^{-2} yr^{-1} (4000 kcal m^{-2} yr^{-1}) are recorded by Marshall (1970) while McIntyre *et al.* (1970) for sandy beach conditions, estimates 50 g C m^{-2} yr^{-1} (500 kcal m^{-2} yr^{-1}). They believe the energy uptake of the meiofauna of intertidal sands to be derived from dissolved organic compounds percolating through the void spaces and utilized first by bacteria which are then eaten by interstitial animals. If this energy route is capable of dealing with large quantities of material, shallow sands may play the part of natural percolating filters in breaking down surplus organic waste. The sand meiofauna indeed seems to be naturally tolerant of high organic loads and of other forms of pollution (Gray, 1971).

There are indications, therefore, that the meiofaunal chain may be particularly important in shallow water areas where the detrital energy component is particularly large. Encouragement of the meiofauna might also help in converting detritus into food in coastal lagoons. In managed prawn fisheries, for example, meiofauna might substitute the trash protein which is normally an expensive element in the industry. A suitably managed meiofaunal chain might similarly improve mullet fisheries (Odum, W. E., 1970). The potential for improvement resulting from more rapid recycling of detritus is well illustrated by the beneficial effects of the introduction of *Nereis diversicolor* into the northern Caspian Sea (Romanova, 1960).

A generation of fundamental study is required to elucidate these processes. In addition to our vast ignorance of the activities of benthic microorganisms, there are two other exciting areas ready for further exploration. First the work of Stephens (1963, 1964) on polychaetes and of the Southwards (1970, 1972a, 1972b) on *Pogonophora* has reopened Pütter's old speculation on direct uptake of dissolved organic nutrients from the environment. These inquiries have been extended to other groups present in deposits (Stephens, 1968), so that the general significance to benthic trophodynamics needs to be assessed. Secondly, the role of the so called "thiobios" in releasing energy in the deeper anaerobic layers of sediment needs investigation (Fenchel, 1969; Fenchel and Riedl, 1970).

MARINE CARNIVORES

I have now reviewed the main types of herbivores responsible for secondary production in the sea, and touched upon the great army of suspension, detritus and deposit feeders which, in company with other benthic organisms, clear up the fall of dead and decaying food that reaches the sea bed. As yet no account has been taken of the terminal carnivores which, as has been shown, dominate the sea as plants dominate the land.

Long before the advent of IBP, fish occupied the central pivot of marine biological research, and some effort was also devoted to the economically important mammals. But the bulk of this vast literature is less concerned with the place of terminal carnivores in the marine ecosystem than with the part their exploitation plays in the human economy. It is only fairly recently that studies of fish nutrition have been started and these have tended to be concentrated on freshwater fish which are easier to deal with.

Table 9 lists the uniformly high values of assimilation efficiency for a number of fish, freshwater and marine. Like most carnivores, fish are excellent assimilators, particularly of protein, with efficiencies of about 90%. Similarly, they are efficient converters of assimilated energy into flesh, but the values of growth efficiency are of course dependent on age, as is indicated by Table 10 for the Pacific sardine. Pandian (1967) gives, for two species of fish, the exponential relationship of consumption, assimilation, respiration and growth to weight, W. Both fish give similar results which indicate that the first three variables rise as $W^{0.75}$ whereas growth increases only as $W^{0.5}$. Hence the growth efficiency P/C must fall with increasing weight as $W^{-0.25}$. Lasker's data (Table 10) gives an identical relationship up to the third year of life but older fish put less energy into production than would be predicted by a $W^{-0.25}$ law. Nevertheless, the above relationship, coupled with population statistics, might make it possible to predict the food consumption of fish

TABLE 9 Assimilation Efficiency in Fish

Species	Assimilation Efficiency (%)	Notes	Reference
Salvelinus fontinalis	90.3	Recalculated by Pandian (1967)	Job (1960)
Cyprinus carpio	89		Ivlev (1939)
Megalops cyprinoides	91.5	cal wet oxidation	Pandian (1967)
	97.2	Protein nitrogen	Pandian (1967)
Ophiocephalus striatus	90.6	cal wet oxidation	Pandian (1967)
	97.1	Protein nitrogen	Pandian (1967)
Lepomis sp.	96–98	Protein nitrogen	Gerking (1952, 1954)
Epinephelus guttatus	96.0	Protein nitrogen	Menzel (1960)
Pleuronectes platessa	92	calories	Birkett (1970)
Sardinops caerulea	90.3	Dry wt (recalculated)	Lasker (1970)

from values of annual fish production. Growth efficiencies rather higher than those given by Lasker for the active pelagic sardine are recorded for the more lethargic flat fish; 14% for plaice (Peterson, 1918), 20% for 0 group plaice, flounder and turbot (Müller, 1969) and 36% for young plaice and dab under optimal conditions (Edwards and Steele, 1968; Edwards *et al.*, 1969). However, the relation between consumption and production for wild populations of fish is likely to be considerably lower than growth efficiencies measured in the laboratory since, in nature, growth will depend very much on feeding rate or "ration" and on searching activity. Mann (1965) gives values in the region of 0.06 for populations of river fish. A factor of 10 to 15 is perhaps appropriate for converting fish production into annual food requirement.

Fortunately, fish yields are known from many parts of the world, and if these yields are sustainable and not a drain on the capital resources of the stock, they can give some indication of total fish production, and hence of the significance of fish in the ecosystem. Gulland (1967, 1970) reviews this problem and arrives at a figure for the North Sea of 0.2 and 0.6 g C m^{-2} yr^{-1} for demersal and pelagic fish respectively.

TABLE 10 Effect of Age on Conversion Efficiency (P/C) for *Sardinops caerulea*. Growth and Reproductive Output Included in Production; Assimilation Efficiency Assumed 90 Percent throughout Adult Life (Lasker, 1962, 1970)

Age	Weight[a]	Food	Conversion Ratio or Growth Efficiency P/C
Embryonic		Yolk	62.0
0–1	<20 g	Artemia	16.6
1–2	20–54 g	Artemia	9.5
2–3	54–106 g	Artemia	6.75
3–4	106–140 g	Artemia	3.8
4–5	140–165 g	Artemia	2.7
5–6	165–180 g	Artemia	1.9

[a]Weight given in gonad-free, fat-free values.

Steele (1974) assumed that fishing accounted for 80% mortality among demersal fish—the North Sea being a very heavily trawled area—and 50% among pelagic fish. From total yields of 0.93 and 2.04 million tons from an area estimated at 5×10^5 km^2, employing Winberg's (1956) approximation (1 g wet wt. fish = 1 kcal) he obtained productions of 2.5 and 8 kcal m^{-2} yr^{-1}, which are close enough to Gulland's estimates.

FOOD CHAIN EFFICIENCY AND FISHERIES YIELD

For the North Sea it is possible to construct an approximate scheme for the main energy chain, which may be typical of that of a cool temperature marine ecosystem. The representation (Figure 2) was inspired by Steele's penetrating analysis of this ecosystem (Steele, 1974) though the treatment here differs in detail.

Figure 2 is divided into a pelagic food chain on the left, based on primary production by microalgae and a benthic food chain on the right, based on detrital energy reaching the sea floor. The upper part of the figure takes primary production as its starting point in the pelagic system and shows the various sources of energy contributing to detritus. The secondary production by pelagic herbivores is shown to reach a value of 175 and 180 kcal m^{-2} yr^{-1} by two independent routes. The supposed conversion of detrital energy by macrobenthic organisms is based on a very crudely approximated biomass of 20 kcal m^{-2}, which is the order of magnitude found by Russian workers for the benthic biomass of fertile temperate seas. It will be seen that there is only just sufficient detritus to supply the macrobenthos, and therefore the energy available for meiobenthic and microbenthic activity is only a small fraction of the total.

The lower part of Figure 2 starts with the known fishery yields and leads to values, first of total pelagic and demersal fish production as outlined above, and thence to values of the energy consumed by fish populations. Between the upper and lower sections of the figure the links are virtually unknown. However, it can be seen that the energy made

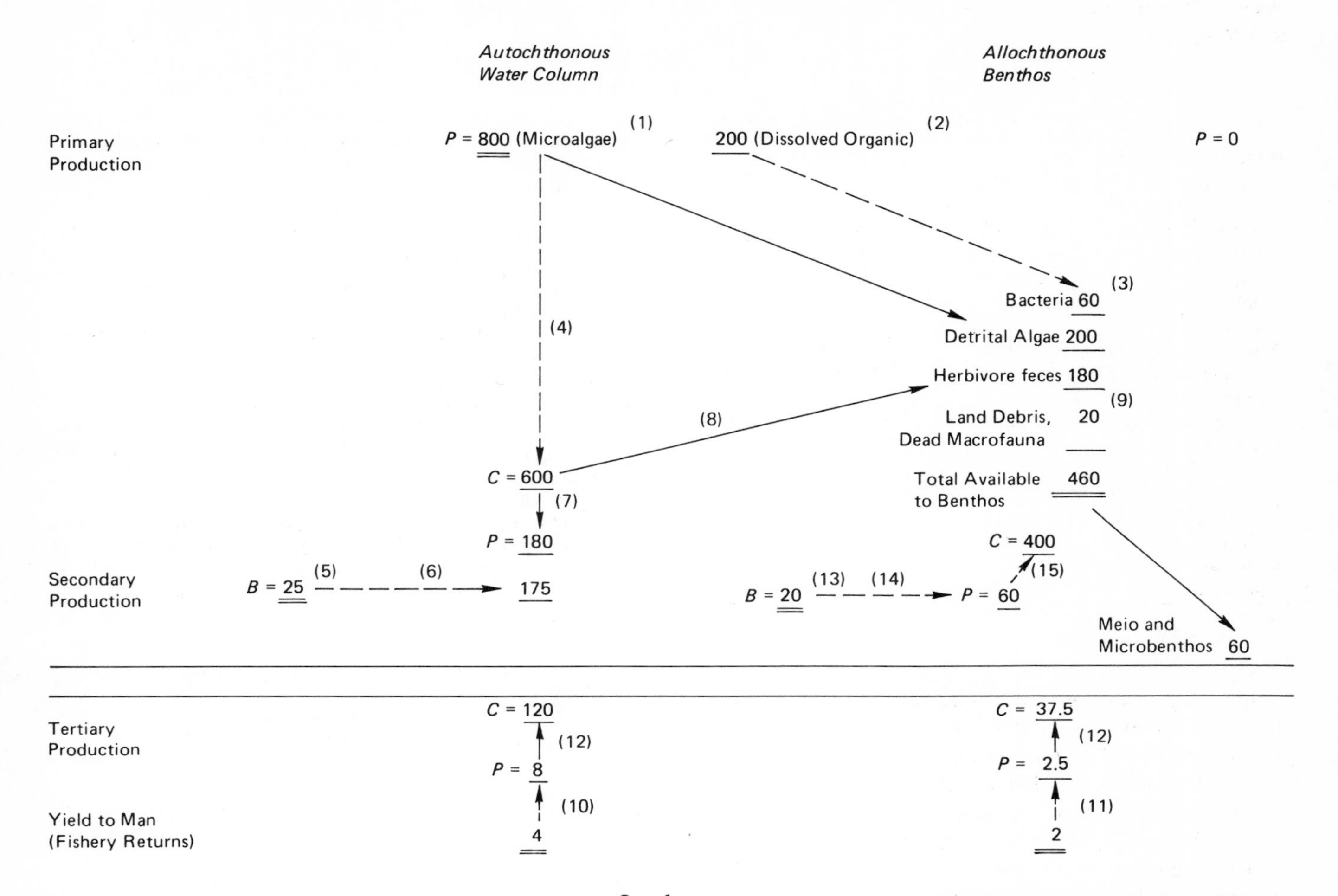

FIGURE 2 Energy flow hypothesis for the North Sea (in kcal m^{-2} yr^{-1}). In the above figure energy flow between trophic levels is linked by arrows, but the direction of the arrow represents the logical derivation of the values given, not the direction of energy flow, which is always from top to bottom of the page. The double underlined values are based on observations, the continuous interconnecting lines are steps based on observed relationships, while the dotted interconnecting lines indicate more speculative relationships. Numbers in parentheses explain the relations observed or assumptions made.

(1) Production values based on Steemen Nielson (1958), Steele (1956), Steele and Baird (1961).

(2) Assumption that 20% of primary production is released as dissolved organic matter (Parsons and Seki, 1971).

(3) Assumption that 30% of dissolved photosynthetic material incorporated in benthic bacteria, or in bacteria which are attached to particles and enter the benthos.

(4) Assumption that 75% of microalgae eaten by herbivores.

(5) 4 X dry weight values of zooplankton from Adams, quoted by Steele (1974).

(6) Production/biomass (P/B) ratio assumed to be 7 for Calanus in natural environment.

(7) P/C or growth efficiency assumed at 30%.

(8) Assimilation efficiency assumed 70%, feces 30% of consumption. (Table 3).

(9) Pure assumption, inserted for completeness.

(10) Pelagic fishing mortality assumed 50% of total.

(11) Demersal fishing mortality assumed 80% of total.

(12) Growth efficiency of fish population P/C taken as 1/15 (Mann, 1965).

(13) Benthic biomass taken as 100 g m^{-2} = 20 kcal m^{-2}, average figure from Russian work (Table 5).

(14) P/B for benthic invertebrates taken as 3.0 (Table 7).

(15) Growth efficiency assumed to be 15%.

TABLE 11 The Contribution of Different Types of Sea Area to Fish Production (Ryther, 1969). Ocean Area 361 X 10^6 km^2

	Type of Sea Area		
	Upwelling	Coastal	Oceanic
Percentage of ocean area, (a)	0.1%	9.9%	90%
Primary production kcal m^{-2} yr^{-1}, (P_1)	3,000	1,000	500
Ecological efficiency per trophic level, (e)	0.20	0.15	0.10
Number of trophic levels from plant production to fish production, (n)	1.5	3	5
Overall efficiency (e^n)	9×10^{-2}	3.4×10^{-3}	10×10^{-5}
Fish production kcal m^{-2} yr^{-1} (P_1e^n)	270	3.4	0.005
Total fish production 10^6 tons/year (Using Winberg's transformation 1 kcal = 1 g fish flesh) (3.61 a P_1e^n)	98	122	1.6
Percentage of fish production	44%	54%	0.7%

available by herbivores and by macrobenthic producers are not greatly in excess of the needs of the fish populations that are supported by them. Hence no great part of the energy flow passes outside the main food chain. There appears to be an excess of only about 50–60% of the energy used in fish production which could be available to other pelagic and benthic carnivores. The above figures are, of course, exceedingly speculative, but they indicate that the temperate shallow water marine ecosystem is efficient in producing mainly organisms useful to man. The food chain described would have a total efficiency of $\frac{10.5}{1000}$ with *two* trophic levels, or 10% ecological efficiency at *each* stage.

Less is known of other marine ecosystems. Ryther (1969) analyzed the probable relationships in three main types of open water situations. He supposed that the greater the fertility of the water and the greater its primary productivity, the fewer would be the trophic levels in the food chain and the more efficient the transfer between each pair of levels. Thus, in rich upwelling areas (Table 11) fish such as the anchovy may feed in part directly on the phytoplankton. The ecological efficiency of each step may be as high as 20%. Conversely in clear oceanic waters there may be multiple trophic levels between plants and terminal carnivores and each trophic step may be only 10% efficient. The result of this high gearing is that the upwelling areas account for a very much greater potential of world fish production than either their area or their rate of primary production would indicate. Conversely, the deep oceanic areas with clear infertile water, despite their huge extent, seem unlikely to provide a fishery yield comparable with that of the coastal and upwelling regions.

REFERENCES

Allen, J. A., and H. L. Sanders. 1966. Adaptations to abyssal life as shown by the bivalve *Abra profundorum* (Smith). Deep-Sea Res. 13:1175–1184.

Atkins, W. R. G. 1926. The phosphate content of sea water in relation to the growth of algal plankton. J. Mar. Biol. Ass. U.K. 14:447–467.

Belyaev, G. M., and N. G. Vindogradova. 1961. An investigation of the Java Trench deep sea bottom fauna. Okeanologiya Akad. Nauk. USSR 1:125–132.

Birkett, L. 1970. Experimental determination of food conversion and its application to ecology, p. 261–264. *In* J. H. Steele (ed.) Marine food chains. Oliver and Boyd, Edinburgh.

Bray, J. R. 1964. Primary consumption in three forest canopies. Ecology 45:165–167.

Bray, J. R., D. B. Lawrence, and L. C. Pearson. 1959. Primary production in some Minnesota terrestrial communities. Oikos 10:38–49.

Burkholder, P. R., L. M. Burkholder, and J. A. Rivero. 1959. Some chemical constituents of turtle grass, *Thalassia testudinum*. Bull. Torrey Bot. Club 86:88–93.

Carefoot, T. H. 1967a. Growth and nutrition of three species of opisthobranch molluscs. Comp. Biochem. Physiol. 21:627–652.

Carefoot, T. H. 1967b. Growth and nutrition of *Aplysia punctata* feeding on a variety of marine algae. J. Mar. Biol. Ass. U.K. 47:565–589.

Cassie, R. M. 1962. Frequency distribution models in the ecology of plankton and other organisms. J. Anim. Ecol. 31:65–92.

Cassie, R. M. 1963. Multivariate analysis in the interpretation of numerical plankton data. N.Z. J. Sci. 6:36–59.

Chukhchin, V. D. 1963. Quantitative distribution of benthos in the eastern part of the Mediterranean Sea. Trudy Sevastopol. Biol. Sta. 16:215–233.

Conover, R. J. 1966. Assimilation of organic matter by zooplankton. Limnol. Oceanogr. 11:338–345.

Cooper, L. H. N. 1933. Chemical constituents of biological importance in the English Channel, Nov. 1930 to Jan. 1932. Part 1. Phosphate, silicate, nitrate and ammonia. J. Mar. Biol. Ass. U.K. 18:677–728.

Corner, E. D. S. 1961. On the nutrition and metabolism of zooplankton. I. Preliminary observations on the feeding of the marine copepod *Calanus helgolandicus* (Claus). J. Mar. Biol. Ass. U.K. 41:5–16.

Corner, E. D. S., C. B. Cowey, and S. M. Marshall. 1967. On the nutrition and metabolism of zooplankton. V. Feeding efficiency of *Calanus finmarchicus*. J. Mar. Biol. Ass. U.K. 42:259–270.

Crisp, D. J. 1964. Grazing in terrestrial and marine environments. Introduction ix–xvi. Blackwell, Oxford.

Crisp, D. J. In press. Energy relations of marine invertebrate larvae. Symp. of Marine Invertebrate Larvae, Roving, 1973.

Crisp, D. J., and D. A. Ritz. 1967. Temperature acclimation in barnacles. J. Exp. Mar. Biol. Ecol. 1:236–256.

Cushing, D. H. 1955. Production and a pelagic fishery. Fish. Invest. II, 18. No. 7. H.M.S.O., London.

Cushing, D. H. 1959. On the nature of production in the sea. Fishery Investigation, Ser. II, 22(6):1–40.

Cushing, D. H. 1964. The work of grazing in the sea, p. 207–225. *In* D. J. Crisp (ed.) Grazing in terrestrial and marine environments. Blackwell, Oxford.

Cushing, D. H., and D. S. Tungate. 1963. Studies of a *Calanus* patch.

I. The identification of a *Calanus* patch. J. Mar. Biol. Ass. U.K. 43:327–337.

Cushing, D. H., and T. Vucetic. 1963. Studies on a *Calanus* patch. III. The quantity of food eaten by *Calanus finmarchicus*. J. Mar. Biol. Ass. U.K. 43:349–371.

Edwards, R. R. C., and J. H. Steele. 1968. The ecology of O group plaice and common dabs in Loch Eive. I. Population and food. J. Exp. Mar. Biol. Ecol. 2:215–238.

Edwards, R. R. C., D. M. Finlayson, and J. H. Steele. 1969. The ecology of O group plaice and dabs in Loch Eive. II. Experimental studies of metabolism. J. Exp. Mar. Biol. Ecol. 3:1–17.

Fenchel, T. 1969. The ecology of marine microbenthos. IV. Structure and function of the benthic ecosystem, its chemical and physical factors and the microfauna communities with special reference to the ciliated protozoa. Ophelia 6:1–182.

Fenchel, T. M., and R. J. Reidl. 1970. The sulphide system; a new biotic community underneath the oxidized layer of marine sand bottoms. Mar. Biol. 7:255–258.

Filatova, Z. A. 1960. On the quantitative distribution of the bottom fauna in the central Pacific. Trudy Inst. Okeanol. Akad. Nauk. CCCP 41:215–233.

Filatova, Z. A., and R. J. Levenstein. 1961. Quantitative distribution of the deep sea bottom found in the northern-eastern Pacific. Trudy Inst. Okeanol. Akad. Nauk. USSR 45:190–213.

Finenko, Z. Z., and V. E. Zaika. 1970. Particulate organic matter and its role in the productivity of the sea, p. 32–44. *In* J. H. Steele (ed.) Marine food chains. Oliver and Boyd, Edinburgh.

Fogg, G. E. 1971. Extra cellular products of algae in freshwater. Arch. Hydrobiol. Beih. Ergebn. Limnol. 5:1–25.

Gerking, S. D. 1952. The protein metabolism of the sunfishes of different ages. Phys. Zool. 25:358–372.

Gerking, S. D. 1954. The food turnover of a bluegill population. Ecology 35:490–498.

Giltay, F. 1898. Uber die vegitabilische Stoffbildung in den tropen und in mitteleuropa. Ann. Jard. Bot. Buitenz. 15:43–72.

Golley, F. B. 1960. Energy dynamics of a food chain of an old field community. Ecol. Monogr. 30:187–206.

Goluake, C. G., W. J. Oswald, H. K. Gea, and B. B. Cook. 1960. Production of low cost algal protein, p. 174–184. *In* Kachroo (ed.) Symp. Algology. New Delhi, India.

Grahame, J. W. 1970. Energy flow in *Littorina* species with special reference to energy expenditure in reproduction. Unpublished thesis. U. of Wales.

Gray, J. S. 1971. The effects of pollution on sand meiofauna communities. Thalassia Jugoslavica 7:79–86.

Greeze, V. N. 1970. The biomass and production of different trophic levels in the pelagic communities of south seas, p. 458–467. *In* J. H. Steele (ed.) Marine food chains. Oliver and Boyd, Edinburgh.

Gulland, J. A. 1967. Area reviews of the world's oceans: Northeast Atlantic. F.A.O. Fish. Circ. 109:Rev. 1. 40 p.

Gulland, J. A. 1970. Food chain studies and some problems in world fisheries, p. 296–316. *In* J. H. Steele (ed.) Marine Food Chains. Oliver and Boyd, Edinburgh.

Harvey, H. W. 1928. Biological chemistry and physics of sea water. Cambridge Comp. Physiol. Cambridge University Press. 194 p.

Harvey, H. W. 1950. On the production of living matter in the sea off Plymouth. J. Mar. Biol. Ass. U.K. 29:97–143.

Heywood, J., and R. W. Edwards. 1962. Some aspects of the ecology of *Potamopyrgus jenkinsi* Smith. J. Anim. Ecol. 31:239–250.

Hughes, R. N. 1970. An energy budget for a tidal flat population of the bivalve *Scrobicularia plana* (da Costa). J. Anim. Ecol. 39:357–381.

Ignatiades, L. 1973. Studies on the factors affecting the release of organic matter by *Skeletonema costatum* (Greville) Cleve in field conditions. J. Mar. Biol. Ass. U.K. 53:923–935.

Ivlev, V. S. 1939. Balance of energy in carps. Zool. Zh. 18:303–318.

Job, S. V. 1960. Growth and calorific approximation in the speckled trout, *Salvelinus fontinalis*. Indian J. Fish 7:129–136.

Johnston, C. S. 1969. The ecological distribution and primary production of macrophytic marine algae in the Eastern Canaries. Int. Rev. Ges. Hydrobiol. 54:473–490.

Johnstone, James. 1908. Conditions of life in the sea. Cambridge Biological Series, Cambridge University Press. 322 p.

Kaczmarek, W. 1967. Elements of organization in the energy flow of forest ecosystems, p. 663–678. *In* K. Petrusevicz (ed.) Secondary Productivity of Terrestrial Ecosystems. Inst. of Ecology, Polish Academy of Sciences, Warsaw. Symposium of IBP (PT) vol. 2.

Kanwisher, J. W. 1966. Photosynthesis and respiration in some seaweeds, p. 407–420. *In* H. Barnes (ed.) Some contemporary studies in marine science. Allen and Unwin, London.

Khailov, K. M., and Z. P. Burlakova. 1969. Release of dissolved organic matter by marine seaweeds and distribution of their total organic production to inshore communities. Limnol. Oceanogr. 14:521–527.

Khailov, K. M., and Z. Z. Finenko. 1970. Organic macromolecular compounds dissolved in sea water and their inclusion in food chains. *In* J. H. Steele (ed.) Marine food chains. Oliver and Boyd, Edinburgh.

Kiriva, M. C., and T. F. Shapova. 1939. Benthic vegetation of the North Eastern part of the Caspian Sea. Bulletin of the Moscow Society of Naturalists, Biology Division 48(2–3):3–14.

Kitazawa, Y. 1967. Community metabolism of soil invertebrates in forest ecosystems of Japan, p. 649–661. *In* K. Petrusevicz (ed.) Inst. Ecology Polish Acad. Sci., Warsaw. Symposium of IBP (PT) 2:649–661.

Kohn, A. J., and P. Halfrich. 1957. Primary organic productivity of a Hawaiian coral reef. Limnol. Oceanogr. 2:241–251.

Kuensler, E. J. 1961. Structure and energy flow of a mussel population in a Georgia salt marsh. Limnol. Oceanogr. 6:191–204.

Kuznetsov, A. P. 1963. Benthic invertebrate fauna of the Kamchatka waters of the Pacific Ocean and northern Kurile Islands. Trudy Inst. Okeanol. Akad. Nauk. USSR 271 p.

Kutnetsov, A. P. 1964. Distribution of the sea bottom fauna in the western part of the Bering Sea and trophic zonation. Trudy Inst. Okeanol. Akad. Nauk. USSR 69:98–177.

Lamprey, A. F. 1964. Estimation of the large mammal densities, biomass and energy exchange in the Tarangire Game Reserve and the Masai in Tanganyika. East African Wildlife Jour. 2:1–46.

Lasker, R. 1962. Efficiency and role of yolk utilization by developing embryos and larvae of the Pacific sardine *Sardinops caerulea*. J. Fish. Res. Bd. Can. 19:867–875.

Lasker, R. 1966. Feeding growth respiration and carbon utilization of a Euphausiid crustacean. J. Fish. Res. Bd. Can. 23:1291–1317.

Lasker, R. 1970. Utilization of zooplankton energy by a Pacific sardine population in the California current, p. 265–284. *In* J. H. Steele (ed.) Marine food chains. Oliver and Boyd, Edinburgh.

Lewis, J. K., G. M. Van Dyne, L. R. Albee, and F. W. Whetzal. 1956. Intensity of grazing; its effect on livestock and forage production. Bull. 459 D.S. State College Agric. Exp. Stn. Brookings. 44 p.

Lund, J. W. G. 1950. Studies on *Asterionella formosa* Hass II. Nutrient depletion and the spring maximum. J. Ecol. 38:1–35.

Lund, J. W. G. 1970. Primary production. Water treatment and exam. 19:332–358.

MacFarlane, C. 1952. A survey of certain seaweeds of commercial importance in Southwest Nova Scotia. J. Fish. Res. Bd. Can. 30:78–97.

Mann, K. H. 1965. Energy transformations by a population of fish in the River Thames. J. Anim. Ecol. 34:253–275.
Mann, K. H. 1972. Ecological energetics of the seaweed zone in a marine bay on the Atlantic Coast of Canada. I. Zonation and biomass of seaweeds. Mar. Biol. 12:1–10.
Marshall, N. 1970. Food transfer through the lower trophic levels of the benthic environment, p. 52–66. *In* J. H. Steele (ed.) Marine food chains. Oliver and Boyd, Edinburgh.
Marshall, S. M., and A. P. Orr. 1927. The relation of the plankton to some chemical and physical factors in the Clyde sea area. J. Mar. Biol. Ass. U.K. 14:837–868.
Marshall, S. M., and A. P. Orr. 1930. A study of the spring diatom increase in Loch Striven. J. Mar. Biol. Ass. U.K. 16:853–878.
Marshall, S. M., and A. P. Orr. 1955. On the biology of *Calanus finmarchicus.* VIII. Food uptake, assimilation, and excretion in adult and stage V *Calanus.* J. Mar. Biol. Ass. U.K. 34:495–529.
Marshall, S. M., and A. P. Orr. 1964. Grazing by copepods in the sea, p. 227–238. *In* D. J. Crisp (ed.) Symp. Brit. Ecol. Soc. Blackwell, Oxford.
McFadyen, A. 1964. Energy flow in ecosystems and its exploitation by grazing, p. 3–20. *In* D. J. Crisp (ed.) Grazing in terrestrial and marine environments. Brit. Ecol. Soc. Symp. Blackwell, Oxford.
McIntyre, A. D., A. L. S. Munro, and J. H. Steele. 1970. Energy flow in a sand ecosystem, p. 19–31. *In* J. H. Steele (ed.) Marine food chains. Oliver and Boyd, Edinburgh.
Menzel, D. W. 1960. Utilization of food by a Bermuda reef fish, *Epinephelus guttatus.* J. Cons. Perm. Expl. Mar. 25:216–272.
Menzel, D. W., and J. H. Ryther. 1961. Zooplankton in the Sargasso Sea off Bermuda and its relation to organic production. J. Cons. Perm. Int. Expl. Mar. 26:250–258.
Menzel. D. W., and J. H. Ryther. 1961. Annual variations in primary production of the Sargasso Sea of Bermuda. Deep-Sea Res. 7:282–288.
Menzel, D. W., J. H. Ryther, E. M. Hulbert, C. J. Lorenzen, and N. Corwin. 1971. Production and utilization of organic matter in Peru coastal current. Invest. Pesq. 35(1):43–59.
Moore, H. B., L. T. Davies, T. H. Frazer, R. H. Gore, and N. R. Lopez. 1968. Some biomass figures from a tidal flat in Biscayne Bay, Florida. Bull. Mar. Sci. 18:261–279.
Muller, A. 1969. Korpergewicht und Gewichtzunahme junger Plattfische in Nord–und Ostsee. Ber. Dtsch. Wiss. Komm. Meeresforsch. 20.
Mullin, M. M. 1969. Production of zooplankton in the ocean: the present status and problems. Oceanogr. Mar. Biol. Ann. Rev. 7:293–314.
Mullin, M. M., and E. R. Brookes. 1967. Laboratory culture, growth rate and feeding behaviour of a planktonic marine copepod. Limnol. Oceanogr. 12:657–666.
Mullin, M. M., and E. R. Brookes. 1970. Growth and metabolism of two planktonic marine copepods as influenced by temperature and type of food, p. 74–95. *In* J. H. Steele (ed.) Marine food chains. Oliver and Boyd, Edinburgh.
North, W. J. 1954. Size distribution, erosive activities and gross metabolic efficiency of the marine intertidal snails *Littorina planaxis* and *L. scutulate.* Biol. Bull. Wood's Hole 106:185–197.
Odum, E. P. 1959. Fundamentals of Ecology. (2nd ed.) Saunders, Philadelphia, Penna. 546 p.
Odum, E. P. 1960. The role of tidal marshes and streams in estuarine production. 19th Ann. meeting. Atlantic States Marine Fisheries Commission, S. Carolina. Sept. 1959.
Odum, E. P., and A. E. Smalley. 1959. Comparison population energy flow of a herbivorous and a deposit-feeding invertebrate in a salt marsh ecosystem. Proc. Nat. Acad. Sci. 45:617–622.
Odum, E. P., C. E. Connell, and L. B. Davenport. 1962. Population energy flow of three primary consumer components of old field ecosystems. Ecology 43:88–96.
Odum, H. T., and E. P. Odum. 1955. Trophic structure and productivity of a windwood coral reef community on Enuvetok Atoll. Ecol. Monogr. 25:291–320.
Odum, H. T. 1956. Primary productivity in flowing waters. Limnol. Oceanogr. 1:102–117.
Odum, W. E. 1970. Utilization of the direct grazing and plant detritus food chains by the striped mullet *Mugil cephalus,* p. 222–240. *In* J. H. Steele (ed.) Marine food chains. Oliver and Boyd, Edinburgh.
Ovington, J. D. 1956. The form, weights and productivity of tree species grown in close stands. New. Phytol. 55:289–304.
Ovington, J. D. 1957. Dry matter production by *Pinus sylvestris* L. Ann. Bot. Lond. (N.S.) 21:287–314.
Ovington, J. D., and D. Heitkamp. 1960. The accumulation of energy in forest plantations in Britain. J. Ecol. 48:639–646.
Ovington, J. D., and H. A. I. Madgewick. 1959. The growth and composition of natural stands of birch. Plant and Soil 10:271–283.
Paine, R. T. 1965. Natural history limiting factors and energetics of the opisthobranch *Navanax inermis.* Ecol. 46:603–609.
Pandian, T. J. 1967. Intake digestion absorption and conversion of food in the fishes *Megalops cyprinoides* and *Ophiocephalus striatus.* Mar. Biol. I:16–32.
Parsons, T. R., and R. J. LeBrasseur. 1970. The availability of food to different trophic levels in the marine food chain, p. 325–343. *In* J. H. Steele (ed.) Marine food chains. Oliver and Boyd, Edinburgh.
Parsons, T. R., and H. Seki. 1971. Importance and general implications of organic matter in aquatic environments, p. 1–27. *In* D. W. Hood (ed.) Organic matter in natural waters. Inst. Mar. Sci., U. of Alaska. Publ. No. 1.
Pavlova, E. V. 1967. Food consumption and energy metabolism of Cladocera populations in the Black Sea, p. 65–85. *In* Structure and Dynamics of Aquatic Communities and Populations. Akad. Nauk. USSR. Inst. Biol. Southern seas. Kiev.
Peterson, C. G. J. 1918. The sea bottom and its production of fish food. Rep. Den. Biol. Stn. 25:1–62.
Petipa, T. S. 1966. Oxygen consumption and food requirements in copepods *Arcartia clausi* Giesbr, and *A. latisetosa* Kritcz. Zool. Zh. 43(3):363–370.
Petipa, T. S., E. V. Pavlova, and G. N. Mironov. 1970. The food web structure, utilization and transport of energy by trophic levels in the plankton communities, p. 142–167. *In* J. H. Steele (ed.) Marine food chains. Oliver and Boyd, Edinburgh.
Petrides, G. A., and W. G. Swank. 1965. Estimating the productivity energy relations of an African elephant population. Proc. 9th Grasslands Congress, Sao Paulo, Brazil.
Qasim, S. Z. 1970. Some problems related to the food chain in a tropical estuary, p. 45–51. *In* J. H. Steele (ed.) Marine food chains. Oliver and Boyd, Edinburgh.
Qasim, S. Z., and P. M. A. Bhattashiri. 1971. Primary production of a seagrass bed on Kavaratti Atoll (Laccadives). Hydrobiologia 38:29–38.
Reeve, M. R. 1970. The biology of Chaetognatha. I. Quantitative aspects of growth and egg production in *Sagitta,* p. 168–189. *In* J. H. Steele (ed.) Marine food chains. Oliver and Boyd, Edinburgh.
Reichle, D. E., and D. A. Crossley. 1967. Investigation on heterotrophic productivity in forest insect communities, p. 499–518. *In* K. Petrusevicz (ed.) Secondary Productivity of Terrestrial Ecosystems. Inst. of Ecology, Polish Acad. Sci., Warsaw. Symposium of IBP (PT) vol. 2.
Riley, G. A. 1956. Oceanography of Long Island Sound 1952–1954.

IX. Production and utilization of organic matter. Bull. Bingham Oceanogr. Coll. Yale U. 15:324–344.

Romanova, N. 1960. Benthos distribution in the Central and Southern Caspian. Zool. J. USSR 39:6.

Ryther, J. H. 1969. Relationship of photosynthesis to fish production in the sea. Science 166:72–76.

Sanders, H. L. 1956. Oceanography of Long Island Sound 1952–54. X. The biology of marine bottom communities. Bull. Bingham Oceanogr. Coll. 15:345–414.

Sanders, H. L., R. R. Hessler, and G. R. Hampson. 1965. An introduction to the study of deep-sea benthic faunal assemblages along the Gay Head–Bermuda transect. Deep-Sea Res. 12:845–867.

Sieburth, J. M., and A. Jensen. 1969. Studies on algal substances in the sea. III. The production of extra-cellular organic matter by littoral marine algae. J. Exp. Mar. Biol. Ecol. 3:290–309.

Slobodkin, L. B. 1962. Energy in animal ecology, p. 69–101. *In* J. B. Cragg (ed.) Advances in Ecological Research. Acad. Press, New York and London.

Smalley, A. E. 1960. Energy flow of a salt marsh grasshopper population. Ecology 41:785–790.

Southward, A. J. 1956. The population balance between limpets and seaweeds on wave beaten rocky shores. Rep. Mar. Biol. Stn. Port Erin for 1955. No. 68.

Southward, A. J., and E. C. Southward. 1970. Observations on the role of dissolved organic compounds in the nutrition of benthic invertebrates. Experiments on three species of Pogonophora. Sarsia 45:69–96.

Southward, A. J., and E. C. Southward. 1972a. Observations on the role of dissolved organic compounds in the nutrition of benthic invertebrates. II. Uptake by other animals living in the same habitat as pogonophores and by some littoral Polychaeta. Sarsia 48:61–68.

Southward, A. J., and E. C. Southward. 1972b. Observations on the role of dissolved organic compounds in the nutrition of marine invertebrates. III. Uptake in relation to organic content of the habitat.

Steele, J. H. 1956. Plant production on the Fladen ground. J. Mar. Biol. Ass. U.K. 35:1–33.

Steele, J. H. 1965. Some problems in the study of marine resources. Int. Comm. N.W. Atlantic Fish. Spec. Pub. 6:463–476.

Steele, J. H. 1974. The structure of a marine ecosystem. Harvard Univ. Press. 125 p.

Steele, J. H., and I. E. Baird. 1961. Relations between primary production, chlorophyll and particulate carbon. Limnol. Oceanogr. 6:68–78.

Steemen Nielson, E. 1958. A survey of recent Danish measurements of the organic productivity of the sea. Rapp. p.v. Cons. Int. Explor. Mer. 144:38–46.

Stephens, G. C. 1963. Uptake of organic material by aquatic invertebrates II. Accumulation of amino acids by the bamboo worm *Clymenella torquata*. Comp. Biochem. Physiol. 10:191–202.

Stephens, G. C. 1964. Uptake of organic material by marine invertebrates III. Uptake of glycine by brackish water annelids. Biol. Bull. Mar. Lab. Wood's Hole 126:150–162.

Stephens, G. C. 1968. Dissolved organic matter as a potential source of nutrition for marine organisms. Am. Zool. 8:95–106.

Teal, J. M. 1972. Energy flow in the salt marsh ecosystem of Georgia. Ecology 43:614–624.

Trevallion, A., R. R. C. Edwards, and J. H. Steele. 1970. Dynamics of a benthic bivalve, p. 285–295. *In* J. H. Steele (ed.) Marine food chains. Oliver and Boyd, Edinburgh.

Trojan, P. 1967. Investigations on production of cultivated fields, p. 545–562. *In* K. Petrusevicz (ed.) Secondary Productivity of Terrestrial Ecosystems. Inst. of Ecology, Polish Acad. Sci., Warsaw. Symposium of IBP (PT) vol. 2.

Varley, G. C. 1967. The effects of grazing by animals on plant productivity, p. 773–778. *In* K. Petrusevicz (ed.) Secondary Productivity of Terrestrial Ecosystems. Inst. of Ecology, Polish Acad. Sci., Warsaw. Symposium of IBP (PT) vol. 2.

Vinogradova, N. G. 1962. Some problems of the study of deep sea fauna. J. Oceanogr. Soc. Japan 20th Anniversary Vol. 724–741 p.

Weigert, R. G. 1964. Population energetics of meadow spittlebugs (*Philaenus spumarius* L.) as affected by migration and habitat. Ecol. Monogr. 34:217–241.

Weigert, R. G. 1965. Energy dynamics of the grasshopper populations in old field and alfalfa field ecosystems. Oikos 16:161–176.

Weigert, R. G., and F. C. Evans. 1967. Investigations of secondary productivity in grasslands, p. 499–578. *In* K. Petrusevicz (ed.) Secondary Productivity of Terrestrial Ecosystems. Inst. of Ecology, Polish Acad. Sci., Warsaw. Symposium of IBP (PT) vol. 2.

Westlake, D. F. 1963. Comparisons of plant productivity. Biol. Rev. 38:385–425.

Widdows, J., and B. L. Bayne. 1971. Temperature acclimation of *Mytilus edulis* with reference to its energy budget. J. Mar. Biol. Ass. U.K. 51:827–843.

Wieser, W., and J. Kanwisher. 1961. Ecological and physiological studies of marine nematodes from a salt marsh near Wood's Hole, Massachusetts. Limnol. Oceanogr. 6:262–270.

Wigley, R. L., and A. D. McIntyre. 1964. Some quantitative comparisons of offshore meiobenthos and macrobenthos south of Martha's Vineyard. Limnol. Oceanogr. 9:485–493.

Winberg, G. C. 1956. Rate of metabolism and food requirements of fish. Fish. Res. Bd. Can. Trans. Ser. 194.

Zatsepin, V. I. 1970. On the significance of various ecological groups of animals in the bottom communities of the Greenland, Norwegian and the Barents Sea, p. 207–221. *In* J. H. Steele (ed.) Marine food chains. Oliver and Boyd, Edinburgh.

Zenkevitch, L. 1963. The biology of the seas of the USSR. Tr. S. Botcherskeya. George Allen and Unwin, London. 955 p.

DECOMPOSITION OF ALLOCHTHONOUS ORGANIC MATTER AND SECONDARY PRODUCTION IN STREAM ECOSYSTEMS

N. K. KAUSHIK

The extent of secondary production in most ecosystems depends largely on indigenous primary production. Although the rate of primary production may be greater in streams than that in various lentic environments (Odum, 1956; Tominago and Ichimura, 1966), its magnitude is limited. Some of the important factors that adversely affect primary production in streams are: unstable and sandy beds restricting suitable substrate for phytobenthos; torrential currents having devastating effects on phytoplankton; and turbulence, turbidity and shading reducing or eliminating sunlight. That indigenously produced organic matter is generally not sufficient to sustain secondary production normally encountered in streams becomes obvious if comparison is made between the rates of gross primary production (P) and community respiration (R). It has been shown (Table 1) that in normal streams the P/R ratio is nearly always less than one and has thus indicated that the bulk of the plant food material, which forms the basis of the secondary production is not synthesized indigenously. Therefore most streams are heterotrophic, and obviously allochthonous organic matter is important for secondary production.

DEGREE OF ALLOTROPHY IN STREAMS

The role of allochthonous organic matter in stream economy was suggested as early as 1912 by Thienemann and since then has been discussed by some limnologists (Hynes, 1963, 1970). However, the measurement of the degree of allotrophy has been attempted only recently and by a few stream ecologists. Teal (1957) attributed 76 percent of the energy at primary producer level in Root Spring, Massachusetts, to allochthonous material. In a Georgian stream, 66 percent of the net primary consumer productivity was found to have its origin outside the stream (Nelson and Scott, 1962). Similarly, in Linesville Creek, Pennsylvania, it has been shown (Cummins *et al.*, 1966) that the biomass of primary macroconsumers was almost the same as that of detrital macroconsumers. A study on Bear Brook, New Hampshire (Fisher and Likens, 1972), the only study that has itemized the nature of input in a stream, has shown that allochthonous organic matter accounted for 99.8 percent of the energy available at primary producer level, and that 52.9 percent of this was in the form of particulate organic litter and the rest as dissolved organic matter.

ALLOCHTHONOUS ORGANIC MATTER AS A FOOD SOURCE

A perusal of the food habits of stream animals shows that members of all the important benthic groups consume plant detritus which in streams primarily consists of decomposed leaves and other plant debris of terrestrial origin. This has been shown for members of Plecoptera, Trichoptera, Ephemeroptera, Diptera, Amphipoda and Isopoda—the groups that comprise the bulk of the biomass of secondary producers in streams. Studies on food habits of a broad spectrum of benthic groups (Minckley, 1963; Minshall, 1967; Hynes, 1961) amply show the importance of allochthonous organic matter. This is further substantiated by the fact that there is a strong correlation between the distribution of detritus and that of many members of stream benthos (Egglishaw, 1964, 1969). Even in a large river, like

TABLE 1 Data from Various Authors Showing P/R Ratios for Some Lotic Environments

Author	Year	Location of Study	P/R Ratio ($g\ O_2/m^2/day$)
Odum[a]	1956	Itchen River (England)	0.1 to 1.1
		Lark River (England)	0.01 to 1.1
		White River, Indiana	
		(a) Zone of recovery from pollution	3.2
		(b) Near pollution outfall	0.008
		Potomac Estuary	0.66
Hoskin	1959	Neuse River system (8 streams) and other streams in N. Carolina	0.2 to 0.7
Nelson and Scott	1962	Middle Oconee River, Georgia	0.1 to 0.26
Duffer and Dorris	1966	Blue River, Oklahoma	mostly 0.4 to 0.6
Tominago and Ichimura	1966	Arakawa River, Japan	0.5 to 0.8

[a]Includes results of other studies cited by the author.

the Missouri, 54 percent of the organic matter ingested by fish has been attributed to terrestrial sources (Berner, 1951).

From the foregoing it is evident that secondary production in streams should largely depend on assimilability of allochthonous organic matter by stream animals. However, this is an area of work that has received negligible attention from aquatic ecologists. Hargrave (1970) has shown that *Hyalella azteca*, an amphipod, is unable to digest cellulose and lignin-like substances, two major constituents of leaf litter. Assimilation efficiency of *H. azteca* for elm leaves was shown to be only about 5 percent. The overall assimilation efficiency of nymphs of *Pteronarcys scotti* (Plecoptera), when maintained on a diet of leaf litter, is only 10.8 percent (McDiffett, 1970). This clearly indicates that most of the particulate allochthonous organic matter is not readily available to stream detritivores. In this context it is also relevant that many aquatic animals lack cellulase activity (Bjarnov, 1972); and that the turnover of sediment by deposit-feeding invertebrates is very rapid (Gordon, 1966), indicating that only a small fraction of the total organic matter is directly available as food. Dissolved organic matter, another major component of allochthonous input, may not be of much consequence in secondary production in streams unless converted into particulate organic matter.

It has been speculated for many years that the detrital energy source becomes available to detritivores through the microbes involved in its decomposition. At a later point in this paper the importance of microbes in the nutrition of stream benthos will be discussed in more detail. It should suffice to mention here that factors controlling microbial activity during detrital decomposition also control the availability of food to detritivores and, hence, secondary production. Therefore, it is essential to understand factors influencing decomposition of allochthonous plant tissue, and those that control conversion of dissolved organic matter into particulate matter.

CONVERSION OF DISSOLVED ORGANICS INTO PARTICULATE ORGANIC MATTER

Processes involved in the formation of particles from leaf leachates, a substantial source of energy in woodland streams, have been investigated recently by Lush (1970), and the factors outlined may be equally applicable to dissolved materials from the other sources. Leaves of two species of maple, *Acer saccharrum* and *A. saccharinum*, were placed in water and leached on a rotary shaker. To measure the rate of precipitation, aliquots of the leachates were removed at different time intervals and filtered onto a 0.45 μ preweighed glass-fiber filter and reweighed. Results (Figure 1), shown as amount of precipitate as a percentage of the initial dry weight of leaves, clearly indicate that the formation of precipitate depends upon leaf species or, in other words, the nature of dissolved organics. Lush (1970) showed that such abiotic factors as turbulence, freezing and the pH of water may also control the amount of material that is eventually precipitated. The initial pH of the water determines not only the amount of dissolved organic matter that leaches out of leaves and the amount that precipitates, but also the size of the particles that are formed. While in acid waters the precipitation is much delayed and an abundant number of large particles are formed while neutral and alkaline waters result in smaller particles. In general, particles resulting because of abiotic factors are smaller

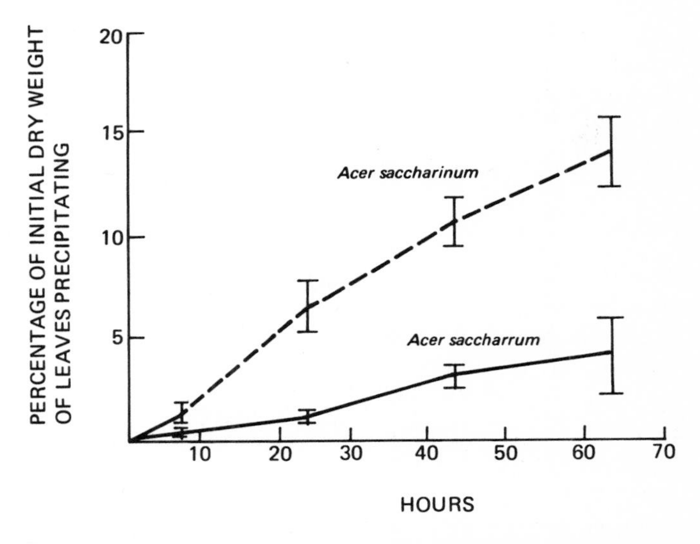

FIGURE 1 Percentage of the dry weight of leaves that precipitates in water as a function of time (Lush, 1970).

than 60 μ in diameter. A few days after formation these precipitates are colonized by microorganisms, both bacteria and fungi. Clumps resulting after biotic activity are larger in size and may grow to a few millimeters, depending upon the turbulence. There can be little doubt that these particles are potential food for stream benthos (Egglishaw, 1969; Brown, 1961). Since formation of the particles is influenced by the nature of riparian vegetation and quality of the water, it is obvious that these factors control the availability of food and hence secondary production in streams.

DECOMPOSITION OF PARTICULATE PLANT ORGANIC MATTER

Because of its importance in nutrient recycling, decomposition of plant tissues, including leaf litter, has been studied in detail by agronomists and woodland ecologists. It has been shown that the processes involved are extremely complicated. In contrast, aquatic ecologists have paid much less attention to these aspects especially in regard to the stream situation, where, as already mentioned, these processes are of utmost importance. In recent years it has been shown that leaves of elm, maple and willow, when placed in lotic environments, disappear much faster than those of oak and beech (Kaushik and Hynes, 1971; Mathews and Kowalczewski, 1969). These dissimilar rates of decomposition of various types of leaf may perhaps be important in that they ensure a food supply in streams for a longer period.

Obviously decomposition depends on the nature of plant tissues that enter streams. Woodland ecologists have suggested that the rate of decomposition is controlled by such intrinsic factors as pH, water-soluble substances, C:N ratio, calcium and nitrogen contents, total ash and constitutents like lignin and tannin. Temperature is another obvious but important factor controlling decomposition. Preleached autumn-shed elm, alder, oak, beech and maple leaves, when incubated (Kaushik and Hynes, 1971) in stream water kept at 10°C and at 20 to 22°C, showed faster rates of decay at higher temperatures (Figure 2a, b). Similar results were also obtained when these leaves were placed in two southern Ontario streams. The effect of temperature on the rate of decomposition of plant tissue in soil has been studied extensively and it has been shown that, although at higher temperatures breakdown of cellulose and hemicellulose is accelerated, the effect is especially marked on lignins (Waksman and Gerretsen, 1931). Beech and oak leaves have a high lignin content, and higher temperatures possibly accelerate decomposition of such plant tissues.

Decomposition of allochthonous plant tissue also depends upon the quality of stream water. Autumn-shed leaves of elm were separately incubated (Kaushik and Hynes, 1971) in stream water enriched with a nitrogen source only, or with both nitrogen and phosphorous sources, or in water without added nutrients (control). The rate of decay increased in the presence of an added nitrogen source and was further accentuated when phosphorus was also added (Figure 2a, b). Similar results were obtained with alder, oak, beech and maple leaves. Possibly other nutrients are also involved. Egglishaw (1968) has shown that breakdown of rice grains in various Scottish streams was faster in those with

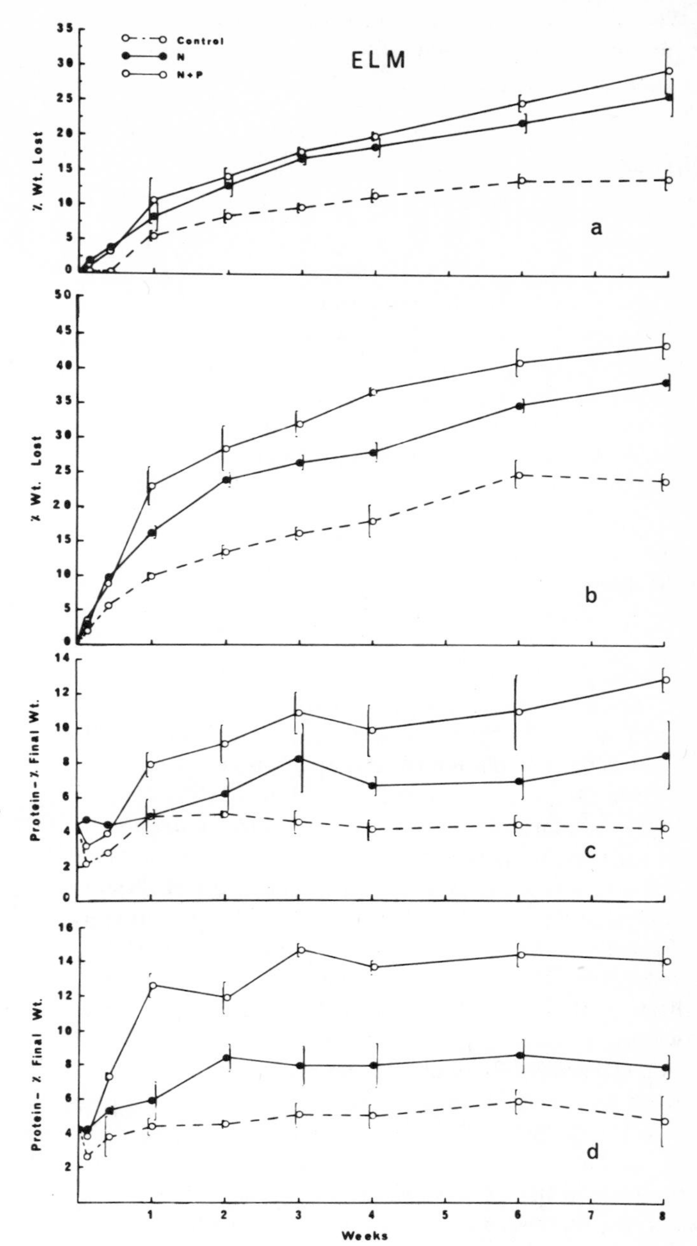

FIGURE 2 Percentage weight loss and protein content elm (*Ulmus*) leaves kept for various time intervals in stream water enriched with only nitrogen or with nitrogen and phosphorus. Mean values ± 95% confidence limits. (a) and (c) at 10 °C and (b) and (d) at 20 to 22 °C. (Kaushik and Hynes, 1971).

higher calcium concentration. These results clearly show that dissolved nutrients in stream water are important in controlling microbial breakdown of detritus.

DECOMPOSITION AND QUALITY OF FOOD

Woodland ecologists (e.g., Bocock, 1964) have found that leaves decomposing on forest floors show an increase in nitrogen content. Similar observations have been recorded for various leaves decomposing in aquatic environments (Mathews and Kowalczewski, 1969; Kaushik and Hynes, 1971) (Figure 3a, b). To determine whether this increase represented an increase in the absolute quantity of nitrogen, Kaushik and Hynes (1971) incubated leaves for various time periods in stream water with and without added nutrients. The amount of nitrogen found in the leaves at the end of each sampling time was calculated as a percentage of the initial weight. The leaves kept in enriched waters showed a significant increase in the absolute quantity of nitrogen; this increase was more pronounced when both the nutrients were added (Figure 3c, d). Uptake of exogenous nitrogen was confined to a certain level that was attained in one to two weeks depending upon temperature; once this level was reached only small further changes occurred. The level of N uptake also depended upon leaf species, e.g., elm and maple gained more nitrogen that did oak or beech.

Since the increase in nitrogen content of decomposing leaves is mostly in the form of microbial protein (Figure 2c, d), this implies that the quality of food available to benthos largely depends upon the capacity of plant tissue to support microbial populations. Thus, a stream receiving elm and maple leaves probably provides better quality food than one that receives oak and becch. It may be noted that during decomposition of allochthonous organic matter there is only a slight change in the caloric value (Kaushik and Hynes, 1971; Mathews and Kowalczewski, 1969) but because of increased protein content, decomposed organic matter is a food of better quality for stream benthos. Since assimilation efficiencies are determined partly by food quality (Boyd and Goodyear, 1971) it is evident that microbial decomposition of organic matter should increase assimilation efficiencies by detritivores and this can lead to higher secondary production.

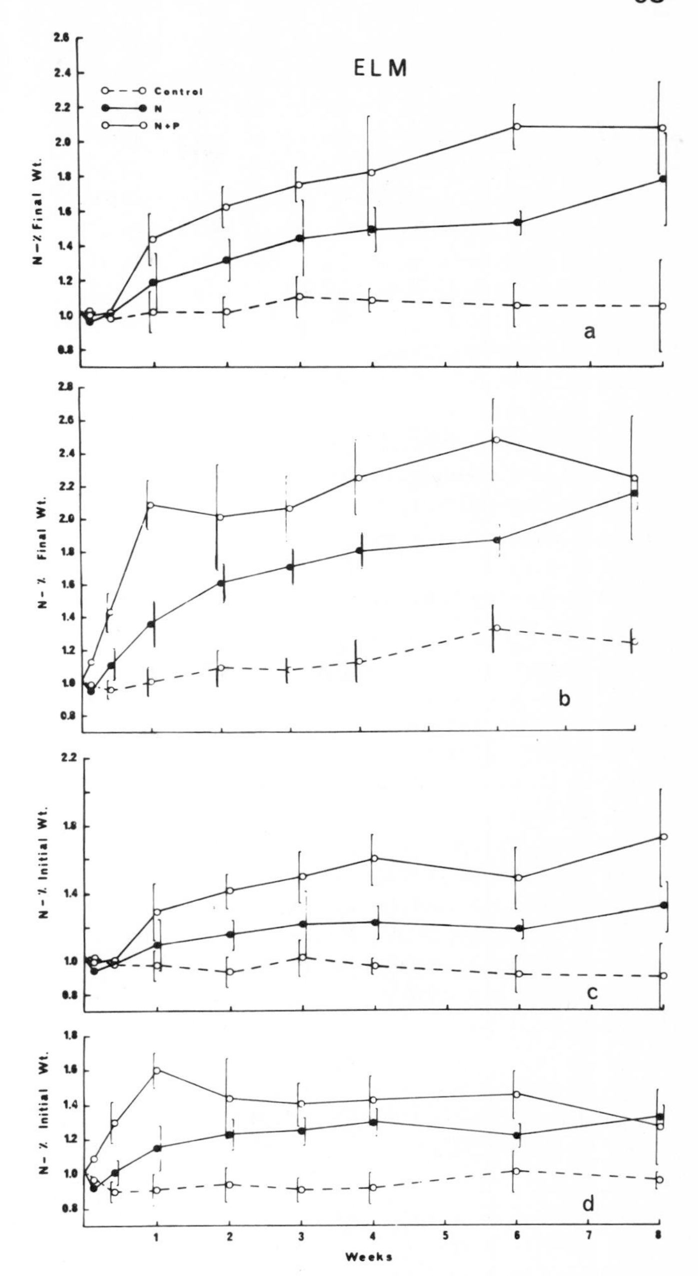

FIGURE 3 Nitrogen content of elm leaves kept for various time intervals in stream water enriched with only nitrogen or with nitrogen and phosphorus. Mean values ± 95% confidence limits. (a) and (c) at 10 °C and (b) and (d) at 20 to 22 °C (Kaushik and Hynes, 1971).

DECOMPOSITION AND FOOD PREFERENCE

Observations, similar to those of woodland biologists, that many invertebrates prefer to feed on certain types of leaves have recently been recorded by aquatic biologists. Elm, maple, ash and alder leaves are preferred by various organisms to those of beech and oak (Dölling, 1962; Wallace *et al*., 1970; Kaushik and Hynes, 1971; Bärlocher and Kendrick, 1973). Many stream invertebrates have been shown to prefer decomposed leaves that support microbial growth over those that are freshly fallen and lack microbial growth (Kaushik and Hynes, 1971). An interesting study on the role of fungi in food preference by *Gammarus pseudolimnaeus* has recently been carried out by Bärlocher and Kendrick (1973). They showed that given a choice between maple leaf discs and colonies of different hyphomycetes, originally isolated from decomposing leaves, the animals preferred fungi.

Amongst the fungi again they found an order of preference. Food selection in *Gammarus* is not only influenced by the type of leaf but also by the fungus it supports. It appears that leaves like elm, maple and ash are preferred by stream animals perhaps because they, in comparison with beech and oak, are better substrates for preferred fungal types.

Hynes (1961, 1963) has shown that streams generally support a larger biomass of benthic organisms in winter. The reason could possibly be that allochthonous organic matter becomes available as food during these months when the water temperature in most temperate streams is very low. Bärlocher and Kendrick (1973) have shown that at winter temperatures aquatic hyphomycetes (*Tetracladium, Tricladium, Anguillospora*) are able to colonize leaves whereas at higher temperatures terrestrial hyphomycetes become important in leaf degradation in streams. Thus, it appears that microbial decomposition of leaf litter, one of the major sources of allochthonous input in streams, is dominated during initial phases by fungi, especially aquatic hyphomycetes.

DECOMPOSITION AND SECONDARY PRODUCTION

Allochthonous organic matter supports a substantial biomass of stream invertebrates and initially this organic matter is not very attractive food for animals. However, its nutritive quality becomes enhanced because of microbial decomposition, and detritivores have the ability to select and feed on such plant tissues that support microbes. It has long been believed that lake detritivores derived their nourishment from bacteria engaged in the decomposition of detritus rather than directly from organic detritus. The ability of chironomid larvae to grow on filter paper to which suitable bacteria had been added (Ivlev, 1945) and of *Simulium* larvae to grow on a bacterial suspension (Fredeen, 1960, 1963) clearly show the nutritional value of microbes in the food of stream invertebrates. Recently it has been shown that the amphipod, *Hyalella azteca*, is able to assimilate 60 to 90 percent of the ingested bacterial biomass (Hargrave, 1970). Similarly, many soil invertebrates are capable of growing on microbial diets (Burges and Raw, 1967). Selective feeding on fungus by soil-dwelling members of Enchytraeidae has been reported by O'Connor (1967). Under laboratory conditions *Tomocerus* is reported to feed selectively on fungal spores. Mycelia of *Trichoderma* and *Phoma* also serve as a diet for Collembola and they show very high growth rates when fed on these fungi (Hale, 1967).

The processes that control decomposition of allochthonous organic matter also control the buildup of microbial populations, and hence the protein content and the quality of the food that ultimately becomes available to stream benthos. Since allochthonous organic matter supports a major part of the biomass of stream invertebrates, it is conceivable that their production should mainly be governed by the processes involved in the decomposition of allochthonous organic matter. Although no study has attempted to directly elucidate this relationship, results from various publications show the importance of some of the processes governing decomposition, and of the quantity of allochthonous input in relation to secondary production.

Jewell (1927) observed that prairie streams, being almost devoid of imported organic detritus, support a sparse fauna as compared with savannah and woodland streams. Streams in heavily wooded areas in Sweden have been reported to have more invertebrates than are found in similar but open streams in Wales (Babcock, 1953). In Morgan Creek in Kentucky, Minshall (1968) observed that the stretch of the creek passing through an area cleared of forest had decreased species diversity and fewer numbers of individuals.

Not only is quantity of allochthonous input important but even quality can affect secondary production. Replacement of deciduous forests with coniferous woodland (*Thuja* and *Picea*) has been shown (Huet, 1951) to have caused reduction in benthic fauna of nearby streams. Although Huet attributed this to the changed water pH and the toxins released by needles, Kendrick (personal communication) speculates that the result was because of comparatively slower decomposition of the coniferous needles. Fisher (1971) and Triska (1970) have shown that most of the particulate organic matter that enters a small stream is processed in the stream. This presumably depends upon the nature of particulate organic matter. Coniferous needles, having a slow decomposition rate, need more time for processing and eventually consumption by benthos. Thus, there is greater chance that needles will be lost downstream depriving benthos of food and causing decreased production.

Perhaps the only study that has attempted to show a correlation between the decomposition of organic matter and the standing stock of invertebrates in streams is that by Egglishaw (1968). He recorded that in nine Scottish Highland streams with varying calcium concentrations, the higher the concentration of Ca^{++} and HCO_3^- ions, the faster the rate of decomposition of dead plant tissue. The streams with faster rates of decomposition showed larger standing stock of many invertebrates and presumably had higher productivity.

It is evident from the preceding that the factors involved in the decomposition of allochthonous organic matter in streams can have important ecological implications. Other factors being equal, secondary and tertiary production in streams depends upon the quality of organic matter. Therefore the manipulations of the right kinds of detritus and microbes could lead to better fish production. Since microbes and detritus are primarily responsible for trapping nutrients, the more efficiently this is done and incorporated into removable secondary and tertiary production the smaller the quantity of nutrients continuing downstream, thus slowing lake eutrophication.

ACKNOWLEDGMENTS

I am grateful to Dr. Lush for allowing the use of unpublished information from his thesis and for critically going through the manuscript.

REFERENCES

Babcock, R. M. 1953. Studies of the benthic fauna in tributaries of the Kavlinge River, Southern Sweden. Rep. Inst. Freshwater Res. Drottningholm 35:21–37.

Bärlocher, F., and B. Kendrick. 1973. Fungi and food preference of *Gammarus pseudolimnaeus.* Arch. Hydrobiol. 72(4):501–516.

Berner, L. M. 1951. Limnology of the lower Missouri river. Ecol. 32:1–12.

Bjarnov, N. 1972. Carbohydrases in *Chironomus, Gammarus* and some Trichoptera larvae. Oikos 23:261–263.

Bocock, K. L. 1964. Changes in the amounts of dry matter, nitrogen, carbon and energy in decomposing woodland leaf litter in relation to the activities of the soil fauna. J. Ecol. 52:273–284.

Boyd, C. E., and C. P. Goodyear. 1971. Nutritive quality of food in ecological systems. Arch. Hydrobiol. 69:256–270.

Brown, D. S. 1961. The food of the larvae of *Chloeon dipterum* L. and *Baetis rhodani* (Pictet) (Insecta Ephemeroptera). J. Anim. Ecol. 30:55–75.

Burges, A., and F. Raw (ed.) 1967. Soil Biology. Academic Press. New York. 532 p.

Cummins, K. W., W. P. Coffman, and P. A. Roff. 1966. Trophic relations in a small woodland stream. Verh. int. Verein. theor. angew. Limnol. 16:627–638.

Dölling, L. 1962. Der Anteil der Tierwelt an der Bildung von Unterwasserböden. Verh. Zool. Bot. Ges. Wien 101–102:50–85.

Duffer, W. R., and T. C. Dorris. 1966. Primary productivity in a southern Great Plains stream. Limnol. Oceanogr. 11:143–151.

Egglishaw, H. J. 1964. The distributional relationships between the bottom fauna and plant detritus in streams. J. Anim. Ecol. 33:463–476.

Egglishaw, H. J. 1968. The quantitative relationship between bottom fauna and plant detritus in streams of different calcium concentration. J. Appl. Ecol. 5:731–740.

Egglishaw, H. J. 1969. The distribution of benthic invertebrates in fast-flowing streams. J. Anim. Ecol. 38:19–33.

Fisher, S. G. 1971. Annual energy budget of a small forest stream ecosystem: Bear Brook, West Thornton, New Hampshire. Ph.D. dissertation. Dartmouth College, Hanover, New Hampshire. 97 p.

Fisher, S. G., and G. E. Likens. 1972. Stream ecosystem: Organic energy budget. BioSci. 22:33–35.

Fredeen, F. J. H. 1960. Bacteria as a source of food for blackfly larvae. Nature Lond. 187:963.

Fredeen, J. H. 1963. Bacteria as food for blackfly larvae (Diptera: Simuliidae) in laboratory cultures and in natural streams. Can. J. Zool. 42:527–548.

Gordon, D. C. 1966. The effect of the deposit feeding *polychaete Pectinaria* gouldii on the intertidal sediments of Barnstable Harbor. Limnol. Oceanogr. 11:327–332.

Hale, W. G. 1967. Collembola, p. 397–411. *In* A. Burges and F. Raw (ed.) Soil biology. Academic Press, New York. 532 p.

Hargrave, B. T. 1970. The utilization of benthic microflora by *Hyalella azteca* (Amphipoda). J. Anim. Ecol. 39:427–437.

Hoskin, C. M. 1959. Studies of oxygen metabolism of streams of North Carolina. Publ. Inst. Mar. Sci. Univ. of Texas. 6:186–192.

Huet, M. 1951. Novicité des boisements en Epicéas (*Picea excelsa* Link.) Verh. int. Verein. theor. angew. Limnol. 11:198–200.

Hynes, H. B. N. 1961. The invertebrate fauna of a Welsh mountain stream. Arch. Hydrobiol. 57:344–388.

Hynes, H. B. N. 1963. Imported organic matter and secondary productivity in streams. Proc. XVI. Int. Congr. Zool., Washington. 4:324–329.

Hynes, H. B. N. 1970. The ecology of running waters. Liverpool Univ. Press. 555 p.

Ivlev, V. S. 1945. The biological productivity of waters. (In Russian) Translated by W. E. Ricker, 1966. J. Fish. Res. Bd. Can. 23(11): 1727–1759.

Jewell, M. E. 1927. Aquatic biology of the prairie. Ecology 7:289–298.

Kaushik, N. K., and H. B. N. Hynes. 1971. The fate of the dead leaves that fall into streams. Arch. Hydrobiol. 68(4):465–515.

Lush, D. L. 1970. Dissolved organic matter in streams. M. S. thesis. Univ. of Waterloo. 78 p.

Mathews, C. P., and A. Kowalczewski. 1969. The disappearance of leaf litter and its contribution to production in the River Thames. J. Ecol. 57:543–552.

McDiffett, W. F. 1970. The transformation of energy by a stream detritivore, *Pteronarcys scotti* (Plecoptera). Ecology 51:975–988.

Minckley, W. L. 1963. The ecology of a spring stream Doe Run, Meade County, Kentucky. Wildlife Monogr. Chestertown 11:124.

Minshall, G. W. 1967. Role of allochthonous detritus in the trophic structure of a woodland springbrook community. Ecology 48: 139–149.

Minshall, G. W. 1968. Community dynamics of the benthic fauna in a woodland springbook. Hydrobiologia 32:305–339.

Nelson, D. J., and D. C. Scott. 1962. Role of detritus in the productivity of a rock-outcrop community in a Piedmont stream. Limnol. Oceanogr. 7:396–413.

O'Connor, F. B. 1967. The Enchytraeidae, p. 213–258. *In* A. Burges and F. Raw (ed.) Soil biology. Academic Press, New York. 532 p.

Odum, H. T. 1956. Primary production in flowing waters. Limnol. Oceanogr. 1:102–117.

Teal, J. M. 1957. Community metabolism in a temperate cold spring. Ecol. Monogr. 27:283–302.

Thienemann, A. 1912. Der Bergbach des Sauerlandes. Internat. Rev. ges. Hydrobiol. Hydrogr. Suppl. 4,2,1. 125 p.

Tominago, H., and S. Ichimura. 1966. Ecological studies on the organic matter production in a mountain river ecosystem. Bot. Mag. Tokyo 79:815–829.

Triska, F. J. 1970. Seasonal distribution of aquatic hyphomycetes in relation to the disappearance of leaf litter from a woodland stream. Ph.D. dissertation. Univ. of Pittsburgh. 189 p.

Waksman, S. A., and F. G. Gerretsen. 1931. Influence of temperature and moisture and extent of decomposition of plant residues by microorganisms. Ecology 12:33–60.

Wallace, J. B., W. R. Woodall, and F. F. Sherberger. 1970. Breakdown of leaves by feeding of *Peltoperla maria* nymphs (Plecoptera:Peltoperlidae). Ann. Entomol. Soc. Amer. 63:562–567.

NUTRIENT CYCLING IN FRESHWATER ECOSYSTEMS

D. W. SCHINDLER, D. R. S. LEAN, and E. J. FEE

ABSTRACT

Flaws in logic or methodology have often caused erroneous conclusions to be drawn from chemical experiments in fresh water. Some examples follow.

The acid-molybdate method was found to greatly overestimate phosphate concentrations in many lakes. Phosphorus was found to be hydrolyzed from organic substances in filtered lake water by the acid molybdate reagent.

The ^{14}C bottle bioassay technique is often misapplied to management questions. It was demonstrated that while lake 227 became eutrophic as the result of additions of phosphorus and nitrogen bioassays indicated that carbon was limiting throughout much of a typical day. Carbon limitation was the result of, rather than the cause of, eutrophication. In spite of low carbon concentrations, enough CO_2 was able to invade the lake from the atmosphere to allow algal blooms to develop. Comparisons with data from the Laurentian Great Lakes indicate that there is no possibility of carbon limitation being of any significance to eutrophication management in those waters.

INTRODUCTION

The futility of the traditional approach to nutrient studies in fresh water, i.e., the measurement of nutrient concentrations, has been soundly demonstrated during the past decade while IBP studies have been in progress. It is paradoxical that biological investigators, while attempting to make precise measurements of rates of movement of one element, carbon, through biotic components of the aquatic ecosystem, have used only concentrations, i.e., pool sizes, of other nutrients to characterize acquatic ecosystems. To use such information to characterize the role of nutrients is equivalent to using algal standing crop as an index of biological productivity. In some circumstances it may give useful information about the dynamics of the system, or it may tell nothing at all.

Several new approaches to nutrient cycling have arisen during the time of the IBP studies. Some of these appeared early enough to be incorporated into programs; other techniques show promise, but have not yet been fully tested; still others have been overlooked nearly completely. The purpose of this paper will be to examine some of the approaches to aquatic chemistry, and to express some opinions about the viability of these approaches as applied to ecosystem productivity.

Some excellent examples of the impotence of the traditional approach to nutrient chemistry have been afforded by the recent controversy over limiting nutrients (Likens, 1972). Although most limnologists believe phosphorus to be the element limiting phytoplankton production and abundance in the majority of freshwater ecosystems (Hutchinson, 1957) and also responsible for the majority of eutrophication problems (Vollenweider, 1968; Anonymous, 1969), their evidence does not appear to have convinced legislators and industrialists in the United States. It is paradoxical that antagonists in the controversy have often used the same data to support their arguments, e.g., the concentration of molybdate-reactive phosphorus. Much of the following information bears on the problem of interpretation of such data, since it is believed that many of the same errors have been made in attempts to interpret the effects of nutrients on biological productivity in freshwater ecosystems.

NUTRIENT BUDGETS FOR LAKES

Phosphorus

The monograph by Vollenweider (1968) presents strong evidence that the quantity of nutrient supplied to a lake, and not nutrient concentration, is the factor of importance in assessing eutrophication. A later theoretical paper (Vollenweider, 1969) elaborates the dynamic aspects of this conclusion. The result has been a convincing condemnation of phosphorus as the primary villain in eutrophication. Sadly, there are a large number of persons who do not recognize the superiority of this approach over older, static concepts like Sawyer's (1947) Limits. More about errors in concentration-based approaches will be presented below.

For studies described to date, estimates of phosphorous input have ranged from 0.016 g P/m^2 year (ultraoligotrophic Char Lake in the high Arctic) (Schindler *et al.*, 1974b) to nearly 20 g P/m^2 year (hypereutrophic Greifensee, Switzerland) (Pleisch, 1970). Corresponding figures for nitrogen range from 0.1 to 45 g N/m^2 year. No input data are available for the extremely productive lakes of central Africa and North America, but hydrological and geological difficulties may render this impossible. The addition of phosphorus and nitrogen appear to cause eutrophication under any climatic regime, even in the high Arctic (Schindler *et al.*, 1974a). Predictions relating the trophic status of lakes to nutrient input and mean depth (Vollenweider, 1968; Anonymous, 1969, Vol. 2, fig. 3.3.1) appear to be remarkably accurate, and it would be pointless to discuss them further here.

The warning by Rigler (1966, 1968) that the standard acid-molybdate technique may overestimate orthophosphate concentrations in natural waters by 10 to 100X has largely been overlooked, and most investigators still rely heavily on this technique. More recent work (Chamberlain, 1968; Lean, 1973a, b, c) has demonstrated that colloidal substances in lake water, mostly of apparent molecular weight $>10^7$, play a major role in aquatic phosphorus dynamics. Our recent unpublished work has shown that the acid-molybdate technique hydrolyses phosphate from the above-mentioned colloidal substances, causing considerable overestimates of phosphate in many cases. It has been found that true orthophosphate concentrations may be less than 0.1 μg/liter,* even in eutrophic waters (S. Levine, unpublished data). The relative importance of colloidal material varies greatly from one water body to the next, however, and in some instances reasonably accurate phosphate results may be obtained with the acid-molybdate test.

If the role of the phosphate "pool" in fresh waters is to be evaluated, concentration, temperature and demand by phytoplankton must be considered. We have found rate constants for turnover of phosphate in natural fresh waters ranging over 5 orders of magnitude. On the other hand, if the phosphate demand of the sestonic phosphorous pool, which is a crude measure of phytoplankton demand, is considered, total range is reduced to 3 orders of magnitude, regardless of trophic status of the water body (Table 1).

The extremely small phosphate pool must be rapidly and continuously replenished in order to maintain even short-term stability in most aquatic ecosystems. Recent work has uncovered two important biologically mediated mechanisms. Lean (1973a, 1973b) found that when radioactive phosphorus was added to unfiltered lake water, most was rapidly taken up by phytoplankton and bacteria. Within minutes, some of the radiophosphorus was excreted as PO_4, and some was transferred to the high molecular weight colloid described above. The colloid was not a direct source of phosphorous for phytoplankton. When phytoplankton were removed from the water by membrane filtration there was no movement of phosphate to the colloid, suggesting that the biota, rather than chemical factors, were responsible for the process. Preliminary evidence suggested that a phosphorylated algal or bacterial excretory product, of molecular weight about 250, was produced which bound to the colloidal material, releasing PO_4 from the colloid in the process. The mode of operation of the mechanism, and its evolutionary significance, are obscure, but regeneration of phosphate phosphorus is possible via this pathway, as a supplement to PO_4 excretion. Both freshwater algae and bacteria are capable of releasing such excretory products, as well as orthophosphate (Lean, 1973a,b,c). The seston takes up and releases phosphorus compounds at a very high rate. The direct release of phosphorus from ultraplankton is the most important factor in the rapid turnover of this element (Lean, 1973a).

Peters (1972) found that most of the phosphorus excreted by zooplankton was in true phosphate form (Peters and Lean, 1973). While this is contrary to many of the conclusions of earlier workers, there is a logical explanation. Peters found that excreted orthophosphate was rapidly taken up by bacteria in the culture flasks. Since incubations by earlier authors lasted several hours, it is likely that the bacteria may have taken up the excreted phosphate and then been included in analyses for dissolved organic phosphorus, due to their small size. Haney (1970) found that zooplankton in many fresh waters graze from 10 to over 100 percent of the phytoplankton standing crop per day. If assimilation rates for zooplankton are 10 to 90 percent, as indicated in the literature, zooplankton excretion is another important source of phosphate supply.

Carbon

Cycling of carbon, where important chemical reactions complicate biological processes, is also poorly understood. Once

* True orthophosphate concentrations may be estimated with reasonable accuracy by bioassay (Rigler, 1966), or by isotope partition using sephadex (Lean, 1973a).

TABLE 1 Turnover Times, Rate Constants, and Flux Rates for Phosphorus in a Variety of Natural Waters. All Samples Are from the Upper Euphotic Zone, from June, July, or August. True Orthophosphate Concentrations Were Obtained by Isotope Partition Using Sephadex (Lean, 1973a)

Lake	Characterization	Total Diss. (P, μg/l)	PO_4-P (μg/l)	PO_4 Rate Constant (k, min^{-1})	Flux (μg, PO_4-P/min)	Sestonic (P, μg/l)	Flux (μg, PO_4-P/μg) Sestonic (P/min)	Water Temperature (°C)
239	Precambrian Shield Oligotrophic N. Temperate	10	0.05	-0.12	0.60×10^{-2}	3	0.20×10^{-2}	21
227	Precambrian Shield Artificially eutrophied N. Temperate	8	0.10	-0.26	0.26×10^{-1}	15	0.17×10^{-2}	24
228 (Teggau)	Precambrian Shield Ultra-oligotrophic N. Temperate	7	0.05	-0.013	0.65×10^{-3}	2	0.33×10^{-3}	17
Char Lake	Limestone Ultra-oligotrophic N. Polar	3	0.22	-31×10^{-4}	0.68×10^{-5}	2	0.34×10^{-5}	1.0
Resolute Bay	Marine Bay Ultra-oligotrophic N. Polar	40	37.7	-0.70×10^{-5}	0.26×10^{-3}	1	0.26×10^{-3}	-1.7
Red River	Fertile clay Valley river N. Temperate Turbid, agricultural	158	151	-0.86×10^{-3}	2.6×10^{-4}	50	0.52×10^{-5}	~20

again, proper consideration has not been paid to kinetic aspects of the cycle. Biologists have concentrated their efforts on the measurement of one rate constant, the uptake of dissolved inorganic carbon (DIC) by phytoplankton, and on the measurement of pool sizes, i.e., bicarbonate, sestonic carbon, etc. The mode and rate of replenishment of DIC has barely been considered.

Most of our knowledge of chemical kinetics affecting DIC has come from chemical experiments with pure solutions of inorganic chemicals (Kern, 1960). Recently, complications have been found which may significantly affect chemical equilibria. Ion-pairing has been known to affect critical reaction rates for some time (Wangersky, 1972; Wigley, 1971). At least one author (Wetzel, 1972) has suggested that chemical reactions in natural waters may be slow enough to cause freshwater phytoplankton to be carbon limited.

A second shortcoming of our knowledge of carbon cycling has been the lack of reliable information on invasion of carbon dioxide into natural waters. This admittedly difficult problem has been a major stumbling block in evaluating the claims of several investigators (Lange, 1970; Kuentzel, 1969; Kerr *et al.*, 1970) that the alleviation of natural carbon deficiencies by sewage additions has caused many eutrophication problems. Our studies of nutrient balance in natural oligotrophic and artificially eutrophied experimental lakes have allowed us to obtain some insight into the above problems.

In natural lakes of the Precambrian Shield, and in other areas, the concentration of dissolved CO_2 gas is usually slightly above atmospheric equilibrium, probably due to regeneration of the gas by decomposition of both autochthonous and allochthonous organic matter in the water and sediments. On the other hand, in some highly eutrophic lakes, the partial pressure of CO_2 in the euphotic zone may be far below that of the atmosphere, due to algal demands on the DIC system and the small pools of bicarbonate and carbonate present (Figure 1). Such deficits do not exist in any oligotrophic lake, nor do they exist in most bicarbonate lakes at temperate latitudes, including the St. Lawrence Great Lakes. This lack of a partial pressure deficit indicates that there is no significant demand on the DIC reservoir by phytoplankton, i.e., they are in no way carbon limited.

It is noteworthy that the carbon deficit in Lake 227 is more extreme than in almost any other lake studied, including some where carbon-limitation has been "proved." More detailed information on carbon cycling in this lake is instructive, therefore, since it is possible to obtain information on carbon dynamics under perhaps the most extreme conditions anywhere in nature.

Dissolved inorganic carbon in Lake 227 before fertiliza-

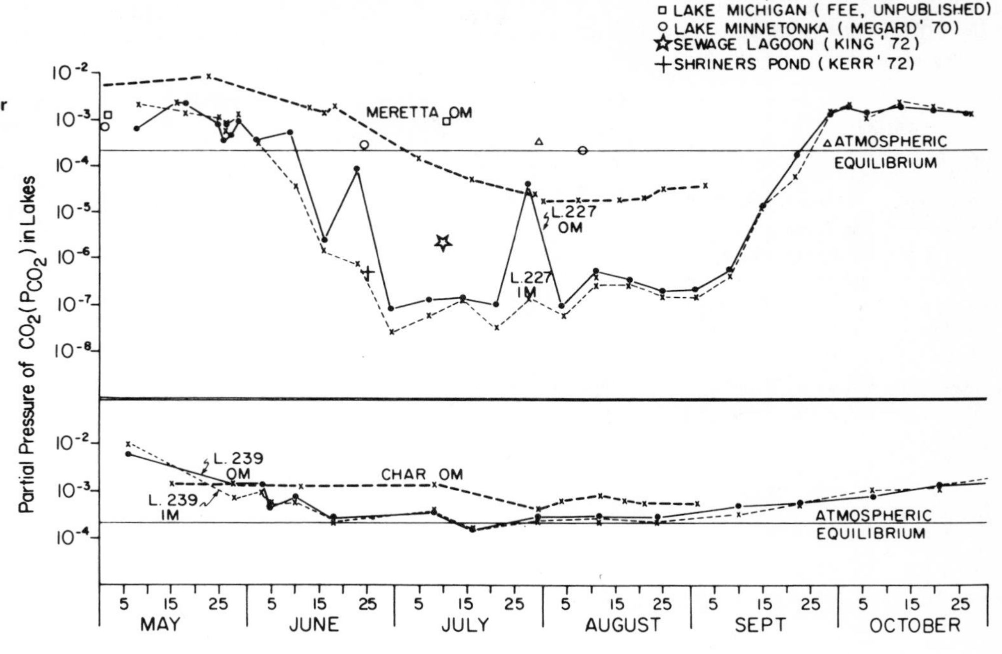

FIGURE 1 Seasonal P_{CO_2} values for a number of eutrophic and oligotrophic lakes. All great lakes are shown on the eutrophic graph. P_{CO_2} is calculated from alkalinity, pH and temperature or from total CO_2, pH and temperature (Lakes 227 and 239 only) using the constants of Garrels and Christ (1965). Char and Meretta Lakes, at 75 °N. lat., are ice-covered for all but a few weeks of the period shown. Other natural oligotrophic lakes examined have annual cycles nearly identical to Lake 239.

tion, and in similar lakes in the Canadian Shield, was usually 50 to 150 μM/liter in midsummer, at pH's of 6.5 to 7.5. A quick calculation reveals that the DIC reservoir is therefore 10 to 40 times smaller than in most lakes of economic importance. Yet by adding phosphate and nitrate to Lake 227, we have been able to increase phytoplankton standing crop (as chlorophyll a) from 1 to 5 to 100 to 300 μg/l, i.e., it is now highly eutrophic.

In its eutrophic state, Lake 227 shows many symptoms of so-called carbon limitation. Euphotic zone pH is over 10 in midsummer, with DIC of about 4 to 10 μM/l at midday. Under such conditions, gaseous CO_2 is a mere 10^{-4} μM/l, six orders of magnitude below atmospheric saturation.* Under such conditions, production is extremely low and in bottle bioassays response to nutrients other than carbon is negligible (Figure 2). These circumstances have been caused by the algal increase due to fertilization of the lake with phosphorus and nitrogen. It is clear that bottle bioassays do not necessarily give useful information about what nutrient has caused eutrophication problems. They merely tell what nutrients are limiting at the time of the assay.

Consideration of diurnal CO_2 and O_2 dynamics in Lake 227 reveals some interesting facts. At dawn the midsummer DIC concentration is usually 50 to 60 μM/liter. This is depleted by algae to 10 μM or less in the next two hours, indicating photosynthetic rates similar to a healthy sewage lagoon (Figure 3). Very little algal "production" takes place during the remainder of the day.

Nighttime regeneration of DIC and consumption of O_2 is equally instructive, particularly if coupled with sealed dark-bottle measurements. The *in situ* change in concentration during night is due to respiration plus invasion from the atmosphere and hypolimnion. That in the sealed bottles is due to respiration alone; therefore invasion may be calculated by difference (Table 2). Assuming that invasion during daylight is equal to that at night, the supply of carbon via this source can be highly significant.

Calculated CO_2 exchange is several times that for oxygen due to the fact that chemical enhancement, i.e., the hydration of invading CO_2, keeps the partial pressure of CO_2 gas in the lake low (Hoover and Berkshire, 1969). The observed enhancement factor agrees reasonably well with theoretically calculated ones (Schindler *et al.*, 1972a; Emerson, 1974), but further work is needed.

The annual carbon budget of Lake 227 is interesting, when compared to that of an oligotrophic Canadian Shield lake. A high proportion of the annual carbon income comes from the air (Table 3). The lake is slowly correcting its own carbon deficit, with an increase in carbon content of the water column averaging 35 percent per year. While limitation of phytoplankton production by carbon is apparently possible in such a lake, algal standing crops are many times desirable levels before the deficiency becomes serious. Moreover, a lake appears to be able to correct its own carbon deficiency from the atmosphere, i.e., the deficiency will become less pronounced from year to year if inputs of phosphorus and nitrogen remain constant.

When $DI^{14}C$ is added to lake water containing phytoplankton, it accumulated on the colloidal fraction much as ^{32}P was

* DIC species and P_{CO_2} are calculated from total DIC, pH and lake temperature, using the constants of Garrels and Christ (1965).

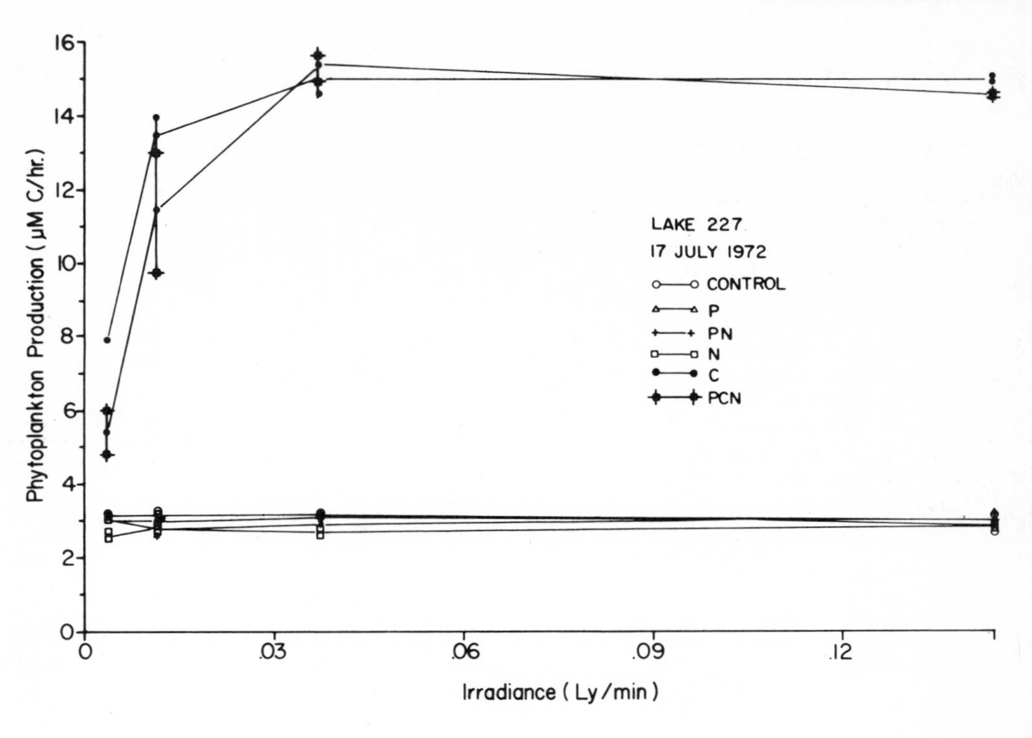

FIGURE 2 Response by phytoplankton taken from Lake 227 (see Figure 1) at midday to a variety of light and nutrient conditions. All samples were run in a light incubator at epilimnion temperature. Additions of nutrients were 5μg/l of PO_4-P, 50μg/l NO_3-N, and 1,000μg/l of HCO_3-C. From Schindler and Fee, 1973.

found to do. Binding of carbon to colloids appears to be much slower than phosphorus, with equilibrium conditions reached after several days, instead of in minutes. So far no investigations of the role of this colloidal fraction in regeneration of DIC have been made, but the fraction cannot be utilized by phytoplankton directly (Lean and Schindler, in prep.). In Canadian Shield waters, colloidally bound ^{14}C appears to be responsible for the correction for filtration error proposed by Arthur and Rigler (1967) (Schindler *et al.*, 1972b).

CARBON LIMITATION OF PRODUCTIVITY

In recent years a claim that carbon added with sewage is responsible for the eutrophication of many natural waters (Kuentzel, 1971) has confused the eutrophication issue, causing delays in critically needed anti-phosphorous legislation. Few aquatic scientists subscribe to the carbon-limitation theory, but most will agree to a platitude like "different nutrients are limiting in different situations." Such unquali-

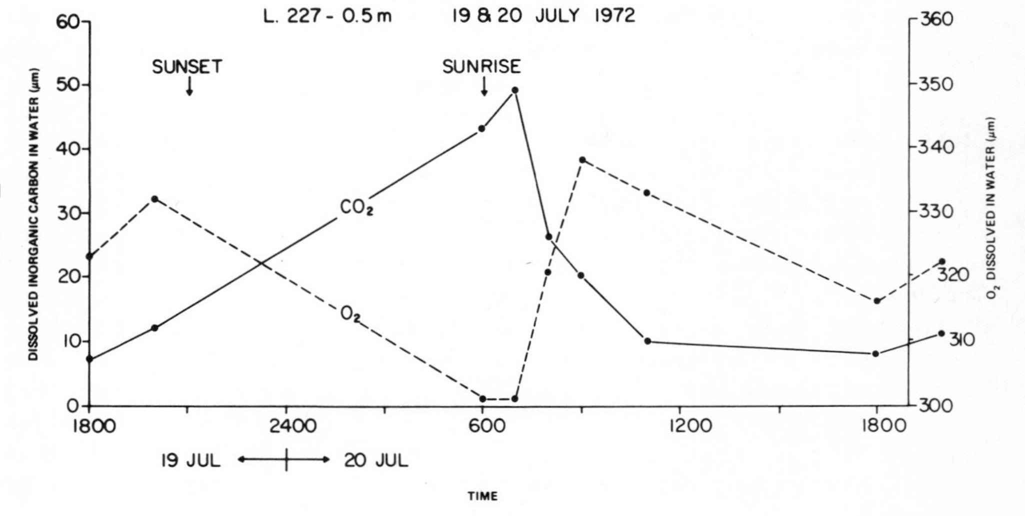

FIGURE 3 Typical diurnal O_2 and CO_2 curves for Lake 227. Depth 0.5 m, 19 July 1972. Note the high early morning photosynthesis. See text and Figure 4 for interpretation of the curves.

TABLE 2 An Example of Components of the Diurnal CO_2 and O_2 Budget for Lake 227. Data are for a Depth of 0.5 m. 31 July–1 August 1972. All Values are in μmoles/liter. Production Figures are Given for CO_2 Only

	CO_2	O_2
1 *In situ* change/hr at night (= invasion + respiration)	+3.04	−1.91
2 Dark bottle incubations (= respiration/hr)	+1.56	−1.72
3 Invasion (+) or Evasion (−)/hr (= ① − ②)	+1.48	−0.19
4 Net observed change in daylight	32	–
5 24 hr invasion	36	–
6 Net production (= ④ + ⑤)	68	–
7 24 hr respiration	37	41
8 Gross production (= ⑥ + ⑦)	105	–

fied statements are totally meaningless when considering causes of eutrophication. It is necessary to specify whether nutrient limitations are causes or results of phytoplankton blooms, and whether they operate five percent of the time or fifty, on a daily as well as an annual scale. Lake 227 offers some excellent illustrative material.

There is no doubt that phosphorus and nitrogen were responsible for the eutrophication of Lake 227, since addition of these two elements causes an enormous increase in phytoplankton despite extremely low DIC. Four years of experiments in large isolated columns (Schindler, unpubl.) have shown that the increase in phytoplankton could not have happened without the phosphorus, but that it would have been smaller if no nitrogen had been added. Carbon, in either inorganic (CO_2 gas or bicarbonate) or organic (sucrose, glucose or acetate) form caused little or no increase in standing crop, whether added alone or along with phosphorus and nitrogen.

Yet our data indicate that at midday in summer photosynthesis is definitely carbon limited, and phytoplankton will not respond to light, or to any nutrient except DIC (Figure 2), even though at dawn the lake exhibited a typical light-controlled response (Figure 4; Fee, unpubl. data), with carbon-fixation of sewage-lagoon proportions (up to 25 μM of C/1 × hour). The total daily production, even though it may take place in only 2 to 3 hours, is far higher than in oligotrophic lakes of the area. As mentioned above, both respiration and atmospheric invasion are important in the diurnal replenishment of CO_2 in the euphotic zone.

Scrutiny of the internal kinetics of the carbon cycle are also of interest. Just after dawn in midsummer, when gross production in excess of 20 μM/1 hour is observed, at a DIC concentration of 50 μM/1 and a pH of 10, dissolved gaseous CO_2 is 7.4×10^{-3} μM/1. If all DIC is supplied to phytoplankton via the gaseous CO_2 pool, this pool must be replenished with a rate constant of 0.75/sec, i.e., its turnover time is 1.33 sec. This is the same magnitude as the slower rate constants for equilibrium reactions in DIC cycle, as found in pure solutions by chemical means. It is therefore unlikely that organic substances slow chemical reactions directly, although under such extreme conditions just collision frequency and membrane efficiency may be important, since there are only 10^7 CO_2 molecules/cell in the euphotic zone.

Demands by phytoplankton for gaseous CO_2 in other productive systems are lower than in Lake 227. For example, phytoplankton in the extremely productive sewage lagoon described by King (1972) would require a rate constant for replenishment of gaseous CO_2 of only 0.06/sec to meet their demands, if calculated in the same manner. If we assume that phytoplankton utilize HCO_3^- directly, i.e., it is not first dehydrated to CO_2, the supply of DIC in Lake 227 is even lower in relation to other lakes. It is clear that in the few natural waters where carbon is "limiting", to phytoplankton, it must be regarded as a symptom, rather than a cause, of eutrophication.

It is obvious that complete access to the carbon, nitrogen

TABLE 3 A Summary of the Phosphorus Budgets for Lake 239 (Oligotrophic) and Lake 227 (Artificially Eutrophied by Addition of PO_4 and NO_3). All Data are in $g/m^2 \times yr$

	Phosphorus		Nitrogen		Carbon	
	L. 239	L. 227	L. 239	L. 227	L. 239	L. 227
Precipitation	0.05	0.05	0.67	0.67	5.31	5.09
Runoff	0.10	0.06	1.03	0.93	38.9	54.12
Atmospheric gases	0.00	0.00	?	?	0.0	22.60
Total natural input	0.15	0.11	1.71	1.60	44.1	81.82
Fertilizer	0.00	0.48	0.00	6.30	0.00	0.00
Total input	0.15	0.59	1.71	7.90	44.1	81.82
Outflow	0.04	0.08	0.81	1.63	28.6	37.45
Retention	0.11	0.50	0.90	6.27	15.5	44.4
% Retention of input	73	85	54	79	35	54

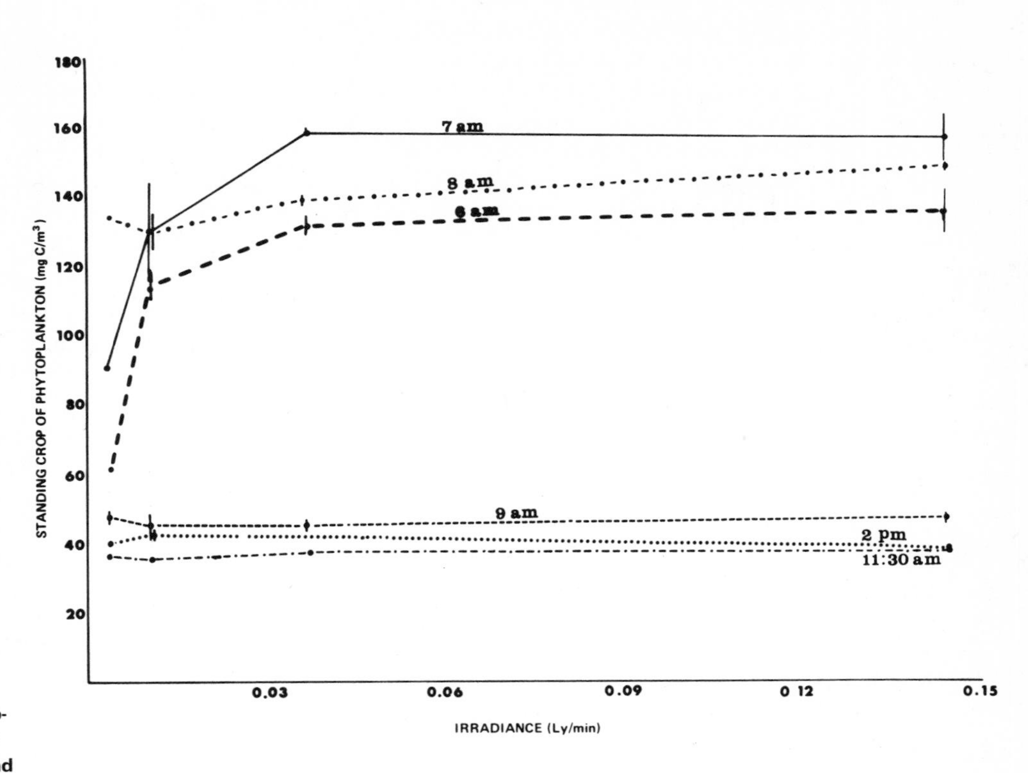

FIGURE 4 Response of Lake 227 phytoplankton taken at different times of day to four different light intensities in an incubator. Concentrations of DIC at the times samples were taken may be obtained from Figure 3. Although phosphate demand by the phytoplankton was uniformly high all day (k = 0.24 to 0.28 for the PO_4-P pool), and the high phytoplankton standing crop at the time of the experiment (170 μg chlorophyll a/l) was due to fertilization of the lake with phosphate and nitrate, there was no detectable short-term increase in carbon fixation due to phosphate additions. In morning (DIC = 50 to 60 μM/l) extremely high photosynthesis rates occurred at high light, and the shape of the light-response curve was typical. By midday (DIC $<$10 μ M/l) photosynthesis was extremely low and controlled by available carbon (see Figure 2). Similar lakes which did not receive phosphorus additions showed the typical light-response curve, but with an asymptote (maximum photosynthetic rate) from 5 to 25 percent that of Lake 227. From Schindler and Fee, 1973.

and oxygen budgets, as in Lake 227, cannot be achieved in great lakes and oceans. Yet estimates of gas exchange for these elements in such waters are important to a number of critical environmental issues besides eutrophication. For example, the rate at which the culturally caused increase in atmospheric CO_2 can be dispersed in the oceans has long-term global implications for climatic change (Broecker *et al.*, 1971).

While gas exchange in such enormous systems has been measured by using bomb-produced and natural ^{14}C (Craig, 1957; Broecker and Olson, 1960) and the naturally produced radioactive gas ^{222}Rn (Broecker, 1965; Broecker *et al.*, 1967; Broecker and Peng, 1971), only recently has it been possible to compare the resulting models to empirically measured chemical enhancement (Schindler *et al.*, 1972a; Emerson, 1974). A reasonably accurate picture of atmosphere-water exchange of carbon and nitrogen in fresh waters should emerge during the next decade.

RETURN OF PHOSPHORUS AND CARBON FROM SEDIMENTS

Understanding of sedimentary "feedback" and its control must be a major goal for limnologists in the next decade. It has been invoked as an important factor in freshwater chemical cycles since the classic work of Mortimer (1941–42), particularly in the control of phosphorous feedback from sediments by precipation with iron (Einsele, 1936; Hutchinson, 1957). Possible chemical interactions at the mud-water interface have been modelled in great detail (Stumm and Leckie, 1971). As late as the mid-1960's, the belief was widespread that sedimentary return would render attempts to rehabilitate culturally eutrophied lakes useless. The high concentrations of phosphorus and other nutrients just above sediments during late summer stagnation was regarded as evidence for "feedback." More recently, the rapid recovery of Lake Washington after diversion of sewage (Edmondson, 1970, 1972) cast considerable doubt on the above belief. In mass-balance studies of Lake Minnetonka, Megard (1970) has shown that although return of phosphorus from sediments occurs, net retention of phosphorus by sediments is extremely high.

In radiotracer experiments where ^{14}C and ^{32}P were allowed to equilibrate with seston and to sediment in *in situ* water columns in Lake 227, no significant return of either radiophosphorus or radiocarbon from sediments has been detected for a period of one year, even under anaerobic conditions (Lean and Schindler, in prep.). In a second experiment, where a ferric hydroxide-phosphate coprecipitate containing both ^{55}Fe and ^{32}P was sprayed over the sediment surface in the anaerobic hypolimnion, there was no significant return of either element. Mass-balance studies of input, output and changes of nutrient in the water column of the entire lake also indicate that sedimentary return is insignificant (Schindler and Lean, in press; Schindler, unpubl.).

Microbial rather than chemical forces appear to regulate movement of phosphorus and carbon at the sediment-water interface in our lakes. Lean and Schindler (in prep.) found high phosphate uptake by seston in anoxic hypolimnetic waters of their isolated column experiments in Lake 227. Once in particulate form, the element was again sedimented. Further studies by Schindler and Frost (unpubl.) have shown that (1) biological processes for sedimentation of phosphorus and carbon are extremely efficient when oxidizing conditions are present at the mud-water interface, and (2) that sedimentation efficiency of phosphorus and carbon is much reduced in the absence of the sediment biota (Figure 5).

On the other hand, it is clear from the study of Burns and Ross (1972) that significant return of phosphorus from sediments does occur in Lake Erie. The reason for these differences is obscure, but since the molybdate-test was employed in the above work, the amount of true orthophosphate returned is unknown.

OVERVIEW

It is not difficult to see why freshwater ecosystems appear complex, and why relationships of biological dynamics to chemical phenomena appear to be obscure. The traditional approach to freshwater chemistry has been a static one, where chemical concentrations serve purely as background information—a sort of decorative tapestry upon which limnologists have displayed the so-called "dynamic" processes, such as energy flow. These processes have been of interest for the past three decades. The small amount of information which we have summarized here demonstrates the sterility of such an approach.

The flux rates of nutrient elements are extremely flexible, and it is possible that small nutrient pools turning over rapidly furnish the same supply of nutrient as large pools turning over slowly. The measurement of pool size, i.e., concentration alone, may therefore be completely misleading when roles of nutrients are assessed.

Modern techniques, such as radiochemistry, allow many important chemical mechanisms and rates to be examined without interfering with ecosystem function, in much the same manner that ^{14}C is used in the study of primary production. Only by using such approaches can similarities

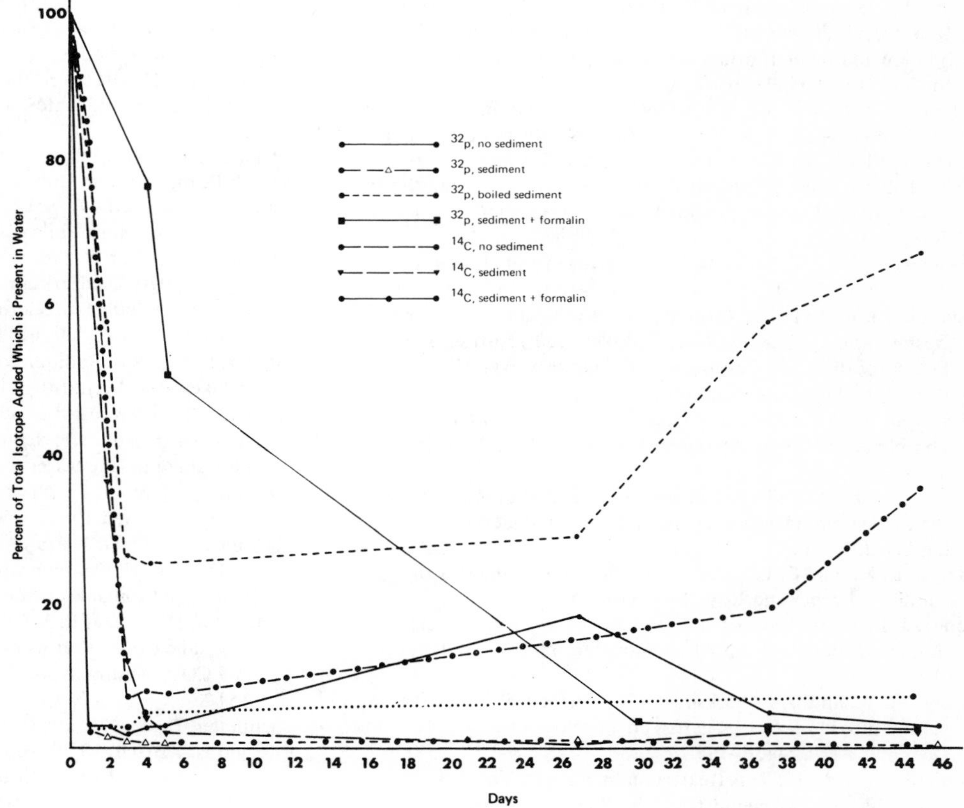

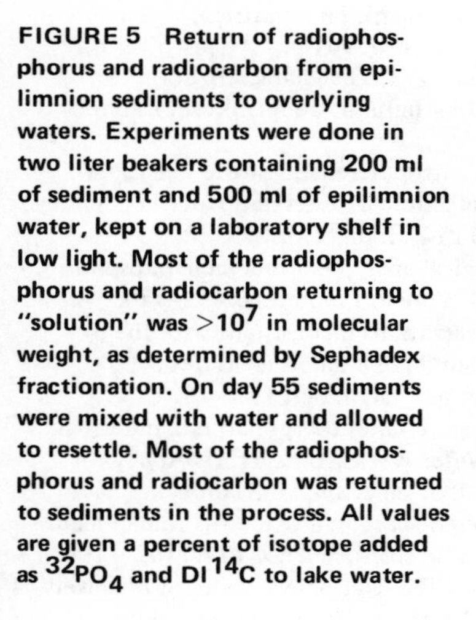
FIGURE 5 Return of radiophosphorus and radiocarbon from epilimnion sediments to overlying waters. Experiments were done in two liter beakers containing 200 ml of sediment and 500 ml of epilimnion water, kept on a laboratory shelf in low light. Most of the radiophosphorus and radiocarbon returning to "solution" was >10^7 in molecular weight, as determined by Sephadex fractionation. On day 55 sediments were mixed with water and allowed to resettle. Most of the radiophosphorus and radiocarbon was returned to sediments in the process. All values are given a percent of isotope added as $^{32}PO_4$ and DI ^{14}C to lake water.

or differences in ecosystem function be accurately judged. It is believed that some of the apparent overwhelming complexity of ecosystems will disappear when such functional relationships for major nutrient elements have been analyzed.

REFERENCES

Anonymous. 1969. International Joint Commission on the Pollution of Lake Erie, Lake Ontario, and the International Section of the St. Lawrence River. Vol. I. Summary. 151 p. Vol. II. Lake Erie. 316 p. Vol. III. Lake Ontario. 329 p.

Arthur, C. R., and F. H. Rigler. 1967. A possible source of error in the ^{14}C method of measuring primary productivity. Limnol. Oceanogr. 12:121–124.

Broecker, W. S. 1965. The application of natural radon to problems in ocean circulation, p. 116–145. *In* T. Ichiye (ed.) Symposium on Diffusion in Oceans and Fresh Waters. Lamont Geol. Observ., Palisades, New York.

Broecker, W. S., and E. A. Olson. 1960. Radiocarbon from nuclear tests. Science 132:712–721.

Broecker, W. S., and T. H. Peng. 1971. The vertical distribution of radon in the BOMEX area. Earth Planet. Sci. Letters. 11:99–108.

Broecker, W. S., Y. H. Li, and J. Cromwell. 1967. Radium-226 and radon-222: Concentration in Atlantic and Pacific Oceans. Science 158:1307–1310.

Broecker, W. S., Y. H. Li, and T. H. Peng. 1971. Carbon dioxide—man's unseen artifact, p. 287–324. *In* D. H. Hood (ed.) Impingement of Man on the Oceans. John Wiley and Sons, Inc. New York.

Burns, N. M., and C. Ross (ed.) 1972. Oxygen-nutrient relationships within the central basin of Lake Erie, p. 120–126. *In* Project Hypo: An intensive study of the Lake Erie central basin hypolimnion and related surface water phenomena. Canada Centre for Inland Waters, Paper No. 6.

Chamberlain, W. M. 1968. A preliminary investigation of the nature and importance of soluble organic phosphorus in the phosphorus cycle of lakes. Ph.D. dissertation. Univ. of Toronto. 232 p.

Craig, H. 1957. The natural distribution of radiocarbon and the exchange time of carbon dioxide between atmosphere and sea. Tellus 9:1–17.

Edmondson, W. T. 1970. Phosphorus, nitrogen and algae in Lake Washington after diversion of sewage. Science 169:690–691.

Edmondson, W. T. 1972. Nutrients and phytoplankton in Lake Washington, p. 172–188. *In* G. E. Likens (ed.) Nutrients and Eutrophication: the limiting nutrient controversy. Allen Press, Inc., Lawrence, Kansas.

Einsele, W. 1936. Über die Beziehungen des Eisenkreislaufs zum Phosphatkreislauf in eutrophen See. Arch. Hydrobiol. 29: 664–686.

Emerson, S. 1974. Radium-226 and radon-222 as limnologic tracers: the carbon dioxide gas exchange rate. Ph.D. dissertation. Columbia U. 183 p.

Garrels, R. M., and C. L. Christ. 1965. Solutions, minerals and equilibra. Harper and Row, New York. 450 p.

Haney, J. F. 1970. Seasonal and spatial changes in the grazing rate of limnetic zooplankton. Ph.D. dissertation. Univ. of Toronto. 177 p.

Hoover, T. E., and D. C. Berkshire. 1969. Effects of hydration on carbon dioxide exchange across an air-water interface. J. Geophys. Res. 74:456–464.

Hutchinson, G. E. 1957. A treatise on limnology. Vol. 1. Geography, physics and chemistry. John Wiley and Sons, Inc. New York, New York. 1015 p.

Kern, A. N. 1960. The hydration of CO_2. J. Chem. Educ. 37:14–23.

Kerr, P. C., D. F. Paris, and D. L. Brockway. 1970. The interrelationship of carbon and phosphorus in regulating heterotrophic and autotrophic populations in aquatic ecosystems. U.S. Govt. Printing Ofc., Washington, D.C. 53 p.

Kerr, P. C., D. L. Brockway, D. F. Paris, and J. T. Barnett, Jr. 1972. The interrelation of carbon and phosphorus in regulating heterotrophic and autotrophic populations in an aquatic ecosystem, Shriner's Pond, p. 41–57. *In* G. E. Likens (ed.) Nutrients and Eutrophication: the limiting nutrient controversy. Allen Press, Inc., Lawrence, Kansas.

King, D. L. 1972. Carbon limitation in sewage lagoons, p. 98–105. *In* G. E. Likens (ed.) Nutrients and Eutrophication: the limiting nutrient controversy. Allen Press, Inc., Lawrence, Kansas.

Kuentzel, L. E. 1969. Bacteria, carbon dioxide, and algal blooms. J. Water Pollution Control Fed. 42:2035–2051.

Kuentzel, L. E. 1971. Phosphorus and carbon in lake pollution. Environmental Letters 2:101–120.

Lange, W. 1970. Cyanophyta-bacteria systems: effects of added carbon compounds or phosphate on algal growth at low nutrient concentrations. J. Phycol. 6:230–234.

Lean, D. R. S. 1973a. An investigation of phosphorus compartments in lake water. Ph.D. dissertation. Univ. of Toronto. 199 p.

Lean, D. R. S. 1973b. Phosphorus dynamics in lake water. Science 179:678–680.

Lean, D. R. S. 1973c. Movement of phosphorus between its biologically important forms in lake water: a community excretion mechanism. J. Fish. Res. Bd. Can. 30:1525–1536.

Likens, G. E. (ed.) 1972. Nutrients and eutrophication: the limiting nutrient controversy. Allen Press, Inc., Lawrence, Kansas. 328 p.

Megard, R. O. 1970. Lake Minnetonka: nutrients, nutrient abatement, and the photosynthetic system of the phytoplankton. Limnological Research Center, Univ. of Minnesota, Minneapolis. Report No. 7. 210 p.

Mortimer, C. H. 1941–42. The exchange of dissolved substances between mud and water in lakes. J. Ecol. 29:280–329; 30:147–201.

Peters, R. H. 1972. The role of zooplankton in nutrient regeneration. Ph.D. dissertation. Univ. of Toronto. 204 p.

Peters, R. H., and D. R. S. Lean. 1973. The characterization of soluble phosphorus released by limnetic zooplankton. Limnol. Oceanogr. 18:270–279.

Pleisch, P. 1970. Die Herkunft eutrophierenden Stoffe beim Pfäffiker und Greifensee. Vierteljahreschrift der naturforschenden gesellschaft in Zürich. Bd. 115 H. 2 p. 127–229.

Rigler, F. H. 1966. Radiobiological analysis of inorganic phosphorus in lakewater. Verh. Internal. Verein. Limnol. 16:465–470.

Rigler, F. H. 1968. Further observations inconsistent with the hypothesis that the molybdenum blue method measures orthophosphate in lake water. Limnol. Oceanogr. 13:7–13.

Sawyer, C. N. 1947. Fertilization of lakes by agricultural and urban drainage. J. New England Water Works Assn. 61:109–127.

Schindler, D. W., G. J. Brunskill, S. Emerson, W. S. Broecker, and T. H. Peng. 1972a. Atmospheric carbon dioxide: its role in maintaining phytoplankton standing crops. Science 177:1192–1194.

Schindler, D. W., and E. J. Fee. 1973. Diurnal variation of dissolved inorganic carbon and its use in estimating primary production and CO_2 invasion in lake 227. J. Fish. Res. Bd. Can. 30:1501–1510.

Schindler, D. W., J. Kalff, H. E. Welch, G. J. Brunskill, H. Kling, and N. Kritsch. 1974 (a). Eutrophication in the High Arctic: Meretta Lake, Cornwallis Island (75° N. Lat.). J. Fish. Res. Bd. Can. 31(7):647–662.

Schindler, D. W., and D. R. S. Lean. 1974. Biological and chemical mechanisms in eutrophication of freshwater lakes. *In* O. Roels (ed.) Proc. Colloq. on the Hudson Estuary, City Univ. of New York, 1972, pp. 129–135.

Schindler, D. W., R. V. Schmidt, and R. A. Reid. 1972b. Acidification and bubbling as an alternative to filtration in determining of phytoplankton production by the ^{14}C method. J. Fish. Res. Bd. Can. 29:11.

Schindler, D. W., H. E. Welch, J. Kalff, G. J. Brunskill, and N. Kritsch. 1974 (b). The physical and chemical limnology of Char Lake, Cornwallis Island (Lat. 74° 42′ N., Long. 94° 50′ W.) J. Fish. Res. Bd. Can. 31:585–607.

Stumm, W., and J. O. Leckie. 1971. Phosphate exchange with sediments: its role in the productivity of surface waters, p. III-26/16. *In* S. H. Jenkins (ed.) Advances in Water Pollution Research. Pergamon Press, New York.

Vollenweider, R. A. 1968. Scientific fundamentals of the eutrophication of lakes and flowing waters, with particular reference to nitrogen and phosphorus in eutrophication. Organization for Economic Cooperation and Development, Paris, France. DAS/CSI/68.27. 159 p.

Vollenweider, R. A. 1969. Möglichkeiten und Grenzen elementarer Modelle der Stoffbilanz von Seen. Arch. Hydrobiol. 66:1–36.

Wangersky, P. J. 1972. The control of seawater pH by ion pairing. Limnol. Oceanogr. 17:1–6.

Wetzel, R. G. 1972. The role of carbon in hard-water marl lakes, p. 84–104. *In* G. E. Likens (ed.) Nutrients and Eutrophication: The Limiting Nutrient Controversy. Allen Press, Inc. Lawrence, Kansas.

Wigley, T. M. L. 1971. Ion pairing and water quality measurements. Can. J. Earth Sci. 8:468–476.

PRODUCTIVITY AND MINERAL CYCLING IN TROPICAL FORESTS

FRANK B. GOLLEY

ABSTRACT

Limited information on tropical forest ecosystems suggest that biomass of vegetation, percent of the nutrient inventory in the biomass, and rapidity of nutrient cycling is higher compared to temperate forest systems. Mean tropical forest net primary productivity is about 23 mt/ha/yr. Mean tropical grassland productivity is higher and savannah is lower than tropical forest net production. Tropical forest appear to use about 70–80 percent of the gross energy converted in photosynthesis in maintenance of the plant mass and, therefore, forest gross primary production is near 70 mt/ha/yr.

The region of the earth termed "the tropics" includes portions of Asia, Africa, the Americas and Oceania. While the image conveyed to many persons resident in temperate regions is a hot, humid, lowland covered with tall forest, actually, the tropics include a wider range of ecological communities than any other portion of the earth. These tropical communities can be arranged in a matrix with altitude on one axis and quantity and duration of rainfall or evaporation on the other (Figure 1). Forest communities fall in the lower and more humid portions of this matrix, as distinct from savannahs, grasslands, deserts and high altitude vegetation. Forests, themselves, form a continuum ranging from the wet tropical rain forest as found in Quibdo, Colombia, to the dry deciduous forest in the Gangetic Plain of India. Further, in most tropical areas disturbance of the landscape has resulted in widespread second-growth forests of different ages. Research on these various forests is uneven. In some we have considerable detailed information on the structure and function of the ecological system; in others we have merely lists of species and miscellaneous natural history observations.

Review of production and mineral cycling of tropical forests ideally would establish the range of variation in these two functional parameters for each type of forest, the frequency distribution of each parameter and the mean or median condition. Unfortunately the available data are inadequate to develop a review in this depth. Nevertheless, the objective of this paper will be to describe the range of production and mineral cycling in tropical forests, within the context of ecosystem structure and function, as far as existing information will allow. The source of data in summary figures and tables has not been listed; the reader should go to the cited papers for references.

CHARACTERISTICS OF TROPICAL FORESTS

An ecological system has numerous structural and functional characteristics, which, together, explain its behavior over space and time. Structure includes biomass, diversity, and chemical makeup; function includes energy flow, productivity, mineral cycling, phenology, resiliency to stress, and linkages with other systems. Because of the variety of types of tropical forests, and the internal complexity of any single forest, it seems best in an analysis of this type to adopt a strategy which will not only provide a review of our knowledge of the function of tropical forests, but also will consider function in the context of the ecosystem and the biosphere. This strategy adopts a black box approach, and focuses on the tropical forest as a unit of study, with linkages to other ecosystems through the environment (Figure 2). This unit can be analyzed into

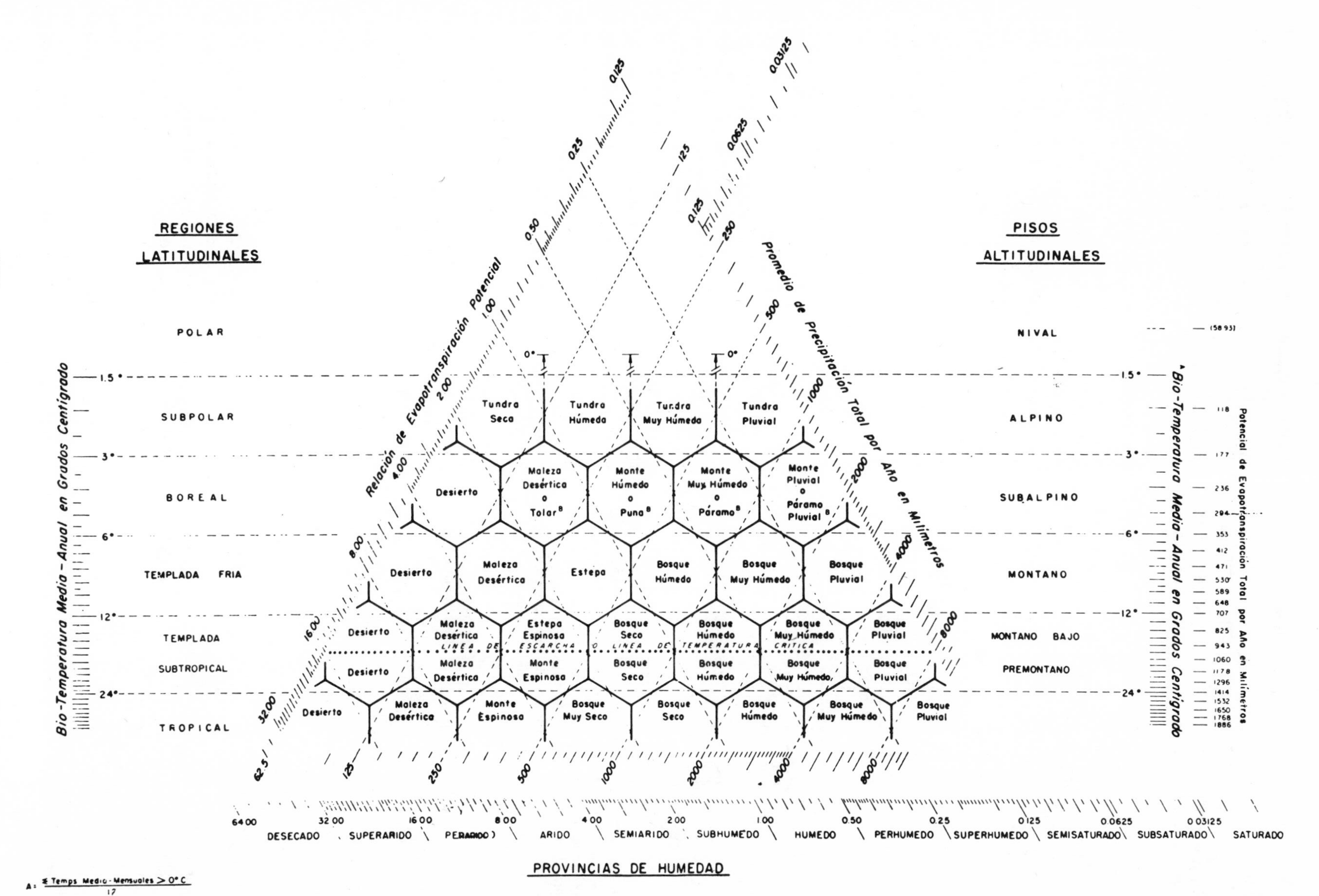

FIGURE 1 Classification of vegetation by temperature, precipitation, and potential evaporation, according to Holdridge (1967).

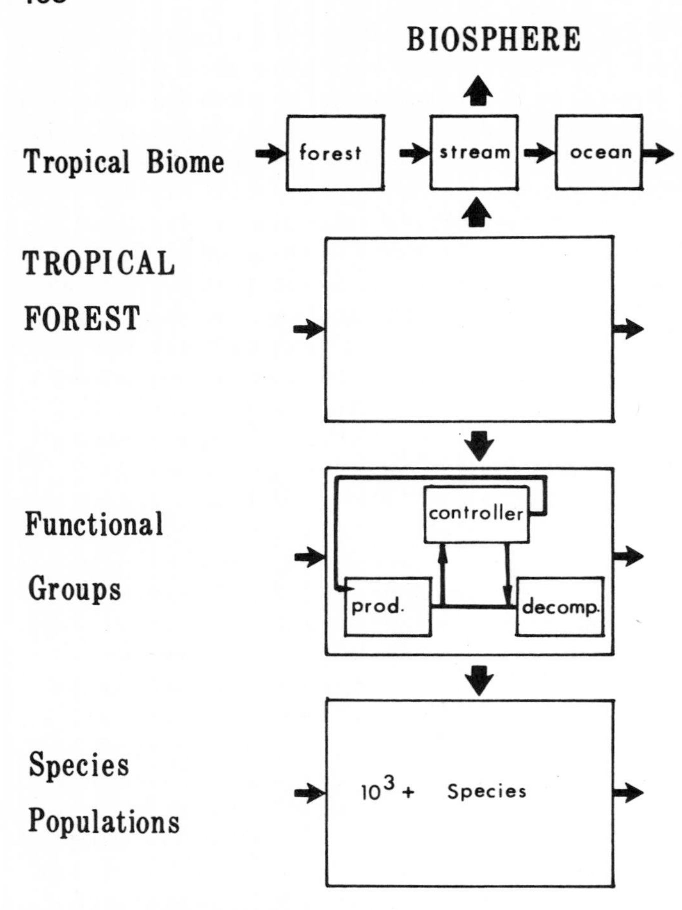

FIGURE 2 Diagram showing the relationships between populations, functional groups, tropical forest ecosystems, and the tropical biome. Beginning at the top of the figure, the forest biome consists of forests linked to streams and to the oceans. The forest ecosystem can be analyzed into functional groups and into species populations depending upon the type of study and the degree of detail needed.

a variety of subunits which interact together to produce the behavior of the whole system.

In this view, tropical forests are made up of a variety of chemical compounds, organized into functional groupings which include producers, decomposers, and controller consumers (Figure 2). These groups are composed, in turn, of species populations, which are the fundamental unit within the community. These groups are linked through the transfer of energy and materials, so that changes in one part are transmitted throughout the system and steady state can be maintained.

With this basic conceptual framework we can conceive of the tropical forest as an ecological unit exposed to the environment and linked to other units through the environment. Two aspects of the environment are of interest in the context of production and mineral cycling. These are the energy input to the system from the sun and the mineral input to the system from rain, dust and the substrate.

Net radiation to the system is the difference between the downward flux from direct and diffuse sunlight and thermal infrared radiation from atmosphere and clouds, and the outward flux from reflected sunlight and thermal, infrared radiation from the ground. The net radiation gain is dissipated by evaporation of moisture, condensation of moisture or conduction and convection by wind. If soil moisture is adequate, evaporation is usually proportional to net radiation. Budyko's (1955, 1956) data mapped by Gates (1962) show that net radiation is higher near the equator and over the oceans. At the equator net radiation ranges from 60 to 140 kcal/cm^2/yr, according to Gates. This energy powers the movement of water through the ecosystem. Since water is the solute for most essential elements, the movement of water powered by net radiation energy is the basic process underlying mineral cycling. As far as I know, Odum (1970) is the only person who has studied this phenomenon on an ecosystem basis in tropical forests.

Vegetation captured through photosynthesis only a few percent of the solar energy incident to the canopy. This photosynthetic energy is employed in the maintenance of the ecological system and, if the system is growing, in the construction of new tissues. Production or productivity in ecosystem terms is the new organic matter grown by a system, or in a steady-state system the amount of matter transferred or available to be transferred to the next functional trophic grouping. Net primary productivity is thus that material available for maintenance of the other nonplant parts of the steady-state system. If the system is not growing there is no net community production; if it is growing there may be net community production. There must always be net primary production, although it becomes smaller in amount as the steady-state condition is established. Thus, the physical-chemical description of the tropical forest ecosystem requires information on the energy input for the work of maintaining the system and on the chemical kinetics of the elements which are essential in the construction of the living system. These two fundamental processes are included within the topics of production and mineral cycling.

PRODUCTIVITY

This review will concentrate on production of vegetation in ecosystems which we assume are at or near steady-state conditions. Few data on true community production in the tropics are known. It is well understood by ecologists that there are two kinds of production of vegetation by the primary producers, as they are often called. The total energy capture by the plants for the work of the ecosystem is the gross primary production. After some of this energy is expended via respiration in the work of maintenance of the vegetation, the remainder is available for other populations in the system. This remaining energy is the net primary production. Since man is one of these "other populations" we are especially interested in net primary production, and we have numerous measures of its magnitude. Gross primary production is less easily measured and the data are fewer.

Tropical forest productivity has been examined in a recent symposium sponsored by the International Society of Tropical Ecology, INTECOL, and the Indian National Science Academy at New Delhi in February 1971 (P. M. Golley and F. B. Golley, 1972). Summarization of the available data on net primary production (Figure 3) suggests that the distribution of data is skewed toward lower productivities, and the mean is about 23 t/ha/yr. Mean tropical grassland net production is slightly higher and savannah lower than forest production. Lieth's map showing the distribution of net primary production over the earth (Figure 4), illustrates the heterogeneity of production in the tropics.

It is not surprising that the frequency distribution of net production is skewed. Production like any other process is limited by environmental factors and we would expect that there will be relatively few ecological situations with optimum production conditions. The expected curve should be skewed to lower productivities with a long tail to the highest production levels. The environmental factors controlling tropical production are described and illustrated in Golley and Lieth (1972).

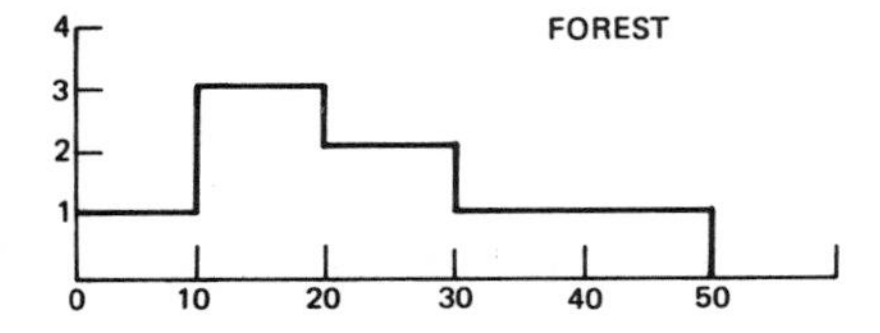

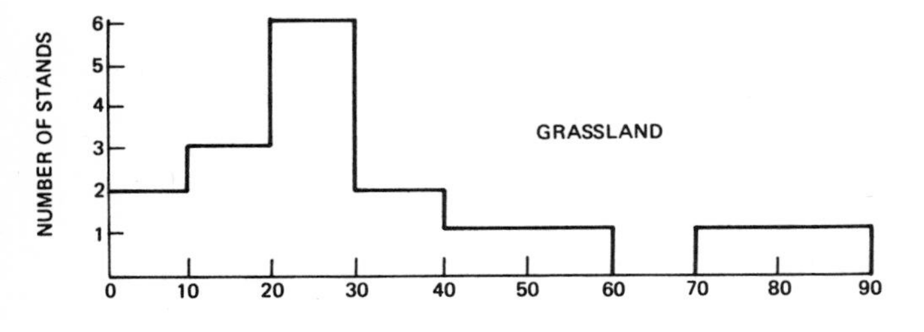

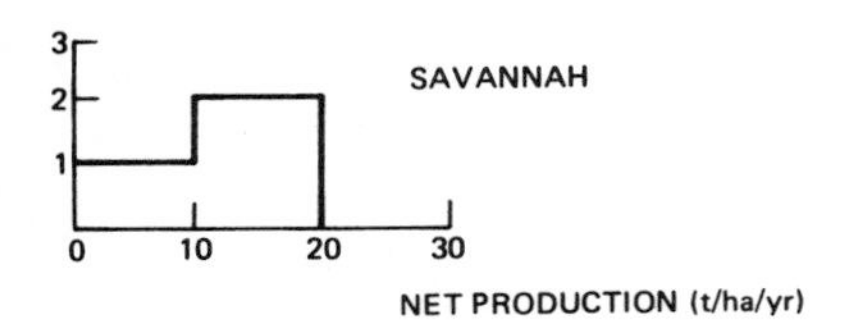

FIGURE 3 Frequency distribution of primary net production of tropical communities in metric tons per hectare per year, after Golley (1972a).

Gross primary production are too few for graphical presentation. At this stage of investigation, about all we can do is correct the net production data by calculating the amount of energy utilized in maintenance of the plant mass. Tropical forests appear to use 70 to 80 percent of the gross energy intake in maintenance, leaving 20 to 30 percent for net production. The frequency distribution of maintenance as a percent of gross production for all vegetation (Figure 5) shows two peaks. One peak represents grassland, herbaceous vegetation and growing plantations of trees; the other peak represents mature tropical and temperate forests. With this correction we can calculate that the average gross primary production of tropical forests is about 67 t/ha/yr or 28×10^3 kcal/m^2/yr (Golley, 1972a). It is not appropriate to put a range on this mean value, since the percentage of gross primary production used in maintenance may vary depending upon growth conditions.

Jordan (1971) also has examined tropical primary productivity from the point of view of the efficiency of production. Efficiency was defined as the sum of energy stored in wood, in leaves, fruit and litter over the total solar energy available to the community. Jordan found that the rate of wood production in intermediate-age stands on mesic sites is similar in tropical and temperate regions; however, the rate of leaf and litter production is higher in the tropics (Table 1). In contrast, efficiency of wood production is higher at higher latitudes. Jordan hypothesizes that where solar energy is abundant, as in the tropics, there has not been selective pressure toward maximization of wood production.

The data on primary production give an estimate of the energy available for work within the tropical forest system. Odum (1970) also has calculated the other ecosystem energy flows for a montane forest in Puerto Rico. Incoming insulation in that forest is 3830 kcal/m^2/day and gross photosynthesis is 131 kcal/m^2/day. Evaporation and transpiration amount to 2975 kcal/m^2/day, which contribute to driving a mineral flow of 0.249 grams/m^2/day from wood to leaves. We would expect a similar range of fluxes in other tropical forests, depending to a large extent on the moisture levels available in the soil and the evapotranspiration rates.

MINERAL CYCLING

There are about 90 chemical elements that might occur naturally in the environment, but not all of these are essential for life or occur commonly in living tissues. The most abundant elements in the biosphere are oxygen, carbon and hydrogen, which occur in amounts greater than 10^4 kg/ha (Deevey, 1970). These elements plus boron, nitrogen, fluorine, sodium, magnesium, silicon, phosphorus, sulfur, chlorine, potassium, calcium, vanadium, chromium, manganese, iron, cobalt, copper, zinc, selenium, molybdenum,

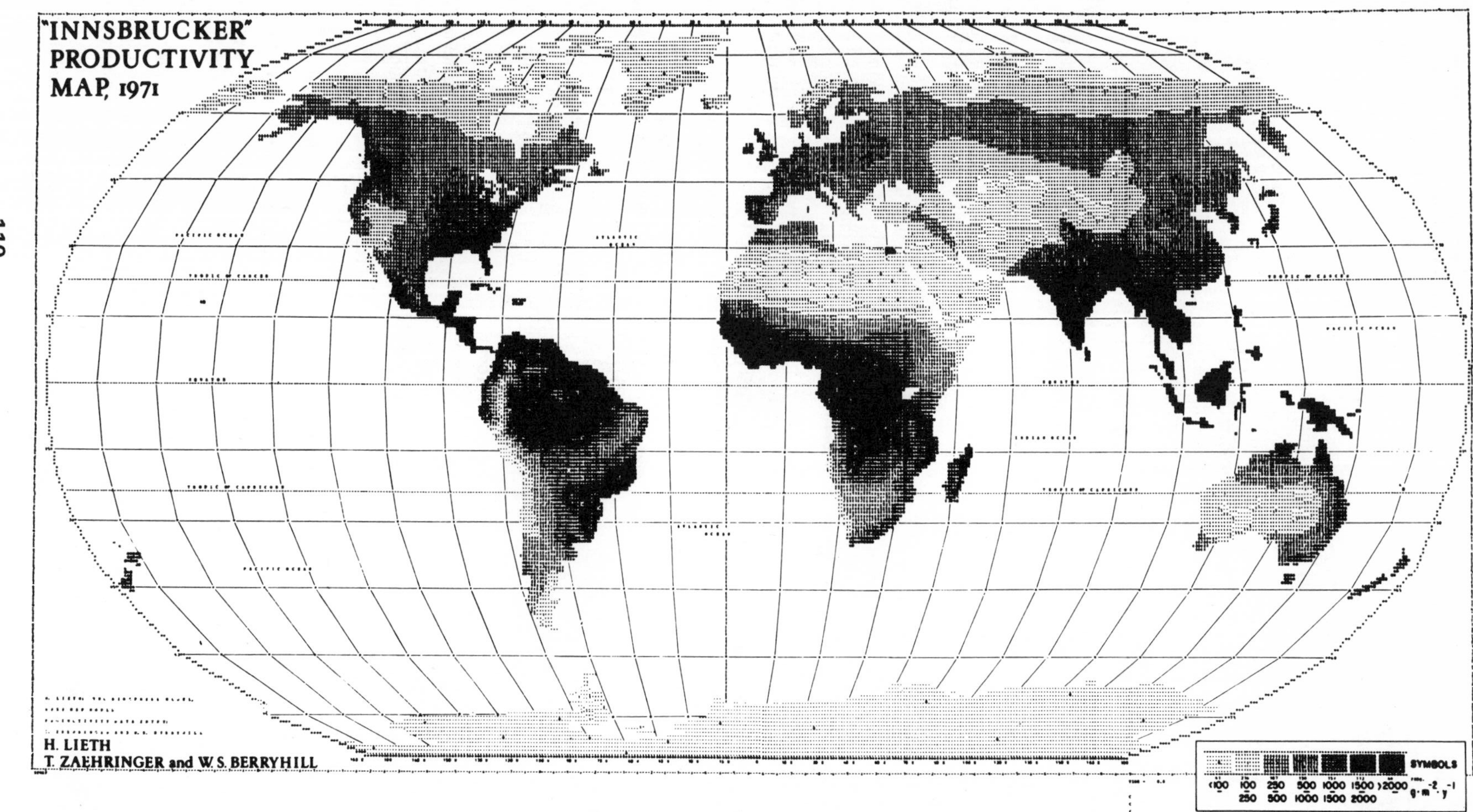

FIGURE 4 "Innsbrucker Productivity Map." This map is a computer simulation of a slightly updated version of an earlier productivity map (Lieth, 1964) for the land areas only. Some unintended alterations in the arctic areas occurred due to the limited number of data values entered for this area. From Golley and Lieth, 1972.

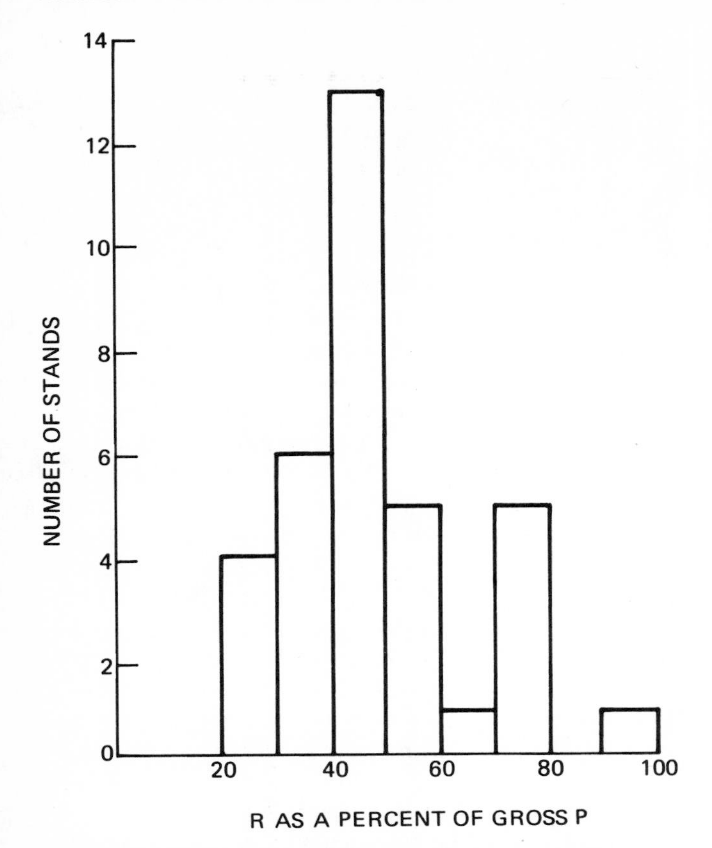

FIGURE 5 The frequency of stands of vegetation with autotrophic respiration (R) as a percent of gross primary production (gross P). All types of vegetation are represented in the figure (from Golley, 1972b).

tin and iodine are essential for life. Nitrogen, calcium, potassium, silicon, magnesium and sulfur occur in quantities of more than 100 kg/ha in the biosphere; the remaining elements usually occur in trace amounts. These elements are accumulated in the living tissue of the community and are cycled between the various species populations and ecosystem components. The movement of elements within the biological part of the ecosystem can be called the biological cycle of elements (Duvigneaud and Denaeyer-de Smet, 1970).

The chemical elements in the biological components are derived from the substrate and the atmosphere. The geological process of erosion and deposition move mineral elements through the lithosphere and hydrosphere. In terrestrial uplands, elements may enter the system through rain or dust and from weathering of soil minerals and are lost to ground water and streams, comprising a geological cycle of elements. The biological cycle forms a shunt on this long term geological process and is characterized by its intensity and rapidity.

The geological cycle is mainly under the influence of water movement and availability, temperature, and topography as well as the parent rock material. High year-around temperatures and abundant water found in certain tropical forest areas result in a relatively rapid geological cycle. It has frequently been suggested in the literature that this geological process is so accelerated in tropical forest environments that the biological cycle has of necessity taken a special form in order to conserve the essential elements for life. Specifically, it has been suggested that nutrients are stored in the biological part of the system where they can be protected from erosive forces, and that the time the elements are in the substrate where they can be leached from the system is minimized by a variety of special biotic adaptations. In this part of the review we will briefly consider the inventory of nutrients in tropical forests, the intensity of the biological and geological cycles, and evaluate the adaptations to preserve nutrients against the geological flux. Only certain macroelements will be discussed here due to limitations of time and space.

Mineral cycling in a tropical forest can be considered in terms of a simple diagram which is merely an expansion of our earlier model (Figure 6). The amount of nutrient in each compartment is determined from its biomass and the concentration of nutrient in that biomass. A large biomass is generally characteristic of tropical forests (Table 2). The quantities of wood, especially, are large in tropical forests and average about 300 t/ha, compared with about 150 t/ha for temperate forests. There also is evidence that the concentration of chemical elements differs from tropical to temperate forests (Figure 7). Rodin and Bazilevich's (1967) excellent summary of data suggests that tropical forests contain in their biomass larger percentages of silicon, magnesium, sulfur and trace elements and smaller quantities of potassium than do temperate forests. The effect of these different quantities of biomass far surpasses the effect of differences

TABLE 1 Comparison of Rate and Efficiency of Production in Different Vegetation (After Jordan, 1971)

	Rate of Production ($g/m^2/yr$)		Efficiency of Production (energy production/total solar energy)	
Forest Type	Wood	Leaves	Wood	Leaves
Tropical forests	350–900	180–2,300	0.28–0.50	0.15–1.15
Temperate forests	410–2,407	270–500	0.42–1.81	0.21–0.44

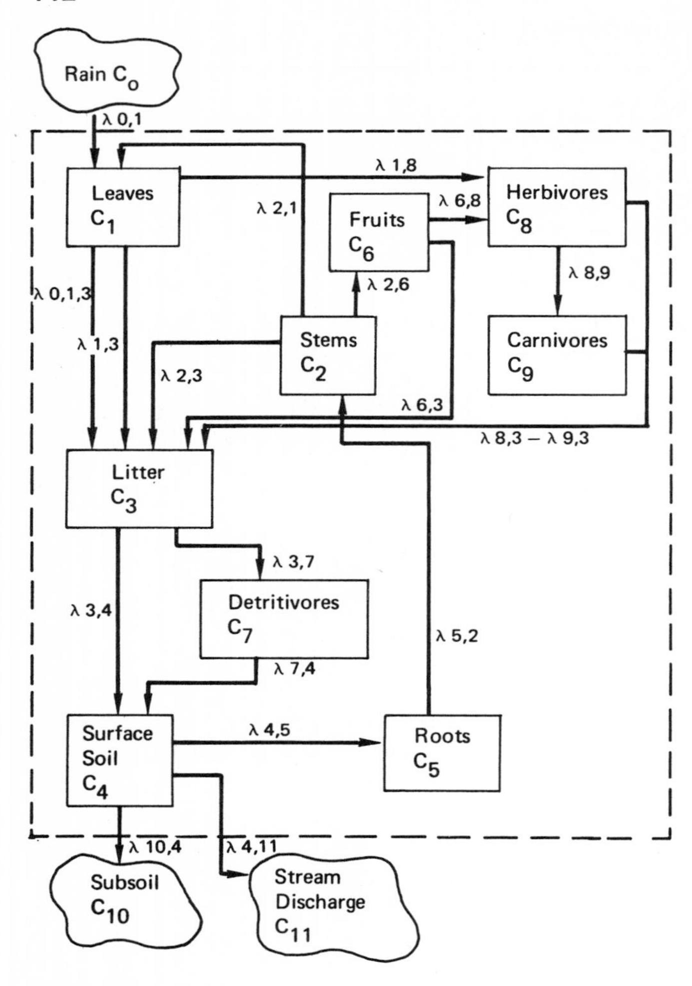

FIGURE 6 Diagram of mineral cycling in a tropical moist forest. The system boundaries are indicated by the dotted lines. System components are identified as boxes, outside sources as amorphous figures. The transfer functions are indicated between components and sources.

in concentration, with the result that tropical forests usually have larger inventories of nutrients in their biomass than do other types of forests.

These between biome comparisons should not conceal the great variation in chemical concentration within the tropical forest. Comparing the averages of Rodin and Bazilevich (1967) with data from forests in Panama (Golley *et al.*, in press), Puerto Rico (Jordan et al., 1972) and the Amazon (Stark, 1971, Table 3 gives an indication of the extent of this variability. These few data show the low nutrient status of the Amazon forest on podzolic sands, which has been so well described by Stark, as compared to the higher levels in Panamanian forests.

If we compare the standing crop of elements in the biological part of the tropical forest ecosystem with that in the active part of the soil, we can judge the role of the vegetation in sequestering nutrients. In five forests in Panama (Golley *et al.*, in press) only phosphorus and potassium are held consistently in large percentages in the vegetation (Table 4). Apparently it is advantageous to the system to concentrate phosphorus and potassium in vegetation because of their mobility, small inventory or both. In the nutrient-poor Amazon podsol sands, Stark (1971) has presented data suggesting that a larger number of elements are stored in the vegetation than in the Panamanian tropical forests.

Detailed information on one important part of the biological cycle, litter fall, is available for a number of tropical forests and serves as a rough index to the rapidity of the cycle. Rather than reproduce these descriptive data, it will be more useful to consider them in terms of the inventory of nutrients available in the vegetation mass. The ratio of element inventory in biomass to annual litterfall is a measure of turnover time. For the macroelements phosphorus, potassium, calcium and magnesium, turnover time is dimensioned in years and is almost always less than 100 years (Table 5) and averages about 20 years. Clearly, the biological cycle is quite rapid for these chemicals.

There are many fewer data on the relation of the biological cycle to the geological cycle in tropical forests. One example from the Tropical Moist forest of Panama will provide some insight into the process. The flux from these ecosystems is represented by discharge to streams; input is by rainfall; the difference between input and output is weathering of soil minerals which recharges the soil inventory. Phosphorus and potassium appear to be in balance in the system (Table 6). The input from rain equals output to streams, and apparently the recharge from weathering is relatively small. Calcium and magnesium recharge, in contrast, must be large to balance input and output of the system. The annual uptake of phosphorus, potassium and calcium by the vegetation exceeds the discharge, while for magnesium the reverse is true. These comparisons, together with those on inventory of nutrients (Table 4), suggest that the supply of potassium and phosphorus is relatively poor and mineral cycling adaptations have developed to conserve these elements. These adaptations include storage in biomass and rapid internal cycling rates. The rate of calcium cycling is rapid, yet this element is abundant in the soil and is not stored at high concentrations in the biomass. For the other elements, the environment appears to provide nearly adequate supplies for system maintenance.

Jordan *et al.* (1972) have found a similar pattern in the Montane forest in Puerto Rico, but Stark (1971) in the Amazon shows a much wider spread of possible nutrient dificiencies. She has postualted a direct nutrient cycling mechanism in the forests growing on podsols in which there is a mycorrhizal connection between dead organic matter and roots allowing the direct transfer of nutrients between them. The inevitable small leakage will eventually result in such poor nutrient conditions that the forest vegetation cannot

TABLE 2 Biomass of Forest Communities (metric tons dry wt/ha)

Forest	Canopy Leaves	Canopy Stems	Understory	Roots	Litter
Tropical forests					
Thailand, rain forest	8.2	360	2.4	33	3.5
Thailand, monsoon forest	3.8	261	2.0	25	–
Thailand, dry evergreen forest	5.6	229	2.9	–	–
Puerto Rico, mangrove	5.4	40	–	5	–
Ghana, moist tropical forest	–	187	–	25	2.3
Congo, secondary forest	6.5	116	–	31	5.6
Puerto Rico, lower montane forest	8.1	269	–	71	19.3
Brazil, mountain evergreen forest	9.1	131	–	33	–
Panama, tropical moist dry	7.3	252	3.9	13	6.2
Panama, tropical moist wet	11.3	355	1.7	10	2.9
Panama, premontane wet	10.5	258	0.8	13	4.8
Panama, riverine	11.3	1,163	1.1	12	14.1
Panama, mangrove	3.5	159	0.1	190	102.1
Temperate forests					
Pine forest	6.4	122	4.0	29	37.5
Coniferous forest	10.4	114	2.0	38	36.6
Deciduous forest	2.6	142	3.0	37	10.5

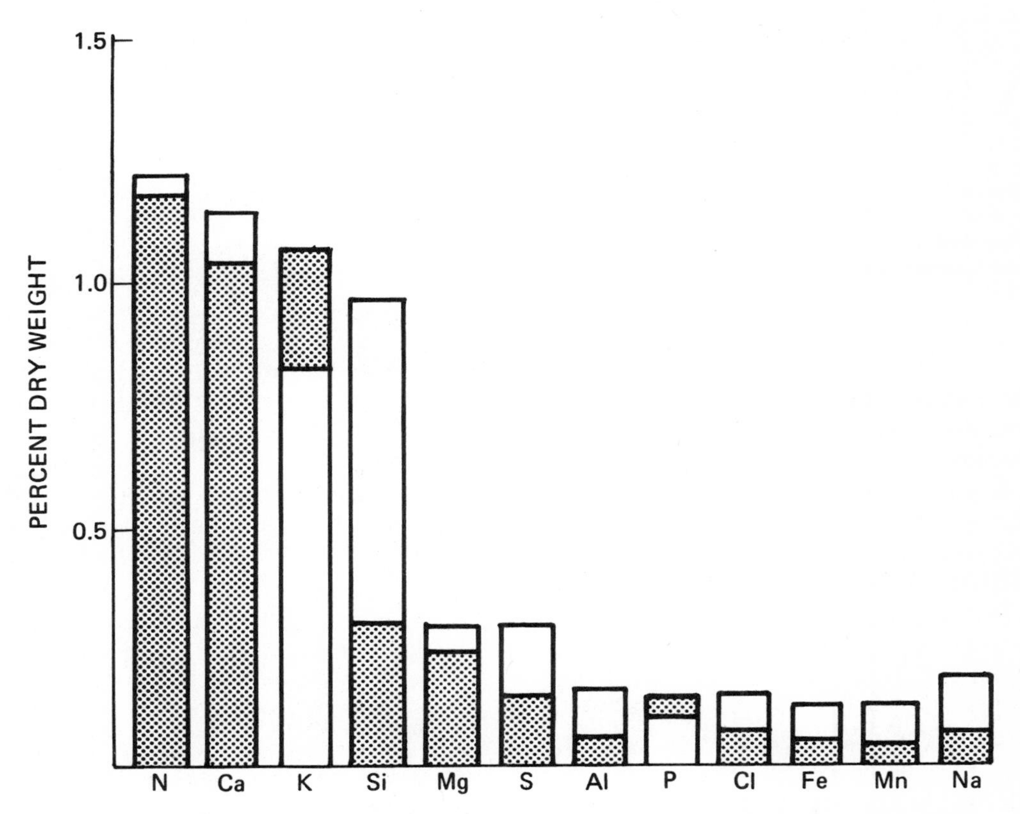

FIGURE 7 Comparison of concentration of selected elements in temperate (black bar) and tropical (white bar) forests. Data from Rodin and Bazilevich (1967).

TABLE 3 Comparison of Element Concentrations in Tropical Forest Vegetation (% dry weight)

	N	P	K	Ca	Mg
Panama	1.40	0.14	1.29	1.42	0.25
Puerto Rico	0.70	0.37	0.51	0.58	0.17
Amazon	1.40	0.13	0.31	0.16	0.16
Overall[a]	1.22	0.10	0.82	1.14	0.28

[a]Nonarithmetic mean.

TABLE 4 Percent of Total Element Inventory in the Vegetation in Panamanian Tropical Forests Represented by Specific Nutrient Elements

Forest	P	K	Ca	Mg	Na
Tropical moist	89	89	12	16	16
Premontane wet	96	85	48	22	47
Riverine	99	97	61	35	14
Mangrove	75	97	32	89	75
Tropical moist second growth (6 yr)	82	58	17	5	2
Average	88	85	34	33	31

TABLE 5 Turnover Time in Years of the Mineral Inventory in Vegetation by Litter Fall in Tropical Forests

	Elements			
Forest	P	K	Ca	Mg
Tropical moist (Panama)	25	37	22	25
Premontane wet (Panama)	9	28	17	10
High forest (Ghana)	12	10	9	6
Deciduous evergreen (Thailand)	5	6	9	8
Montane (Puerto Rico)	–	84	9	10
Bamboo (Burma)	8	3	12	9
Dry deciduous (Thailand)	5	7	12	5
Dry deciduous (India)	38	–	16	–
Dry acacia (Senegal)	27	57	27	9
Riverine (Panama)	94	93	79	45
Mangrove (Panama)	23	8	24	5

TABLE 6 Comparison of Biological and Geological Cycles in a Tropical Moist Forest

	Geological Cycle		Biological Cycle
Element	Input Rain (kg/ha/yr)	Stream Output (kg/ha/yr)	Annual Uptake (kg/ha/yr)
P	1.0	0.7	11
K	9.3	9.5	187
Ca	29	163	270
Mg	5	44	30

be maintained, in which case renewed erosion will permit new nutrient sources to be developed. This very interesting hypothesis should be tested by study of the mineral dynamics on these podsol soils.

CONCLUSIONS

While the data on tropical forest production and mineral cycling are very limited, we have sufficient information to develop hypotheses for the design of the next stage of our work. We speculate that as one moves from cold temperate to moist tropical forest conditions, the biomass of vegetation, the percent of the nutrient inventory in the biomass and the rapidity of the biological portion of the mineral cycles will increase. These patterns have not one single cause, rather they are an expression of the change in environment providing improved growth conditions in the tropics but also more rapid chemical kinetics. The result is that in certain locations (podsols in the Amazon) and for certain essential elements (P and K in Panama) the nutrients may become limited due to the intensity of the geological portion of the mineral cycling process. These situations may result in fragile types of ecosystems, even though they consist of a large mass with high diversity. Disturbance of these systems could require a very long time for recovery, since the rapidity of the chemical kinetics would work against establishment of stability. If this is true, then, recovery would be very expensive in terms of time and money. For this reason alone, it is worthwhile to increase support of tropical forest studies, which should be directed toward establishing the fragility of the system, the steps required for recovery of disturbed systems and the role of the tropical forest system in influencing the geological and hydrological cycles.

REFERENCES

Budyko, M. T. 1955. Atlas of the heat balance. Leningrad. (From Gates, 1962).

Budyko, M. T. 1956. The heat balance of the earth's surface. Translation PB 131692. U.S. Dept. Commerce, Office Tech. Services, Washington, D.C.

Deevy, E. S., Jr. 1970. Mineral cycles. Sci. Am. 223:148–158.

Duvigneaud, P., and S. Denaeyer-De Smet. 1970. Biological cycling of minerals in temperate deciduous forest, p. 199–225. *In* D. E. Reichle (ed.) Analysis of temperate forest ecosystems. Springer-Verlag, New York. 304 p.

Gates, D. M. 1962. Energy exchange in the biosphere. Harper and Row. New York. 151 p.

Golley, F. B. 1972a. Summary, p. 407–413. *In* P. M. Golley and F. B. Golley (compilers). Tropical ecology, with an emphasis on organic production. Athens, Georgia. 418 p.

Golley, F. B. 1972b. Energy flux in ecosystems, p. 69–90. *In* J. S. Wiens (ed.) Ecosystem structure and function. Oregon State Univ. Press, Corvallis. 176 p.

Golley, F. B., and H. Lieth. 1972. Bases of organic production in the tropics, p. 1–26. *In* P. M. Golley and F. B. Golley (compilers).

Tropical ecology, with an emphasis on organic production. Athens, Georgia. 418 p.

Golley, F. B., J. T. McGinnis, R. G. Clements, G. I. Child and M. J. Duever. (In press). Mineral cycling in a tropical moist forest ecosystem. Univ. of Georgia Press. Athens, Georgia.

Golley, P. M., and F. B. Golley (compilers). 1972. Tropical ecology, with an emphasis on organic production. Athens, Georgia. 418 p.

Holdridge, L. R. 1967. Life zone ecology. Tropical Science Center. San Jose, Costa Rica.

Jordan, C. F. 1971. Productivity of a tropical forest and its relation to a world pattern of energy storage. J. Ecol. 59:127–142.

Jordan, C. F., J. R. Kline, and D. S. Sasscer. 1972. Relative stability of mineral cycles in forest ecosystems. Am. Nat. 106:237–253.

Lieth, H. 1964. Versuch einen kartographischen Darstellung der Troduktizitat der Pflanzendecke auf der Erde, p. 72–80. *In* Geographisches Taschenbuch 1964–65. M. Steiner, Wiesbaden.

Odum, H. T. 1970. Rain forest structure and mineral cycling hypothesis, p. H3–52. *In* H. T. Odum and R. F. Pigeon (ed.) A Tropical Rain Forest. A study of irradiation and ecology at El Verde, Puerto Rico. Div. of Tech. Info., U.S. Atomic Energy Commission. (H)219 p.

Rodin, L. E., and N. I. Bazilevich. 1967. Production and mineral cycling in terrestrial vegetation. Oliver and Boyd, Edinburgh. 288 p.

Stark, N. 1971. Nutrient cycling. II. Nutrient distribution in Amazonian vegetation. Tropical Ecology 12:117–201.

ANALYSIS OF CARBON FLOW AND PRODUCTIVITY IN A TEMPERATE DECIDUOUS FOREST ECOSYSTEM*

W. F. HARRIS, P. SOLLINS, N. T. EDWARDS, B. E. DINGER, and H. H. SHUGART

ABSTRACT

Carbon metabolism of a mesic *Liriodendron tulipifera* forest is summarized based on recent estimates of pools of carbon in ecosystem components and annual fluxes of carbon within the system. Estimates of metabolic parameters of the total ecosystem are derived from component processes (e.g., photosynthesis, autotrophic respiration, heterotrophic respiration). Residence time of carbon in the forest ecosystem is comparatively short (10 years) because of the large carbon efflux in respiration, even though some components such as woody biomass and soil organic matter have residence times of 100 to 150 years. Experimental constraints on interpretation of the current summary of ecosystem carbon metabolism are discussed with emphasis on improved measurement technology and the need for similar analyses of ecosystem metabolism for diverse ecosystem types in order to adequately assess the role of the biosphere in regulating global carbon balance.

INTRODUCTION

The International Biological Program has entered Phase III, involving synthesis and exchange of data gathered during the program. Synthesis involves drawing together all the available information in order to answer or clarify specific questions and to determine scientific generalities underlying the functional organization and interactions among organisms, populations and ecosystems. The variety and amount of information available for such synthesis is awesome. Therefore, to realize the potential contribution of the IBP to its primary objective, increased understanding of productivity in nature, Phase III has three tasks: 1) to summarize our understanding of trophic level components of various ecosystems, 2) to complete syntheses which emphasize coupling of trophic levels to evaluate both the total ecosystem behavior and the influence of particular ecosystem processes on the response of the total system and 3) to define areas requiring further research. This last task helps set the stage for the logical development of ecosystem studies to follow.

Ecosystem Analysis In order to integrate the results of large, multidisciplinary research into a holistic ecosystem framework, a formal basis of analysis is needed (Reichle and Auerbach, 1972). Considering the ecosystem as a functional unit composed of a trophic hierarchy dispersed in time and space provides such a framework. Trophic-level analysis of ecosystems relates flows of materials between components to overall metabolism of the system. In turn, fluxes of matter to and from the ecosystem couple it to adjacent systems, providing some insight into the function of larger landscape units. Conceptualization of how the ecosystem functions forms the basis for intensive observation, measurement and manipulation of system parameters. This body of data in turn supports expression of ecosystem behavior as sets of mathematical functions. Together these operations of ecosystem analysis are a powerful technique with which to examine ecosystem properties, derive inferences about ecosystem function, both natural and following perturbation, and define areas of additional research.

Carbon Cycle As an example of ecosystem analysis and its

* Research supported by the Eastern Deciduous Forest Biome, US-IBP (Contribution No. 54) funded by the National Science Foundation under Interagency Agreement AG-199, 40-193-69 with the Oak Ridge National Laboratory, which is operated by Union Carbide Corporation for the U.S. Atomic Energy Commission.

potential for contribution to IBP synthesis, this paper summarizes progress on analysis of the carbon cycle of a temperate deciduous forest (Reichle *et al.*, 1973a). Additional data have been obtained, particularly on below-ground carbon dynamics and autotrophic metabolism, which provide a clearer, but certainly not final, assessment of carbon dynamics in this ecosystem.

A thorough understanding of carbon dynamics in the biosphere is necessitated by the role of carbon in the chemistry of living systems, the coupling of biological systems by carbon through a common atmospheric pool and the interaction of biosphere and atmosphere to regulate global carbon balance. Man's activities (largely combustion of fossil fuel) are resulting in an increase in atmospheric carbon dioxide content of 0.2% per year (~7 ppm by volume). The observed increase in atmospheric carbon dioxide content represents roughly one-half the carbon dioxide emitted in fossil fuel combustion (Anonymous, 1970). Uptake and storage in the biosphere and oceans account for the remainder. Recent discussions (Olson, 1970; Whittaker and Likens, 1973) conclude that the role of terrestrial ecosystems, especially forests, previously has been underestimated in assessment of global productivity and carbon dynamics. Evidence is accumulating that the total impact of man's activities has reduced biological production significantly and therefore altered the capacity of the global terrestrial ecosystem to regulate atmospheric carbon balance (Olson, pp 33–43). However, it has been difficult to assess recent trends of global carbon exchange, much less extrapolate future trends, because of a lack of measurements of carbon storage and flow in different types of terrestrial ecosystems (Olson, 1970). Questions of how terrestrial ecosystem processes influence carbon turnover, its residence time in the biosphere and, in turn, atmospheric carbon balance urgently need our attention.

METHODS AND RESULTS

Annual Budget of Carbon

Site Description The study site is a second-growth, mesophytic deciduous forest on karst topography within the U.S. AEC Reservation at Oak Ridge, Tennessee. The forest is established on a deep, alluvial, silt loam soil (Emory series). Mean annual temperature is 13.3 C, annual precipitation averages 126.5 cm and total short-wave radiation averages 123.5 kcal cm^{-2} yr^{-1} (Sollins *et al.*, 1973). The forest is dominated by yellow poplar (*Liriodendron tulipifera*) interspersed with various oaks (*Quercus velutina, Q. alba*, and *Q. rubra* principally). Understory species include redbud (*Cercis canadensis*), flowering dogwood (*Cornus florida*) and occasional blackgum (*Nyssa sylvatica*). Virginia creeper (*Parthenocissus quinquefolia*), woody hydrangea (*Hydrangea arborescens*) and Christmas fern (*Polystichum acrostichoides*) account for approximately 90% of the ground cover biomass, although many other species are present (Taylor, 1974).

Conceptual Model A conceptualization of the forest as a series of compartments representing structural components in a functional notation is shown in Figure 1. This compartmentalization served as the basis for calculation of a budget which describes amounts of organic matter 1) in each compartment and 2) transferred annually along each pathway. An annual budget synthesis is useful for preliminary descrip-

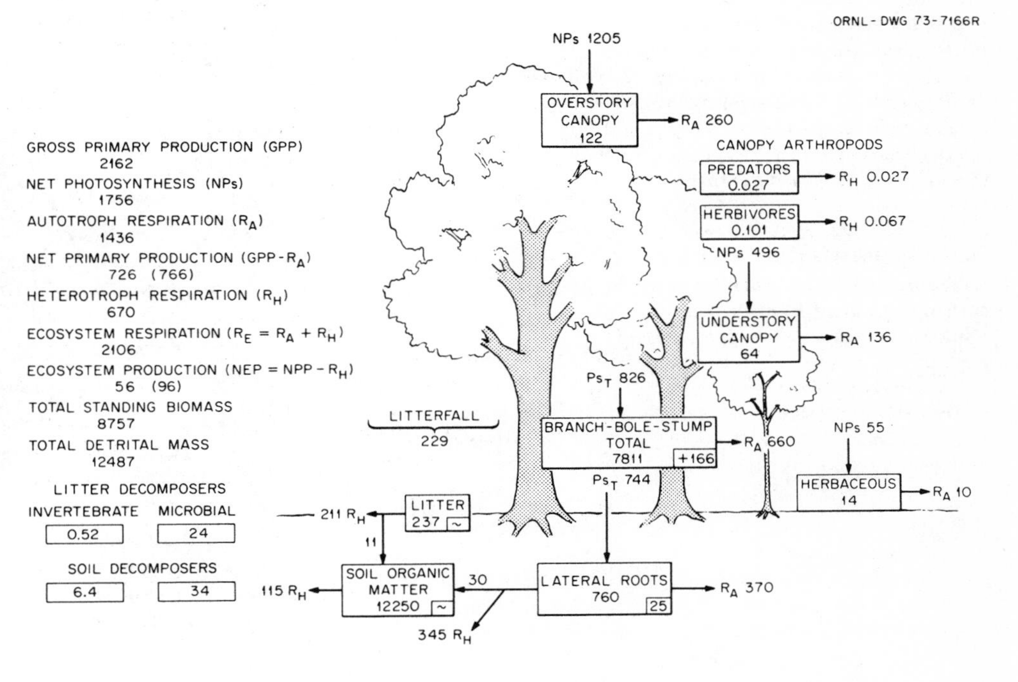

FIGURE 1 Diagram of conceptual model of organic matter/carbon storage and flow in a temperate forest ecosystem. Ps = net photosynthesis; R_A = autotrophic respiration; R_H = heterotrophic respiration; L = losses due to litterfall or root sloughing. Litter and soil decomposers include both microbial and invertebrate organisms. Standing deadwood is included in the branch-bole-stump compartment. Double-headed arrows denote photosynthate translocation pathways.

tions of the system behavior. These budgets also represent initial steps in the construction of dynamic simulation models (Reichle *et al.*, 1973).

Forest Carbon Budget Determination of a forest carbon budget requires use of a broad spectrum of data from physiological analyses to structural and population parameters (Reichle *et al.*, 1973a). Independently derived estimates of transfers using rates determined by diverse methodologies (e.g., harvest/allometric analyses vs. gasometric analysis to estimate production parameters) serve to focus attention on problems of methodology as well as to verify estimates. The time resolution appropriate for particular interpretations helps determine which data sets are emphasized. Short-term environmental influences on physiological processes are not addressed when interpreting data on an annual basis, but information about total system dynamics can be obtained which is not readily apparent when viewed at more detailed levels of resolution.

The carbon budget for the yellow poplar forest is summarized in Figure 2. Carbon determinations were based on flame photometric detection following high temperature pyrolysis and catalytic hydrogenation (Horton *et al.*, 1971). Estimates of autotroph aboveground and central root carbon pools (8.03 kg C m^{-2}) rates of annual, aboveground accumulation (0.166 kg C m^{-2} yr^{-1}) were based on allometric relations of weight of tree components and diameter breast height (Sollins and Anderson, 1971; Harris *et al.*, 1973), periodic inventory of tree diameters (Sollins *et al.*, 1973) and excavation of stump and lateral root biomass (Harris *et al.*, in press).

Underground autotroph carbon pool and associated physiological processes are the least accurately measured and least understood forest ecosystem components. Mean standing pool of lateral root carbon (0.76 kg C m^{-2}) was determined from a series of soil core data collected through the year. Net compartment increment was determined from net seasonal differences in lateral root biomass pools to be ~8% for stands with 5.0 to 10.0 kg C m^{-2}. Turnover of lateral roots was estimated from the summation of net decreases in standing pools determined monthly during 1971-72. Significant death of roots <0.5 cm diam occurred in late spring and late autumn, while root production occurred in late winter and midsummer (Harris *et al.*, in press). Early spring increases in biomass of roots <0.5 cm were accompanied by apparent decreases in biomass of roots >0.5 cm, suggesting growth of the smaller roots at the expense of stored carbohydrates. Radiotracer experiments using late-

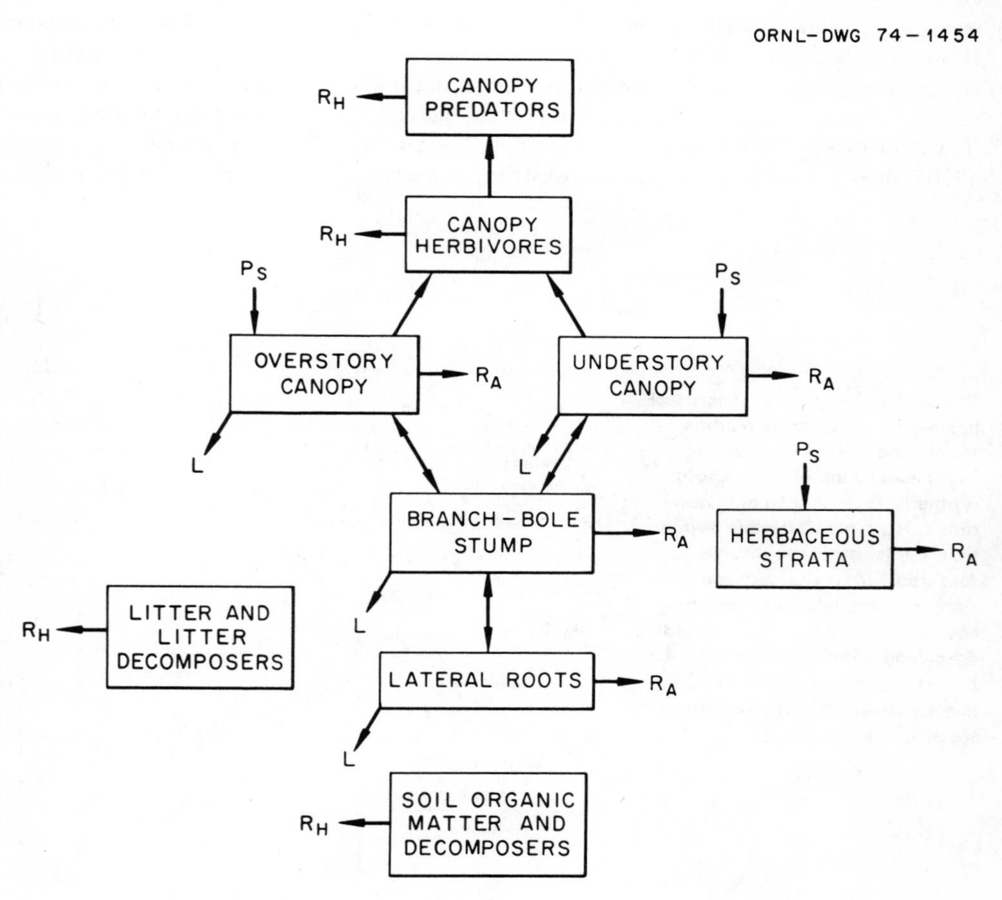

FIGURE 2 Annual carbon cycle in a temperate deciduous forest. Major fluxes in the system are illustrated by arrows. Ps_T denotes translocated photosynthate. Mean annual standing crops are summarized in the center of each box; net annual increment is shown in the lower, right corner. Units of measure are g C m^{-2} yr^{-1} and g C m^{-2}. A summary of ecosystem metabolism (after Woodwell and Botkin, 1970) and standing crops are shown to the right of the figure. Values of NPP and NEP in parentheses are based on harvest/allometric methods; other values are based on gasometric analyses (revised after Reichle *et al.,* 1973a).

fall inoculation of trees with ^{14}C-sucrose revealed a sharp elevation in ^{14}C activity in small roots corresponding to the early-spring period of active growth (Shugart and Harris, unpublished data). Root turnover through death largely occurs from roots <0.5 cm diam. However, cyclic renewal of large support roots >2.5 cm diam has been observed by Kolesnikov (1968) in orchards. Large dead roots attached to living trees also have been observed in our soil monolith analysis of oak-hickory, pine and mesic hardwood forest ecosystems (Harris *et al.*, in press). The rate of cyclic renewal of large roots is unknown. On an annual basis, we have assumed turnover of large roots to be very much slower than that of small roots. Other data on decay rates of various sized roots tend to corroborate this assumption (Harris, unpublished data).

Mean annual standing crop of O_1 and O_2 litter layers was 237 g C m^{-2} based on monthly collections. Soil organic matter decreased from 4.6% (% dry weight of soil) in the upper 10 cm of soil to 1.3% at 21-30 cm depth. The total amount of soil carbon was estimated to be 12.3 kg C m^{-2} to a depth of 75 cm, assuming 58% carbon content of soil organic matter (after Jackson, 1958).

Calculations of the carbon pool present as canopy arthropods utilized weekly measurements of population densities per leaf with conversion to a unit area biomass basis (Reichle and Crossley, 1967). Using the mean carbon content of insect tissue of 45%, carbon pools of canopy herbivores and predators were 101 and 27 mg C m^{-2}, respectively. Litter and soil invertebrates were estimated from population analyses. Litter invertebrates amounted to 520 mg C m^{-2}, while soil invertebrates (primarily earthworms, *Octolasium*) averaged 6.4 g C m^{-2} (Moulder and Reichle, 1972; McBrayer and Reichle, 1971). The mean annual carbon pool in total litter and soil microflora based on ATP analysis was 58 g C m^{-2}, with approximately 65% in fungi and 35% in bacteria (Ausmus, 1973).

Annual Carbon Fluxes Fluxes of carbon in net photosynthesis and autotroph respiration were determined under natural temperature and light conditions by means of gas exchange analysis in controlled environment chambers (Dinger, 1972; Reichle *et al.*, 1973a). Data from several hundred hours of measurement were used to determine average daily and seasonal flux (assuming a 180-day growing season) in contrast to the limited data employed in the earlier summary of carbon metabolism (Reichle *et al.*, 1973a). Converting CO_2 fluxes to carbon resulted in a gross carbon uptake of 2.15 kg C m^{-2} yr^{-1}. This value of total carbon influx is a minimum estimate of true gross photosynthesis in that light respiration of foliage was assumed equal to dark respiration of 0.20 kg C m^{-2} yr^{-1}. Recent measurements of yellow poplar suggest that light respiration is approximately 4 times as great as dark respiration (Richardson *et al.*, 1972).

Respiration of lateral roots was estimated from manometric determinations and monthly biomass density (Reichle *et al.*, 1973a). Total annual carbon efflux from this compartment was 0.392 kg C m^{-2} yr^{-1}. Estimates of shoot respiration are based on data of Woodwell and Botkin (1970). Total carbon loss by branch-bole tissues was 0.660 kg C m^{-2} yr^{-1}. Preliminary analysis of woody shoot respiration of *Liriodendron* suggests good agreement with their data. Foliage respiration evolved at least 0.400 kg C m^{-2} yr^{-1}. Including minor contributions from forest floor autotrophs, total autotrophic respiration was 1.44 kg C m^{-2} yr^{-1}.

Total heterotrophic respiration from the forest floor was measured for 24-h intervals through the year using gas analysis procedures (Edwards and Sollins, 1973). Carbon evolution from the forest floor was 1.04 kg C m^{-2} yr^{-1}. Decomposer respiration from litter was estimated to be 0.21 kg C m^{-2} yr^{-1}, based on monthly estimates of pool size and respiratory flux per unit weight determined manometrically. Respiration of canopy-feeding insects was estimated from body size-metabolism regressions (Reichle, 1971) and mean body size for each age-class for the various insect species. Summed canopy insect respiration (herbivorous and predatory) was 0.094 g C m^{-2} yr^{-1}.

Annual litterfall averaged 229 g C m^{-2}. Leaves accounted for 78% of annual litterfall. Over the eight-year period (1962-1970), tree mortality determined from stand inventory was assumed to have occurred at a uniform rate, 50 g C m^{-2} yr^{-1} (Sollins *et al.*, 1973). The loss of photosynthetic surface area through insect consumption varied by a factor of nearly 2 over a three-year period. Actual foliage consumption varied from 1.9% to 3.4%, while actual reduction in photosynthetic surface due to hole expansion ranged from 5.6% to 10.1% (Reichle *et al.*, 1973b). Using a mean carbon content of leaves of 50%, the carbon flux due to actual consumption was 4.5 g C m^{-2} yr^{-1}.

DISCUSSION

Carbon Budget Comparisons and Implications A total carbon budget provides a basis for estimating ecosystem metabolism (Figure 2). Based on mass balance calculations and field determinations, a lower bound on gross photosynthesis was taken to be 2.15 kg C m^{-2} yr^{-1}, net primary production was estimated at 0.716 to 0.752 kg C m^{-2} yr^{-1} (based on gas analysis and allometric estimates, respectively) and net ecosystem production was 0.046 to 0.082 kg C m^{-2} yr^{-1}. Autotrophic respiration (R_A) was estimated at 1.436 kg C m^{-2} yr^{-1} and heterotrophic respiration (R_H) was 0.67 kg C m^{-2} yr^{-1}. Heterotrophic biomass of <60 g C m^{-2} contributed 31% of ecosystem respiration ($R_A + R_H = 2.11$ kg C m^{-2} yr^{-1}) due almost entirely to decomposer activity.

Analysis of ecosystem metabolism provides insight to overall dynamics and relative importance of the separate components. Another comprehensively analyzed ecosystem is the xeric, oak-pine forest at Brookhaven National Laboratory (Woodwell and Botkin, 1970). While obviously differing structurally from the *Liriodendron* forest (compare total standing crops of autotrophs, Table 1), the two ecosystems had similar net primary production. Relative production of the two systems also is similar. The observed variation between the two forests is within the range observed from year-to-year within the same ecosystem (Harris, unpublished data). Autotrophic respiration (R_A) of the two systems is similar (see Table 1). The differences in metabolism are in heterotrophic respiration (R_H), which for the yellow poplar forest was more than twice that of the more xeric forest at Brookhaven (Table 1). The large R_A and relatively larger R_H of the yellow poplar forest resulted in total ecosystem respiration (R_E) of 2.11 kg C m^{-2} yr^{-1} (2.08 times greater than the oak-pine forest). Basic to these differences is the apportionment of R_E between R_A and R_H; the ratio of R_A/R_H for the oak-pine forest was 2.5 (only 0.68 in the yellow poplar forest). The larger R_H of the yellow poplar forest represents annual decay of 712 g C m^{-2} yr^{-1}, while the oak-pine system loses 360 g C m^{-2} yr^{-1}. Comparison of the ratios of NEP to total standing crop (0.05 for oak-pine and 0.007 for yellow poplar) indicates that the oak-pine system is accumulating carbon 7 times as rapidly as the yellow poplar forest. Thus, while cursory examination of single components of ecosystem metabolism would suggest overall similarity, in fact these two systems are quite different. Ecosystem analysis makes these differences readily apparent. Additionally, the value of R_H in the *Liriodendron* forest confirms the existence of a high root turnover rate; decomposition of 222 g C m^{-2} yr^{-1} from litterfall, dead bole and frass inputs is substantially less than 712 g C m^{-2} yr^{-1} in decomposer respiratory losses. The difference is well within the range of an independent estimate of root turnover, 375 g C m^{-2} yr^{-1}.

Carbon Residence Times Interpretation of the global carbon cycle requires understanding of the ecological factors affecting carbon turnover in local and regional terrestrial ecosystems (Olson, 1970; Anonymous, 1970). Residence times of carbon in components of the *Liriodendron* forest ecosystem vary considerably depending in part upon rates of carbon efflux and carbon pool sizes. Components such as foliage have turnover times of one year or less (Table 2). Our data indicate a surprisingly rapid turnover rate of lateral roots of 0.5 yr^{-1} (a residence time of ~2 years) largely attributable to sloughing of roots <0.5 cm diam. Other components of the forest ecosystem have much longer residence times. The turnover rate (9.3×10^{-3} yr^{-1}) of soil carbon in the *Liriodendron* forest represents a residence time of 107 years, while turnover of carbon to litter through mortality of woody trees is also slow (50 g C m^{-2} yr^{-1}/699 g C m^{-2} = 6.4×10^{-3} yr^{-1}) and represents a residence time of 155 years. The values of carbon residence time agree reasonably well with those reported by SCEP (Anonymous, 1970). The comparatively long residence times of both of these pools suggests that forests have a large capacity to act as a carbon "sink," thus buffering increments to atmospheric CO_2 content, but only for a few decades as decomposition of a larger detrital pool would return increasingly larger amounts of carbon dioxide to the atmosphere.

TABLE 2 Turnover Rates and Residence Times for Carbon in the *Liriodendron* Forest at Oak Ridge, Tennessee. Turnover Rates (yr^{-1}) Are Calculated from Flux (g C m^{-2} yr^{-1}) Divided by Compartment Size (g C m^{-2})

Compartment	Turnover Rate[a] (yr^{-1})	Mean Residence Times (yr)
Rapidly decomposable litterfall	0.89	1.1
Lateral roots[b]	0.49	2.0
Aboveground woody component	0.0064	156.0
Total woody component	0.049	20.0
Soil organic matter	0.00931	107.0
Forest ecosystem	0.10	10.0

[a]Turnover rate calculated from ratio of R_E (total carbon efflux) to total ecosystem carbon pool.
[b]Lateral roots are all roots except the central stump.

TABLE 1 Comparison of Metabolism and Structure of Two Terrestrial Ecosystems. Units of Measure are kg C m^{-2} and kg C m^{-2} yr^{-1} for Compartments and Fluxes Unless Otherwise Noted

Parameter	*Liriodendron* Forest[a]	*Quercus-Pinus* Forest[b]
Total standing crop (TSC)	8.76 kg C m^{-2}	5.96 kg C m^{-2}
Net primary production (NPP)	0.73 kg C m^{-2} yr^{-1}	0.60 kg C m^{-2} yr^{-1}
Relative production (NPP/TSC)	8.3%	10%
Autotroph respiration (R_A)	1.44	0.68
R_A/TSC	0.16	0.11
Heterotroph respiration (R_H)	0.67	0.29
Ecosystem respiration ($R_E=R_A+R_H$)	2.11	1.01
Net ecosystem production (NEP=NPP-R_H)	0.06	0.28
Annual decay	0.70	0.36

[a]Revised after Reichle *et al.*, 1973a.
[b]Woodwell and Botkin, 1970.

When the forest ecosystem is viewed as a single entity, the overall residence time of carbon is only 10 years (2.110 g C m^{-2} yr^{-1}/21,150 g C m^{-2} = 0.10 yr^{-1}). This more rapid turnover is the result of the large respiratory efflux. Viewing the system in this manner detracts from the idea that forests act as long-term carbon "sinks." These data suggest that carbon residence time in ecosystems can be manipulated through changes in ecosystem respiration ($R_E = R_A + R_H$). Thus, maintenance of younger forests with relatively high net primary production but with lower autotroph maintenance respiration and smaller detrital pools could result in increased carbon residence time in the biosphere (compare NEP and annual decay for forests in Table 1). Clearly additional summaries of carbon metabolism of diverse ecosystem types are needed.

Experimental Constraints Estimates of carbon dynamics of ecosystems are extremely sensitive to the accuracy of measurement of component processes. Some ecosystem parameters can be determined directly as well as from measurement of physiological fluxes. The degree to which these independently derived estimates corroborate one another is the only present check on accuracy and precision of our measurement techniques applied to total ecosystems. While estimates of autotrophic processes in our forest seem reasonably well defined by harvest techniques and measurement of physiological fluxes (NPP of 750 and 720 g C m^{-2} yr^{-1}, respectively), the difference between the two estimates yields values of net ecosystem production (NEP) which differ approximately two-fold (Figure 2). Thus a 5% difference in estimates of autotrophic processes is magnified 20-fold in estimation of NEP. Additional summaries from other IBP research will clarify the range of likely estimates, precision and accuracy of methodologies, and perhaps lead to greater standardization of techniques and analysis.

Estimates of carbon residence times, and particularly measurements of carbon efflux and carbon pool sizes, also are subject to close scrutiny. In the present example, a 25% change in estimated carbon efflux ($R_A + R_H$), or total carbon respired by the ecosystem, results in a change of similar magnitude in carbon residence time. Because of the large pool of carbon, variations in estimates of net carbon accumulation (NEP or NPP) have little influence on estimates of residence time. Our data emphasize the importance of acquiring estimates of respiratory (both autotrophic and heterotrophic) fluxes for diverse types of terrestrial ecosystems.

Budgeting of ecosystem resources follows the principle of conservation of mass. Our present summary of heterotrophic metabolism is based on the assumption that total soil carbon efflux is 1.04 kg C m^{-2} yr^{-1}. Estimates of litter and root decay and maintenance of living roots were subtotaled. The difference, 115 g C m^{-2} yr^{-1}, was attributed to the decay of soil organic matter. The accumulation and turnover of soil carbon may be a major factor determining the influence of terrestrial ecosystems on the atmospheric carbon balance. Therefore, direct measurement of, as well as knowledge of the factors controlling, the metabolism of soil carbon are required to assess this important role of terrestrial ecosystems.

Another carbon flux requiring further analysis is the humification of detritus. This flux represents the interface of cycles of carbon and other elements. Measurements of *in situ* humification are at best approximate, and the chemistry involved is only incompletely understood for even the simplest of systems. While methodology and data associated with estimation of component processes are subject to further analysis and interpretation, the fact remains that large variances will be associated with estimates of metabolic parameters of total ecosystems even though component processes may be known within 5% of their real values. Confidence in estimates of ecosystem metabolism initially can be gained through systematic comparison of results of current IBP studies as we await development of improved analytical techniques.

REFERENCES

Anonymous. 1970. Man's impact on the global environment: Assessment and recommendations for action. Report of the study of critical environmental problems (SCEP). The MIT Press. Cambridge, Mass. 319 p.

Ausmus, B. S. 1973. Litter and soil microbial dynamics in a deciduous forest stand. EDFB-IBP-73-10. Oak Ridge National Laboratory, Oak Ridge, Tenn. 184 p.

Dinger, B. E. 1972. Gaseous exchange, forest canopy and meteorology, p. 49–56. *In* Ecological Sciences Division Annual Progress Report for period ending September 30, 1971. ORNL-4759. Oak Ridge National Laboratory, Oak Ridge, Tenn. 188 p.

Edwards, N. T., and P. Sollins. 1973. Continuous measurement of CO_2 evolution from partitioned forest floor components. Ecology 54:406–412.

Harris, W. F., R. A. Goldstein, and G. S. Henderson. 1973. Analysis of forest biomass pools, annual primary production and turnover of biomass for a mixed deciduous forest, p. 43–64. *In* H. E. Young (ed.) IUFRO Symposium: Forest biomass. Univ. Maine Press, Orono.

Harris, W. F., R. S. Kinerson, Jr., and N. T. Edwards. In press. Comparison of belowground biomass of natural deciduous forests and loblolly pine plantations. *In* J. K. Marshall (ed.) The belowground ecosystem: A synthesis of plant associated processes.

Horton, A. P., W. D. Shults, and A. S. Meyer. 1971. Determination of nitrogen sulfur, phosphorus and carbon in solid ecological materials via hydrogenation and element-selective detection. Anal. Letters 4:613–621.

Jackson, M. L. 1958. Soil chemical analysis. Prentice Hall, Inc., Englewood Cliffs, N.J. 498 p.

Kolesnikov, V. A. 1968. Cyclic renewal of roots in fruit trees, p. 102–106. *In* M. S. Ghilarov, V. A. Kovda, L. N. Novichkova-Ivanova, L. E. Rodin, and V. M. Sveshnikova (eds.) Methods of productivity studies in root systems and rhizosphere organisms. International symposium, USSR, August–September 1968. USSR Acad. of Scis. Nauka, Leningrad. (Reprinted by Biddles, Lts., Guildford, U.K. 240 p.)

McBrayer, J. F., and D. E. Reichle. 1971. Trophic structure and feeding rates of forest soil invertebrate populations. Oikos 22:381–388.

Moulder, B. C., and D. E. Reichle. 1972. Significance of spider predation in the energy dynamics of forest floor arthropod communities. Ecol. Monogr. 42:473–498.

Olson, J. S. 1970. Carbon cycles and temperate woodlands, p. 226–241. *In* D. E. Reichle (ed.) Analysis of temperate forest ecosystems. Ecological studies 1. Springer-Verlag, New York, Heidelberg, Berlin.

Reichle, D. E. 1971. Energy and nutrient metabolism of soil and litter invertebrates, p. 465–477. *In* P. Duvigneaud (ed.) Productivity of forest ecosystems. Proceedings, Brussels Symposium, October 1969. UNESCO, Paris.

Reichle, D. E., and S. I. Auerbach. 1972. Analysis of ecosystems, p. 260–281. *In* J. Behnke (ed.) Future directions in the life sciences. Oxford Univ. Press, New York. 502 p.

Reichle, D. E., and D. A. Crossley, Jr. 1967. Investigation of heterotrophic productivity in forest insect communities, p. 563–587. *In* K. Petrusewicz (ed.) Secondary productivity of terrestrial ecosystems, Vol. II. Proceedings of Working Meeting, Jablonna, 1966. Polish Acad. Sci., Warsaw.

Reichle, D. E., B. E. Dinger, N. T. Edwards, W. F. Harris, and P. Sollins. 1973a. Carbon flow and storage in a forest ecosystem, p. 345–365. *In* G. M. Woodwell and E. V. Pecan (eds.) Carbon and the biosphere. Brookhaven Symposium in Biology. AEC CONF-720510.

Reichle, D. E., R. A. Goldstein, R. I. Van Hook, and G. J. Dodson. 1973b. Analysis of insect consumption in a forest canopy. Ecology 54:1076–1084.

Reichle, D. E., R. V. O'Neill, S. V. Kaye, and P. Sollins. 1973. Systems analysis as applied to modeling ecological processes. Oikos 24:337–343.

Richardson, C. J., B. E. Dinger, and W. F. Harris. 1972. The use of stomatal resistance, photopigments, nitrogen, water potential and radiation to estimate net photosynthesis. EDFB-IBP-72-13. Oak Ridge National Laboratory, Oak Ridge, Tenn. 130 p.

Sollins, P., and R. M. Anderson. 1971. Dry-weight and other data for trees and woody shrubs of the southeastern United States. ORNL-IBP-71-6. Oak Ridge National Laboratory, Oak Ridge, Tenn. 80 p.

Sollins, P., D. E. Reichle, and J. S. Olson. 1973. Organic matter budget and model for a southern Appalachian *Liriodendron* forest. EDFB-IBP-73-2. Oak Ridge National Laboratory, Oak Ridge, Tenn. 150 p.

Taylor, F. G., Jr. 1974. Phenodynamics of production in a mesic deciduous forest, p. 337–354. *In* H. Lieth and F. Stearns (eds.) Phenology and seasonality modeling. Springer-Verlag, New York, Heidelberg, Berlin.

Whittaker, R. H., and G. E. Likens. 1973. Carbon in the biota, p. 281–302. *In* G. M. Woodwell and E. V. Pecan (eds.) Carbon and the biosphere. Brookhaven Symposium in Biology. AEC CONF-720510.

Woodwell, G. M., and D. B. Botkin. 1970. Metabolism of terrestrial ecosystems by gas exchange techniques: The Brookhaven approach, p. 73–85. *In* D. E. Reichle (ed.) Analysis of temperate forest ecosystems. Ecological studies 1. Springer-Verlag, New York, Heidelberg, Berlin.

PLANT NUTRIENTS AS LIMITING FACTORS IN ECOSYSTEM DYNAMICS

C. O. TAMM

INTRODUCTION

A satisfactory model of ecosystem functions must enable simulation of not only steady state conditions but also effects of changes in manipulation or other external influences. Therefore, recognition of the various regulatory mechanisms in the system is of even greater importance than accurate description of processes within undisturbed ecosystems. While no one denies that lack of light, water, adequate temperatures, or plant nutrients may cut short the primary production of the ecosystem, thus seriously affecting all organisms depending on primary producers, few investigators have tried to explore the mechanisms involved in the adaptation of plant and animal communities to environments with a poor supply of certain nutrients. Experiences with field crops have initiated innumerable physiological studies on the effect of various plant nutrients on growth and other processes. However, "nutrient deficiency experiments" have become standard demonstration experiments, and the scientific frontier has also moved in the direction of biochemistry and molecular biology in plant nutrition work. A few applied ecologists are still trying to explore the effect of fertilization on the species composition of grass swards, or the role of sewage nutrients in lake eutrophication. Little is known, however, about how changes in nutrient supply affect primary production, particularly in terrestrial ecosystems, and next to nothing is known about the effects on secondary consumers and decomposers.

The concept of limiting factors dates back to Liebig's work of about 1840. His idea was that yield was proportional to the most deficient element, as long as this element was "in minimum." Later experience has shown that Liebig's "law of the minimum" was an oversimplification. Yet there are cases where his theory, as well as Blackman's somewhat similar theory for factors limiting photosynthesis, describes fairly well what happens. The next great step was Mitscherlich's exponential equation for the relationship between added fertilizer and harvest—the "law of diminishing returns." In its original form, this well-known curve accounts only for growth increases up to a certain maximum or "optimum." It is fairly simple, however, to include a factor which causes the harvest to decrease when the independent variable increases above some optimum.

Three-dimensional diagrams are one way of illustrating Mitscherlich curves, when two independent variable are involved, but it is often more convenient to show a series of curves (Figure 1) (Nielsen, 1963). Evidently the growth of a grass sward depends on both nitrogen and irrigation. One consequence of this conclusion is that the response to fertilizer may be very different in different years, owing to changes in precipitation. In other cases, differences in other climatic factors may interact with nutrition, e.g., the length of the frost-free season. Therefore, field experiments with plant nutrients must be repeated, preferably over a series of years, before confidence in the results can be established.

Logistic functions, such as the Mitscherlich equation, are more laborious to work with than are simple parabolas. The latter type of relationship is often satisfactory within a limited range but extrapolation may be very misleading.

Different plant species react differently to environmental influences with plant nutrient supply being no exception. An ecologically important difference is that between plants

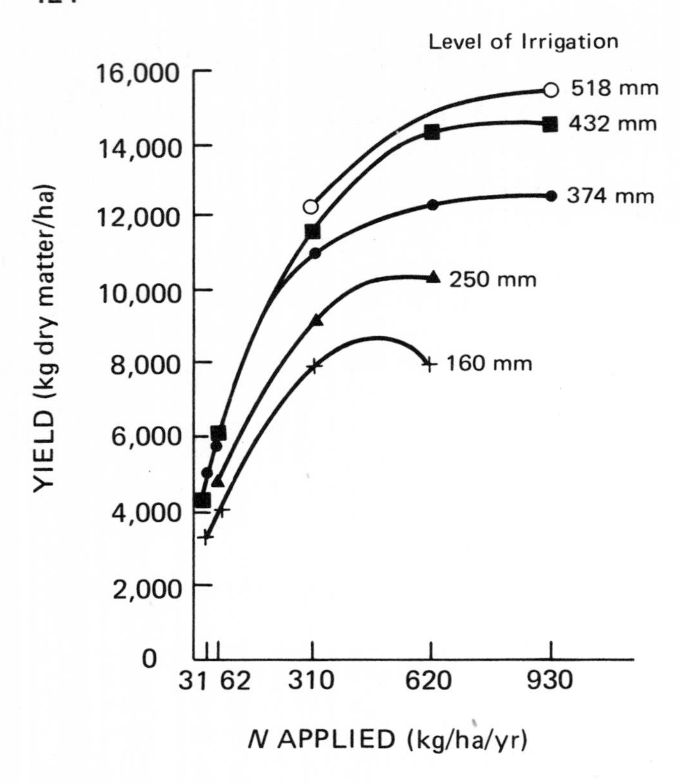

FIGURE 1 The response to fertilizer N of ryegrass grown in covered lysimeters at 5 levels of irrigation (Nielsen, 1963). Data points are averages of five harvests.

having symbiotic fixation of nitrogen and those other plants which do not. This difference is also valid for whole plant communities with and without nitrogen-fixing species. Grass-clover mixtures react less to nitrogen supply than do pure grass swards (Whitehead, 1970).

As far as more natural vegetation is concerned, much less is known about the requirements of different species. A classical paper in forest science is that by Mitchell and Chandler (1939), where a number of Northeastern deciduous trees were classified as nitrogen-demanding, nitrogen-tolerant, or intermediate. Work was done with foliar analysis combined with field experiments in which nitrogen was supplied. A number of New England forest sites were classified with respect to their nitrogen regime. A conclusion from this work (Mitchell and Chandler, 1939) is that the nitrogen regime must be one of the main factors in the competition between tree species. Nitrogen-demanding species will be poor competitors on sites with a low level of available nitrogen. The less exacting species, however, will not be able to take full advantage of a fertile soil.

On the other hand, it does not necessarily follow that nutrient circulation on a particular site must be different in a pure stand of a nitrogen-demanding species from that occurring in a stand of a nitrogen-tolerant species. There are usually some differences, but Ovington (1957, 1958) has shown that the most decisive factor for nutrient uptake by a tree species is its growth energy under the prevailing conditions. This is a result in good agreement with more physiological investigations by Ingestad (1962). It is a well-known fact that nutrient uptake by plant roots does not always reflect the nutrient demand; there is often a luxury consumption of ions in excess. Different plant species have varying abilities to absorb and accumulate different elements—both essential ones and others—but this discrimination is never complete. Therefore, studies of nutrient circulation within an ecosystem usually reveal relatively little about the nutrient status of the various plants within this ecosystem. Full advantage from nutrient circulation models cannot be derived until it has been complemented with information on the extent to which the nutrient demand in the dominant organisms is satisfied (usually the primary producers).

EXAMPLES OF NUTRIENT DEFICIENCIES IN FORESTS AND PEATLANDS

Before proceeding further in the discussion of the mechanisms behind the concept, "limiting factors," comments will be made on some typical cases of deficiency in one or more nutrients. It has long been known that lack of nitrogen may limit growth. Northern coniferous forests on medium and poor sites are one example, as shown first by Hesselman (1937) and Romell and Malmström (1945).

An equally well-known case of deficiency in available nitrogen concerns heathland afforestation in Britian (Zehetmayr, 1960). Growth check in spruce is common there, while other species, such as pines and larches, appear able to obtain more nitrogen from the soil in the early stages of afforestation. The significance of this difference is probably that soil microbiological conditions, including mycorrhiza formation, provide an important regulatory mechanism for nutrient uptake by trees, and that some trees, or their mycorrhizal fungi, are more sensitive than others. In Australia, *Araucaria* species behave in much the same way as spruce in European heathland plantations (Richards and Bevege, 1969).

A number of cases of nutrient deficiency can be related to man's activities. A deficiency in nitrogen in a forest may be the result of litter removal. Such an operation was practiced for centuries in many Central European forests.

Infertile soils, low in both total and available nitrogen, may be the result of repeated or particularly intensive forest fire. In other circumstances, forest fire or prescribed burning may, both in Scandinavia and elsewhere, help to maintain, or even to restore, the fertility of a forest site. This is obviously the case in the southeastern United States, where prescribed burning soon leads to invasion by nitrogen-fixing leguminous plants.

Agricultural use tends to deplete the soil of organic matter. On sandy soils, soil organic matter is often the only cation-absorbing complex available. Therefore when fields are abandoned for cultivation and planted with trees, destruction of the organic matter may lead to a deficiency in potassium, or magnesium, or both.

Drainage of a peatland means a profound change in the nutrient regime. Many peatlands receive some plant nutrients from floodwater during rainy periods. This supply is cut off by ditching, and the water movement in the root-zone is changed from predominantly horizontal to vertical. Potassium deficiency is a common consequence of change in the water and nutrient regime. Deficiency in phosphorus is also common on drained peatlands, where a low content of phosphorus is often a characteristic feature. Lack of phosphorus and potassium on drained peatlands limits not only tree growth but also the production of the ground cover.

The fact that drainage of a peatland may disturb the nutrient status by no means implies that undrained peatlands are normally well supplied with nutrients. Few experiments have been carried out, but there is evidence that deficiency in phosphorus may occur on undrained peatland (Tamm, 1954). Potassium and nitrogen also may well limit production. Gore (1972) has worked out an interesting mathematical model for an ecosystem on blanket peat, dominated by *Eriophorum vaginatum*. The experiment consisted in the removal of vegetation by repeated clipping for several years, and the model suggests that the production of *Eriophorum vaginatum* is limited by available phosphorus.

In the case of grassland, it is quite clear that the application of nutrients may increase productivity. This is well demonstrated by the results from the German IBP project in Solling (Figure 2). There are indications that fertilization with nitrogen may be replaced to some extent by symbiotic, and perhaps also asymbiotic, nitrogen fixation. Production and nitrogen turnover may be slightly lower in a grass/clover mixture without fertilizer nitrogen, than in a pure grass sward receiving large amounts of nitrogen, but so are leaching losses (see Figures 4-7 in Whitehead, 1970). It is hoped that the final synthesis of the IBP grassland work will shed more light on these problems.

HOW LACK OF NUTRIENTS RESTRICTS PRIMARY PRODUCTION

The last section of this paper will be devoted to an admittedly very incomplete discussion of the mechanisms by which nutrient deficiencies may limit ecosystem primary

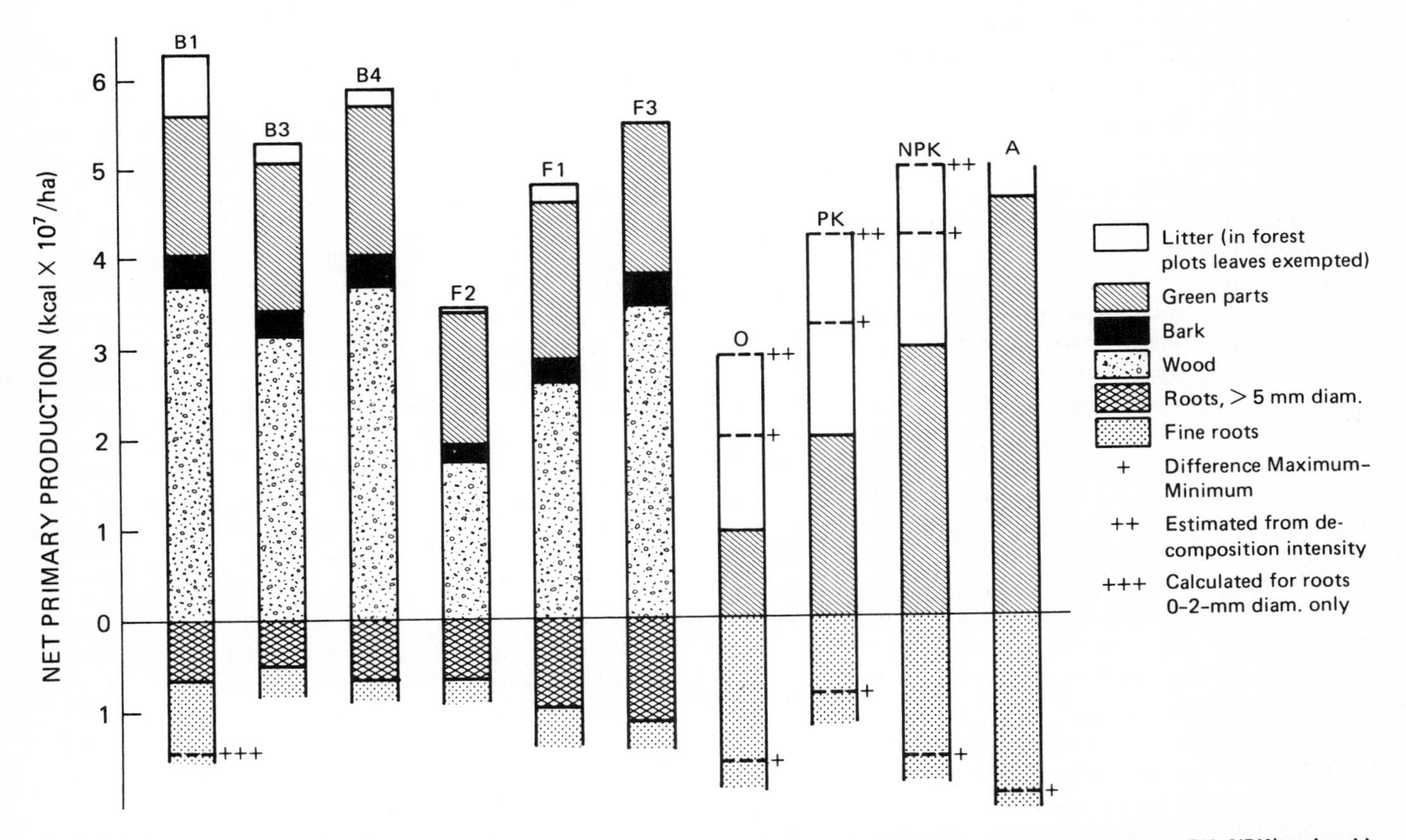

FIGURE 2 Energy fixation in net primary production in beech forest (B), spruce forest (F), grassland (fertilizer regime O, PK, NPK) and arable land (A). Data from Solling project, supplied by M. Runge.

production. Some of this discussion will be based upon work carried out within the Swedish IBP program. This is out of necessity since relatively little experimental work has yet been completed on plant nutrition within the IBP-PT program.

A dosage experiment was started in 1957 with nitrogen supplied annually to a young plantation of Norway spruce on an abandoned field—an experiment that was later incorporated into the Swedish IBP program. Half the number of plots also received a combined P, K fertilizer. This experiment was never a fully integrated ecosystem project, but biomass data are available from two inventories and primary production data, together with much other information, from the final sampling in 1969–1970 (Table 1).

One objective of this experiment was to study what happens in a stand when different nitrogen levels are maintained over extended periods of time. There has been some variation in internal nitrogen level in the spruce, but on the whole the experiment was successful in this respect (Figure 3). At first there was also a positive response to nitrogen, as assumed on the basis of foliar analyses before the start of the experiment. An optimum curve for total stem production versus foliar nitrogen level was obtained, together with some evidence for an interaction between N on the one hand and P, K on the other (Tamm, 1968).

A parallel experiment was laid out in another plantation of the same age, planted on forest land. There was no positive response to nitrogen in this experiment for the first revision period. The reason is no doubt the so-called assart effect: the fertilizing effect of the removal of the old stand. On good sites this effect may be both strong and long-lasting. I advise against the uncritical supply of nitrogen fertilizer to recently established saplings on former forest land. The assart effect, particularly intensive after forest fires, often makes fertilization unnecessary for some time.

Returning to the experiment on the old field, the total amounts of fertilizer added during the 13-year period were very large (Table 2). The originally positive response to nitrogen changed to a negative one—at least at the higher N levels. There remained a significant P, K-N interaction, meaning that P, K alone has no effect, while N, P, K fertilization is consistently better than N alone. At the end of the experiment there was no indication that growth-rate was higher on fertilized plots than on controls. Stem growth was lower at high N (Figure 4), while total primary production appeared to differ little between most treatments (Table 3).

The interesting point here is that needle biomass was much higher on N fertilized plots than on controls at the biomass sampling in 1960 (Table 1). In 1970, however, there was hardly any difference in needle biomass between plus N and minus N. Moreover, the N plots had increased their needle biomass very little from 1960 to 1970, while the minus N plots had increased their needle biomass by more than 50 percent. These data support the assumption that the main effect of the nitrogen fertilization was to allow the spruce to build up a large crown more rapidly (cf. Brix and

TABLE 1 Biomass (September 1960 and May 1970) of a Spruce Stand Planted in 1947. Experiment E1 Hökaberg, Remningstorp

		O	PK	N1	N1PK	N2	N2PK	N4	N4PK
$\sqrt{n} \times H^2$	1956	26	20	20	25	22	23	22	22
Mean height (cm)	1956	186	163	158	181	167	172	165	172
Stem volume (o.b.m^3)	1960	27	20	26	35	31	33	31	33
	1970	143	125	130	171	145	160	137	152
Dry weight (kg/ha)	1960	8,520	6,920	8,740	10,530	9,680	10,760	10,340	10,700
stem wood	1970	40,060	35,380	38,490	49,090	43,520	47,570	41,860	45,740
Stem bark	1960	1,720	1,400	1,760	2,120	1,960	2,170	2,100	2,160
	1970	6,400	5,720	6,260	7,760	6,970	7,560	6,770	7,260
Branches	1960	7,510	6,570	9,200	10,480	9,920	10,880	10,610	10,680
	1970	12,860	11,190	11,020	15,190	12,470	14,490	12,110	13,980
Needles	1960	8,680	7,560	10,840	12,310	11,670	12,790	12,480	12,540
	1970	13,400	11,770	11,580	15,530	12,940	14,900	12,680	14,360
Sum above	1960	26,430	22,450	30,540	35,440	33,230	36,600	35,530	36,080
stumps (kg/ha)	1970	72,720	64,060	67,350	87,570	75,900	84,520	73,420	81,340
Stumps	1970	2,390	2,160	2,380	2,880	2,610	2,820	2,570	2,700
Roots > 5 mm diam.	1970	11,070	9,180	9,850	13,150	10,910	12,130	10,780	12,170
Roots < 5 mm diam.	1970	4,740	4,740	4,740	4,740	4,740	4,740	4,740	4,740
Total biomass	1970	90,920	80,140	84,320	108,340	94,160	104,210	91,510	100,950

NOTE: Weights above stumps estimated by means of measurements on sample trees from all treatments and allometric equations (in case of branches and needles different equations for +N and −N). Stump weights and weights of roots > 5 mm based on only 12 sample trees and common allometric equations. Roots < 5 mm estimated from sample pits; no difference between treatments established, because of wide scattering. Biomass depends both upon experimental treatments and on stand condition at start of experiment, which may be expressed as the so-called Bjørgung index, $\sqrt{n} \times H^2$ (n = number of trees per plot, H = their mean height 1956).

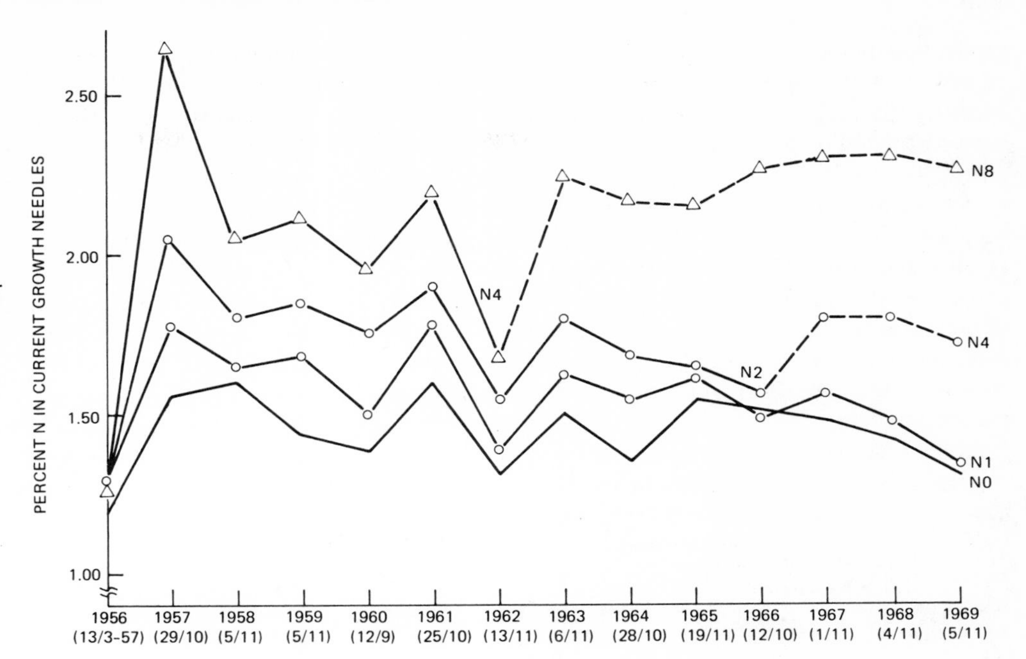

FIGURE 3 Nitrogen contents in exposed current needles of spruce at four different nitrogen regimes (see Table 2). Expt. E1 Hökaberg, S. Sweden (after Tamm, 1968).

Ebell, 1969). The ultimate size of the green crown on this site seems to be determined by other factors than added nitrogen, and therefore the fertilizer effect is not persistent.

The effect of plant nutrient supply as the factor limiting the amount of photosynthetically active organs is probably the most common, and the simplest, mechanism by which this factor limits primary production. In ecosystem modelling it should be observed that this mechanism automatically leads to a lag between nutrient uptake and nutrient effects on primary production. Depending on the plant species and the element examined, the length of the lag may vary, as different species and different growth forms may have varying ability to redistribute a particular element in their tissues.

Another indication of the existence of the same mechanism of growth limitation can be taken from an old Swedish experiment. Here wood ash was used to stimulate forest growth on a drained peatland that remained at a very low productivity without fertilizer. After additions of wood ash in 1918 and 1926, birch stands established themselves. Various amounts of wood ash produced different amounts of forest growth, both total and annual (Figure 5). Initially, there were large differences in nutrient concentrations between treatments, but in time these differences decreased—in the case of potassium almost to nil (Figure 6). Potassium appears to be the element "in minimum," to use Liebig's terminology. Evidently the stand adjusts its growth very closely to the amount of available potassium.

Other mechanisms are also operating, when nutrient supply limits plant growth. It is evident that the chlorotic needles on some sites that are deficient in nitrogen or potassium must have a photosynthetic capacity below normal. Yet it is not too well known to what extent less spectacular deficiencies affect photosynthesis (Keller, 1971; Brix, 1971, 1972).

It has been suggested that a deficiency in potassium affects the water economy of plants (Arland, 1954; Brag, 1972). It is certainly true that deficiency in potassium, or perhaps rather an unfavorable ratio of K/Ca, affects winter survival. Furthermore, the possibility should not be overlooked that a nutritional imbalance may affect the redistribution of

TABLE 2 Nutrients Added, 1957–1969 (kg/ha) in Experiment E1 Hökaberg, Remningstorp

	Treatment							
Element	O	PK	N1	N1PK	N2	N2PK	N4	N4PK
N	–	–	625	625	1,550	1,550	3,900	3,900
P	–	151	–	151	–	151	–	151
K	–	280	–	280	–	280	–	280
Mg	–	200	–	200	–	200	–	200

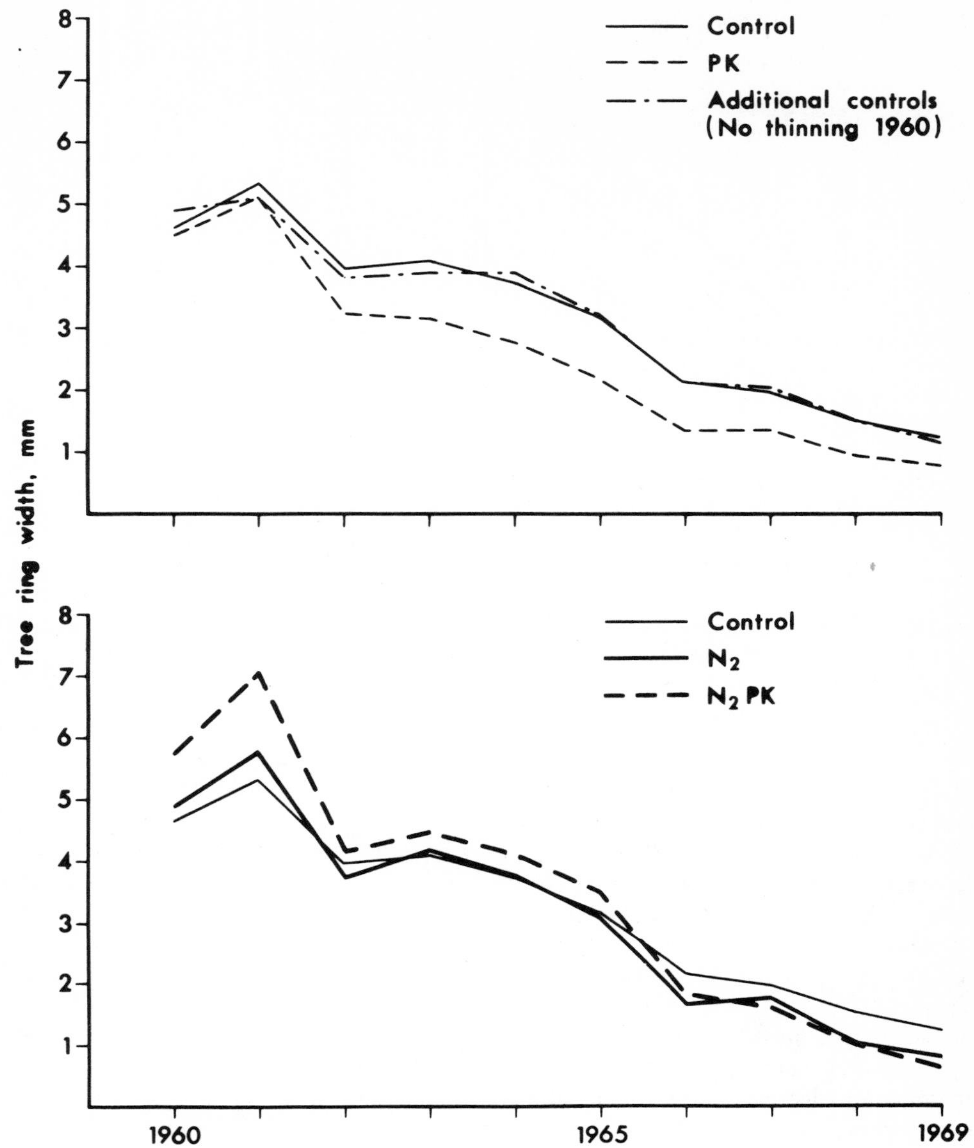

FIGURE 4 Example curves of diameter growth in Experiment E1, Hökaberg.

TABLE 3 Annual Production by the Spruce Stand in Table 1 (kg/ha)

	O	PK	N1	N1PK	N2	N2PK	N4	N4PK
Needles	3,280	2,960	2,920	3,770	3,310	3,630	3,170	3,500
Branches	2,160	1,950	2,020	2,600	2,300	2,520	2,200	2,420
Stem	4,660	4,280	3,990	4,850	4,110	4,300	3,520	3,630
Stumps	240	220	210	250	210	220	190	180
Roots > 5 mm	1,860	1,500	1,800	2,260	2,020	2,050	1,950	2,200
Roots < 5 mm	1,070	1,100	1,100	1,060	1,110	1,060	1,100	1,060
TOTAL PRODUCTION	13,270	12,010	12,040	14,790	13,060	13,780	12,130	12,990
Current needles (1969)	2,600	2,350	2,250	2,910	2,560	2,810	2,450	2,710

NOTE: Stem production and current needles determined on sample trees from all treatments. Ratio between on one hand current needles and on the other hand needle production (=last years needles) and branch production determined on 12 sample trees from treatments O, N1, and N1PK. Stump production assumed to be proportional to stem production, roots < 5 mm to branch production (same ratio biomass:production) and roots > 5 mm to needle production.

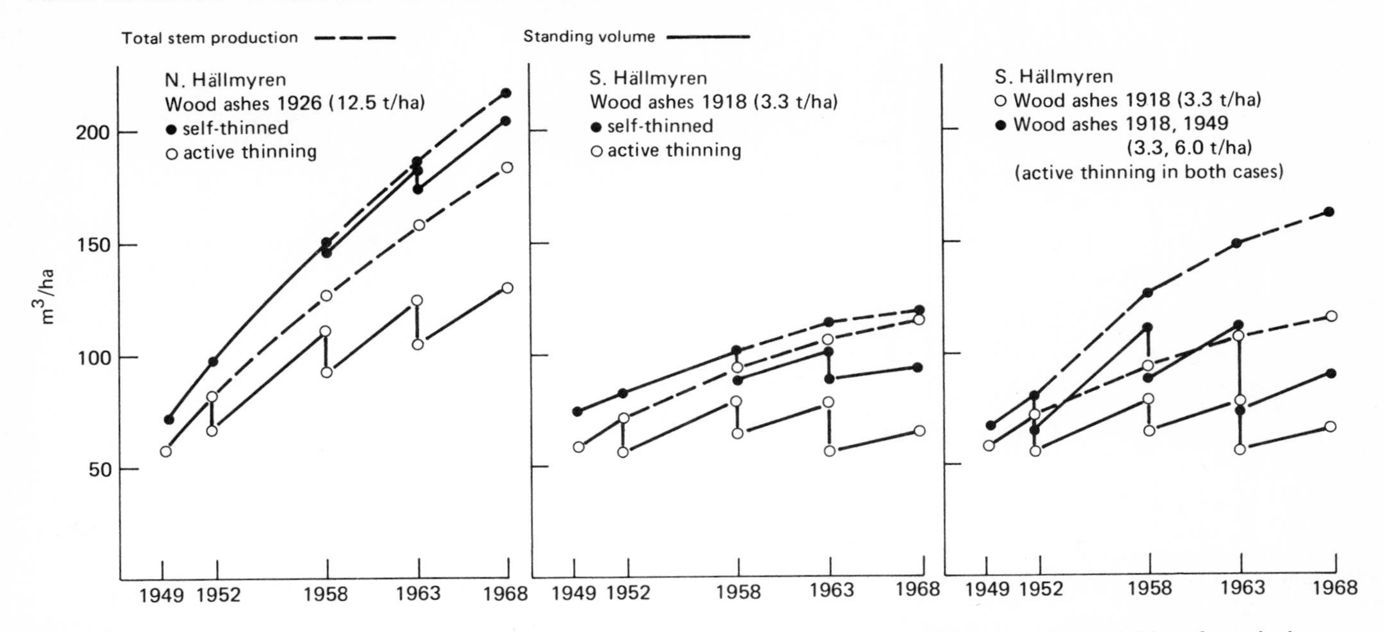

FIGURE 5 Total stem production and standing stem volume in birch stands established on drained peatland after addition of wood ashes. Expt. E16, Robertsfors, N. Sweden.

photosynthetic products and other metabolites in the plant and hence affect growth. In fact, there is some evidence in the spruce experiment described earlier, that nitrogen overdoses affected stem growth more than needle production.

CONCLUSION

It is hoped that the synthesis of the IBP work within the various biomes will supply more examples of the close relationship between nutrient supply and growth. Analysis of extreme situations, due to variations either in soil or in weather, may be helpful in this respect. Still more experimental work will be needed on the mechanisms by which the plant nutrient supply limits growth—both in the form of physiological studies in the laboratory and ecophysiological research in the field.

In conclusion a short presentation will be made of some results from recent experiments started under the auspices of IBP but not yet completed, to illustrate one type of field experiment needed. A strong reaction to nitrogen was obtained in the spruce growing on a poor site (Figure 7). Phosphorous supply also increased growth, although to a much smaller extent than did nitrogen supply. The effect of other plant nutrients still remains obscure. A preliminary optimum curve was obtained for nitrogen (Figure 8), and there has been a dramatic change in production in both spruce and ground vegetation. On the other hand, fertilization caused increases in winter damage and in vertebrate browsing. It is certainly necessary to take the entire ecosystem into account in both planning and interpretation of field experiments.

It makes no difference whether the ecosystem is a forest, a tundra, a grassland, or a desert. The breakthrough for this integrated way of looking at ecological problems is certainly one of the major achievements of the International Biological Program.

SUMMARY

This paper reviews early work in plant nutrition, and describes a few cases when lack of nitrogen, potassium, or phosphorus restricts the growth of certain or all primary producers. While nutrient cycling studies form a well-recognized component of ecosystem analysis projects, it is also necessary to investigate the nutrient status of the dominant primary producers. Without such information it will be impossible to forecast effects of changes in nutrient cycling on productivity. Mechanisms are discussed by which nutrient supply affects primary production. It is concluded that plants often adjust their growth to the amounts of nutrients available. Poor sites, therefore, often have a low amount of photosynthesizing organs (low leaf area index). Lack of nutrients may also affect the efficiency of leaves and the redistribution of phytosynthetic products, and interfer with water regulation mechanisms. The need for further ecophysiological studies is emphasized.

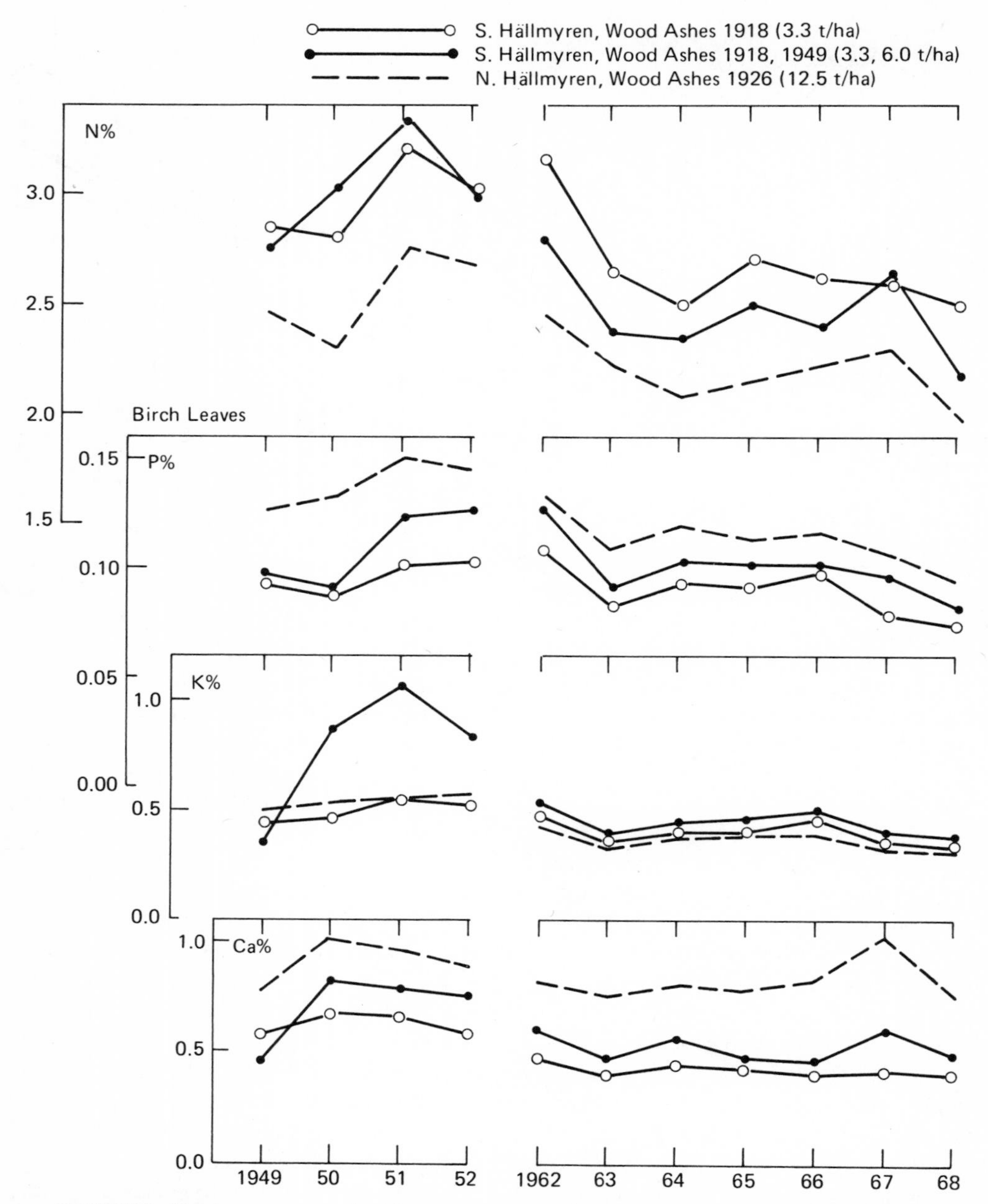

FIGURE 6 Nutrient concentrations (percent dry weight) in birch leaves from the plots in Figure 5.

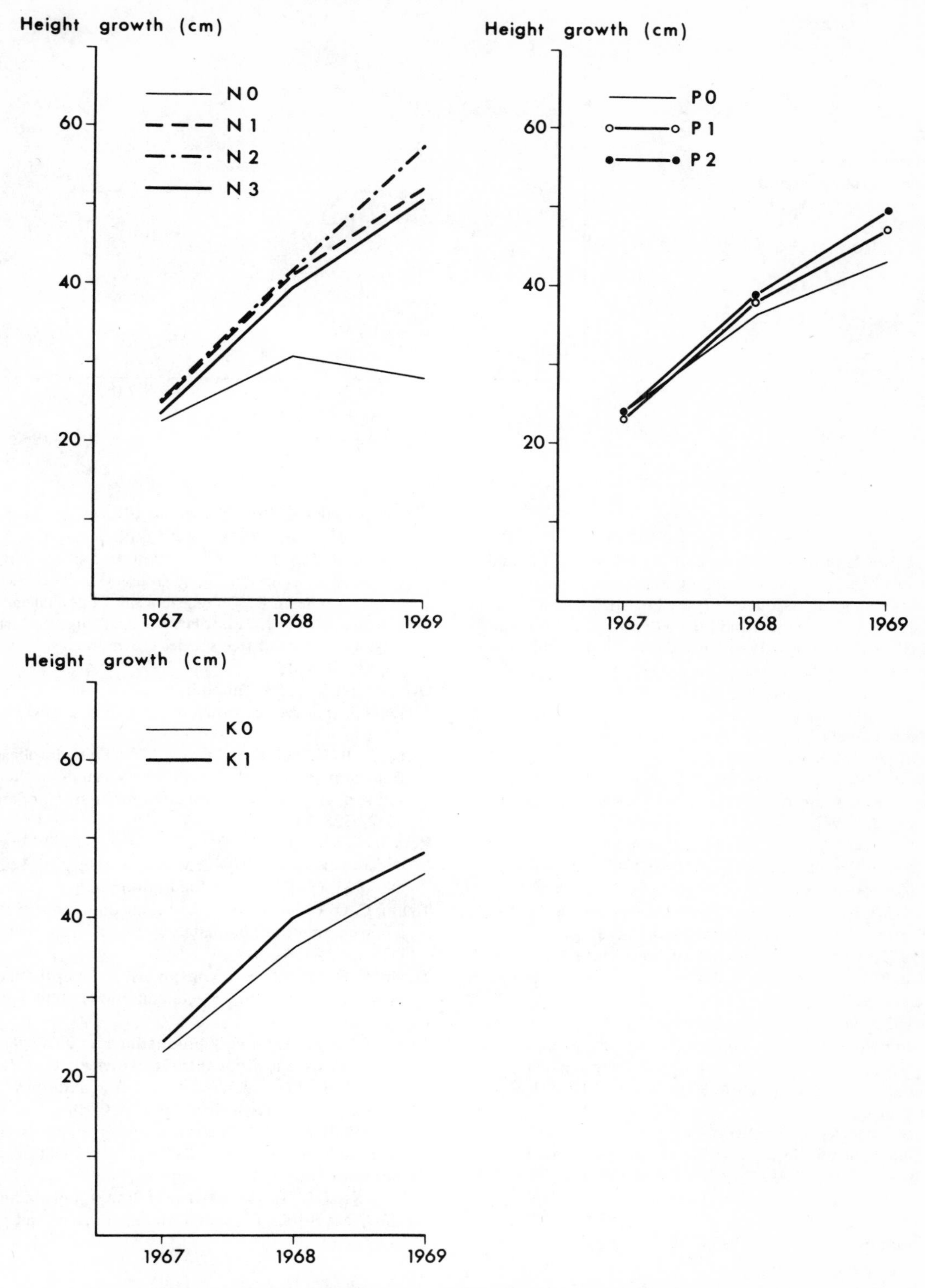

FIGURE 7 Height growth responses in the optimum nutrition experiment E26 A.

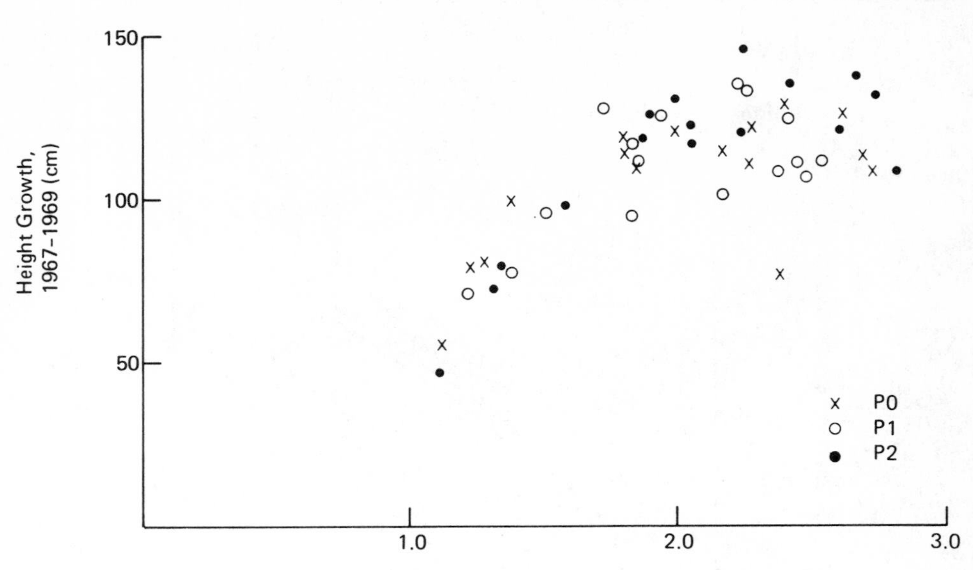

FIGURE 8 Height growth 1967-1969 plotted against foliage nitrogen concentration of spruce fertilized annually from 1967 onwards. Experiment E26 A Stråsan.

REFERENCES

Arland, A. 1954. Die Transpirationsintensität der Pflanzen als Grundlage bei der Ermittlung optimaler acker- und pflanzenbaulicher Kulturmassnahmen unter besonderer Berücksichtigung von Pflanzenanalysen. Plant analysis and fertilizer problems. Colloque organisé par l'I.R.H.O. sous la Présidence de H. Lundegårdh dans le cadre du VIIIe Congrès International de Botanique, p. 35–47. Paris.

Brag, H. 1972. The influence of potassium on the transpiration rate and stomatal opening in *Triticum aestivum* and *Pisum sativum.* Physiol. Plant. 26:250–257.

Brix, H., and L. F. Ebell. 1969. Effects of nitrogen fertilisation on growth, leaf area, and photosynthesis rate in Douglas-fir. Forest Science 15(2):189–196.

Brix, H. 1971. Effects of nitrogen fertilisation on photosynthesis and respiration in Douglas-fir. Forest Science 17(4):407–414.

Brix, H. 1972. Nitrogen fertilisation and water effects on photosynthesis and earlywood-latewood production in Douglas-fir. Canadian Journal of Forest Research 2(4):467–478.

Gore, A. J. P. 1972. A field experiment, a small computer and model simulation, p. 309-325. *In* J. N. R. Jeffers (ed.) Mathematical models in ecology. Symp.

Hesselman, H. 1937. Om humustäckets beroende av beståndets ålder och sammansättning i den nordiska granskogen av blåbärsrik Vaccinium-typ och dess inverkan på skogens föryngring och tillväxt. Medd. St. Skogsförsöksanst. 30:529–668. (Swedish with German summary.)

Ingestad, T. 1962. Macroelement nutrition of pine, spruce, and birch seedlings in nutrient solutions. Medd. St. skogsforskn. Inst. 51(7):1–150.

Keller, Th. 1971. Der Einfluss der Stickstoffernährung auf den Gaswechsel der Fichte. Allg. Forst-u. J. -Ztg. 142(4):89–93.

Mitchell, H. L., and R. F. Chandler. 1939. The nitrogen nutrition and growth of certain deciduous trees of Northeastern United States. The Black Rock Forest. Bull. 11:1–94.

Nielsen, B. F. 1963. Plant production, transpiration ratio and nutrient ration as influenced by interactions between water and nitrogen. Thesis. R. Vet. Agric. Coll. Copenhagen.

Ovington, J. D. 1957. The volatile matter, organic carbon and nitrogen contents of tree species grown in close stands. New Phytol. 56:1–11.

Ovington, J. D. 1958. The sodium, potassium and phosphorus contents of tree species grown in close stands. New Phytol. 57:273–284.

Richards, B. N., and D. I. Bevege. 1969. Critical foliage concentrations of nitrogen and phosphorus as a guide to the nutrient status of Araucaria underplanted to Pinus. Plant and Soil 31(2):328–336.

Romell, L. -G., and C. Malmström. 1945. Henrik Hesselmans tallhedsförsök åren 1922–42. Medd. St. skogsförsöksanst 34(11): 543–625. (Swedish with English summary.)

Tamm, C. O. 1954. Some observations on the nutrient turn-over in a bog community dominated by Eriophorum Vaginatum L. Oikos 5(2):189–194.

Tamm, C. O. 1968. An attempt to assess the optimum nitrogen level in Norway spruce under field conditions. Stud. for. suec. No. 61:1–67.

Tamm, C. O. 1971. Primary production and turn-over in a spruce forest ecosystem with controlled nutrient status (a Swedish IBP project). Systems analysis in northern coniferous forests–IBP Workshop. Bull. from the Ecol. Res. Comm. 14:114–145.

Whitehead, D. C. 1970. The role of nitrogen in grassland productivity. Bull. 48, Commonwealth Bur. Past. and Field Crops. Hurley, Berkshire, England.

Zehetmayr, J. W. L. 1960. Afforestation of upland heaths. Bull. of the Great Britain Forestry Commission. England. 145 p.

MINERAL CYCLING IN TERRESTRIAL ECOSYSTEMS

P. DUVIGNEAUD and S. DENAEYER-DE SMET

INTRODUCTION

In any optimistic or pessimistic perspectives we may have about the future of our planet, botany is rarely considered. Although plants are the absolute base of the alimentation of animals and men, some people may ask themselves "Is botany still alive?" Recently, Epstein (1972) has pointed out a "blind spot" in biological sciences: the *knowledge of the mechanisms involved in plant mineral nutrition.* In our opinion, this gap results from insufficient knowledge of the relations *in nature* between plants and the mineral elements of their abiotic environment.

For a long time the study of the plant-soil relations in nature has been approached by many disciplines but with little interchange among them (botanists, geologists, agronomists and foresters; epidemiologists). Today, a better coordination enables us to gather these investigations in a modern science, called *biogeochemistry*, from which begin to emerge important principles and laws.

One of the most important aspects of biogeochemistry is the study of mineral cycling, because, as well as energy and water, some mineral elements (called nutrients = biologically essential elements) are required in sufficient quantities for sustenance of the ecosystems and their components. Biogeochemistry is of practical interest and application when some elements, essential or not, are artificially incorporated into the ecosystems by human activities and become toxic for living organisms (e.g., S, Zn, Cd and Pb). On that viewpoint, biogeochemistry may contribute substantially to solving the problems resulting from the increasing pollution of the environment.

BIOGEOCHEMISTRY

Most of the investigations performed till now in biogeochemistry have concerned only plant-soil relations, except for health related trace elements, such as Se (see p. 150), and some other elements for which investigations on food-chain relationships began about ten years ago. A general review of this important aspect of biogeochemistry may be found in Hopps and Cannon (1972).

Owing to their more spectacular character, abnormal soils (those affected by "biogeochemical anomalies," such as excess of salt, heavy metals or deficiences in P, Co, Cu, etc.) and the corresponding vegetation have been much more studied than "normal"* soils and vegetation. The study of soils unbalanced by excess or lack of an element, which may cover wide areas, is the basis of very important practical applications, such as the utilization of sea water for the irrigation of desert zones (Boyko, 1966). Data obtained from the study of abnormal soils may lead to generalization and contribute to the establishment of a general theory of mineral nutrition of plants in nature.

Indicators–Accumulators–Concentrators

Some plant species show by their presence particular chemical soil conditions. Exclusive or preferential, these *"indicators"* are generally adapted genetically to their substrate (physiological ecotypes or distinctly differentiated species).

* A soil may be considered as normal when its chemical composition is not an essential limiting factor for plant development.

Indicator species are often called *specialists* (selenophytes, cuprophytes, zincophytes, etc.). In fact in a given region, every species is a specialist of a particular ecological site. Duvigneaud (1946, 1949), Duchaufour (1957), Ellenberg (1954, 1966), and Schlenker (1939, 1960) led to a phytosociological or pedological mapping for several regions of Europe. Much remains to be done in this field.

Exception here has to be made for the weeds which colonize all types of disturbed soils (roadsides, gravelly shores, mine spoil-heaps, gardens, abandoned cultures), and which extend to the tailings of abandoned Cu, Zn and Pb mines and to all kinds of polluted sites (even able to persist at high levels of ionizing radiation). These species which constitute ecosystems in the most polluted areas have been called *generalists* by Woodwell (1970).

The mineral composition of plants depends on many factors: concentration of the chemical elements in the soil, pH, ion antagonism (for their influence on the availability of nutrients for plants), soil moisture and temperature, root morphology (not yet well known), root cations exchange capacity (still discussed, see Wacquant, 1968, 1969; Hemphill, 1972), age of plants, organs (leaves, stems, fruits, roots differ generally in mineral composition) and genetic constitution of the plant species.

Selective Absorption and Accumulation of Chemical Elements by Plant Species

The mineral composition of different plant species growing on the same soil or sampled in their respective ecological optimum conditions may vary widely because of the *selectivity of the absorption and accumulation of ions by plants species* (Hohne, 1962; Duvigneaud and Denaeyer, 1966, 1968, 1970b; Hemphill, 1972). This phenomenon may be masked by influence of the substrate (luxury or poverty consumption) and a detailed nomenclature has been proposed by Duvigneaud and Denaeyer (1970a). The same authors published recently (1973) a paper on the ecology of mineral nutrition of plants in nature.

Plants may be divided into poor and rich species on the basis of foliar analysis and with reference to standard plants of average mineral composition; among the latter, they are *accumulators* of elements present in excess in the soil: 12 to 16 percent Na in halophytes (Denaeyer *et al.*, 1968), 3 to 7 percent S in gypsophytes (Duvigneaud and Denaeyer, 1968), and *concentrators* which absorb very high quantities of elements found in normal concentrations in the soil: 800 ppm F in *Camelia sinensis* (Hemphill, 1972), 5,000 ppm in *Vaccinium myrtillus* (Denaeyer–De Smet, 1966), 845 ppm Co in *Nyssa sylvatica* (Kubota and Allaway, 1971), 10,000 ppm Se in *Astragalus bisulcatus* (Trelease and Trelease, 1939).

Concentrators can play a very important part in the evolution of the ecosystems by accumulating, on the soil surface through litter shedding and decomposition, elements which are dispersed in the soil or formed directly from bed rock. If these elements are little mobile, they form a highly enriched superficial layer and their chemical form may be modified, e.g., mineral Se converted into organic Se by "*Transformators*" plant species (see also p. 150).

The mineral composition of animal species is little known. Edwards *et al.* (1970) and Reichle (1971) who studied the content of Na, Ca and K of 37 kinds of invertebrates in a *Liriodendron* wood (Oak Ridge, Tennessee) have shown a very high specificity: Isopoda and Diplopoda accumulate Ca (103 mg to 546 mg/g ash free dry matter; Lepidoptera accumulate K (50 mg/g) and, to a lesser extent, Na (9.2 mg). Data concerning the mineral composition of Lumbricids and Basidiomycetes are given in Duvigneaud *et al.* (1971). Earthworms accumulate mainly N (8.5 percent dry weight), S (0.6 percent dry weight) and P (0.4-0.8 percent dry weight). The carpophores of *Basidiomyceta* are K accumulators (2–8 percent K dry weight) but do not absorb much Ca (0.06–0.15 percent dry weight) probably not essential for fungi.

The soil–plant–animal relations have been recently overviewed by Horvath (1972) for the most important health related trace elements. He emphasized the importance of accumulator plants which may intoxicate grazing cattle: excess of Mo after intensive N fertilization, Se occurring in native soils, or Cd emmissions originating from Zn industry. Because plants are the main source of the mineral elements required by animals and man, plants are also involved in health diseases because of the deficiency of some elements, e.g., Cu (deficient in very acid-sandy soils, in peat and muck soils, in highly N-fertilized soils, etc, see also p. 151). Horvath (1972) emphasized also the role played by the genetic constitution of animals in their resistance to excesses or deficiencies of nutrients. The competition among mineral elements in relation to digestive absorption by animals has been reviewed by Davis (1972).

MINERAL CYCLING IN TERRESTRIAL ECOSYSTEMS

One of the great contributions of biogeochemistry is the study of the mineral cycling that contributes to our understanding of the functioning of the ecosystems. Following the definitions given by Ovington (1968), mineral cycling involves the *biological cycle* which corresponds to the circulation of elements within ecosystems (mainly between the phytocenosis and soil) and the *geochemical cycle* which means the flux of elements from the abiotic environment into and out of ecosystems. Mineral cycling includes both essential elements (macro- and micronutrients) as well as ballast elements (Al, Si) and toxic elements occurring in nature (Se, Cu, etc.) or incorporated artificially in the ecosystems as a result of human activities (Pb, Cd, Hg, etc.).

BIOLOGICAL CYCLES

In Western Europe, the concept proposed by Albert (in Dengler, 1930) for forest ecosystems is often followed:

absorption = retention + restitution.

Retention means the quantity of elements retained in the annual increment of perennial organs. In forest ecosystems retention is very high and leads into a stocking ("mineralo-mass" sensu Duvigneaud, 1968) of elements in stems, branches and roots of the trees. Sometimes a concentration may occur ($CaCO_3$ in the stems of *Chlorophaea excelsa*, Al in the stems of different tropical species). *Restitution* to the soil of a part of the absorbed elements is carried out by shedding and decomposition of litter (dead leaves, fruits, dead wood, etc.) and by rain leaching of the phytocenosis (for more details see Carlisle *et al.*, 1967; Denaeyer, 1969; Nihlgard, 1972). When litter decomposes slowly, restitution may lead to an accumulation of elements adsorbed or chelated on the humus colloids. Restitution of nutrients may also occur by living roots (desorption, exosmosis, secretion) especially at the end of vegetation period. These problems have been intensively studied by several Soviet authors (Boriskina, Titova, Pogrebniyak, 1955).

Figures 1-3 show different methods for representation of the biological cycle following the classic concept of Dengler (1930). Figure 4 reflects the concepts of Soviet authors. They utilize the notion of "carrying capacity of the biological circulation" which corresponds to the quanti-

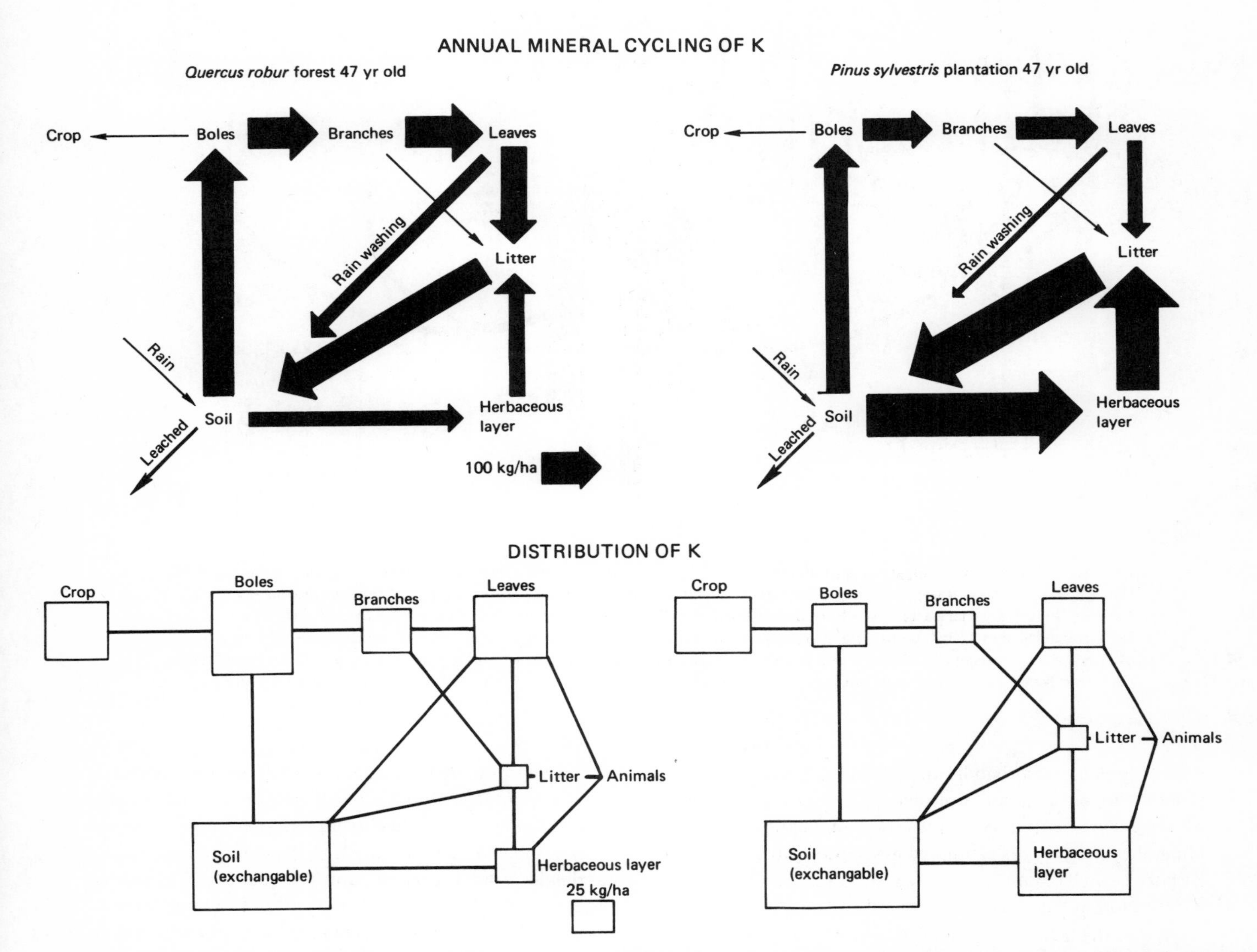

FIGURE 1 Annual biological cycle of K in a 47-year-old *Quercus robur* forest and in a *Pinus sylvestris* plantation of the same age (after Ovington, personal communication).

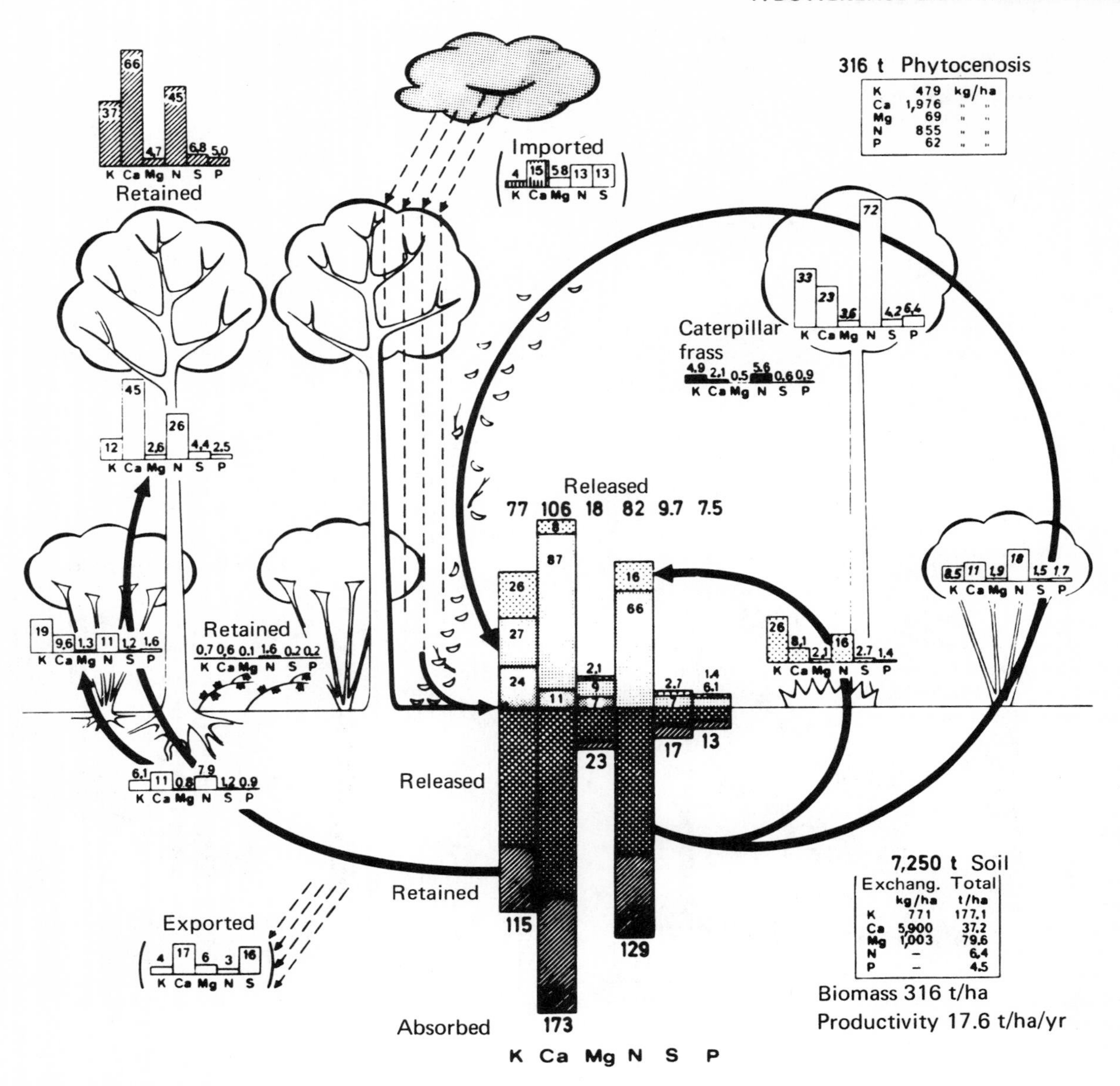

FIGURE 2 Annual biological cycle of elements in a 117-year-old oak-hazel forest (*Querceto-Coryletum*) in Belgium (Ferage near Houye). *Retained* in increment of trees and shrubs, and in the perennial parts of the herbaceous layer. *Released* by dead aerial parts of the herbaceous layer (lightly stippled), by tree and shrub litter (moderately stippled), by throughfall and stemflow (highly stippled). *Absorbed* = retained + released. *Imported* by incident rainfall (values of K and Ca at Ferage; Mg, N and S at a nearby site (Dourbes, National Institute for Epidemiology and Public Health). *Exported* by drainage water (extrapolated, after Likens *et al.*, 1971). Italics: Mineral content of green leaves (July).

ties of elements absorbed by the net primary productivity (NPP = ligneous increment plus green leaves) and the notion of "true increment" (NPP minus total annual litterfall). Mineral cycling is thus estimated as follows: Absorption in NPP (measured) minus Restitution by total litter (measured) = Retention in true increment (calculated).

Due to the lack of sufficient data, Soviet authors do not take in account restitution by rainleaching of the canopy and stemflow. Following the Soviet concept, in a climax forest the values obtained for retention may be negative if the quantity of dead wood is larger than the ligneous increment. For meadow ecosystems Soviet authors simplify the biological cycle by considering that the elements released by aerial dead parts correspond to the quantities of elements absorbed annually by the living parts. (However, the knowledge of the restitution of elements by dead roots may be necessary because of the high productivity of these organs.) Scandinavian authors also estimate annual absorption through

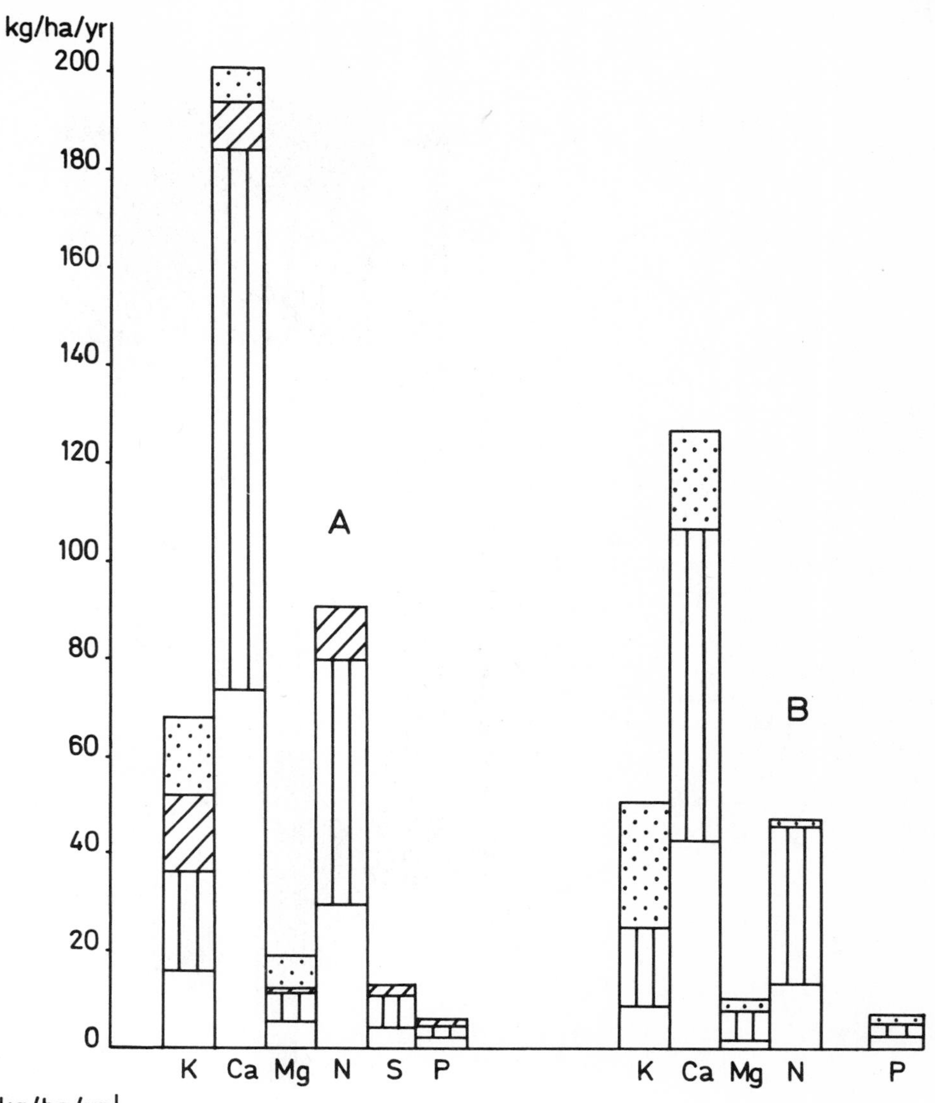

FIGURE 3 Simplified graphic representation of the annual biological cycle of nutrients in several forest ecosystems of Western Europe.

A. *Quercetum mixtum* at Virelles, Belgium (Duvigneaud and Denaeyer, 1970b), productivity = 14.6 t/ha/yr;

B. *Quercetum ilicis* at Montpellier, France (Lossaint and Rapp, 1971), productivity = 6.5–7 t/ha/yr;

C. *Fagetum nudum* at Mirwart, Belgium (Denaeyer and Duvigneaud, 1972), productivity = 14.8 t/ha/yr;

D. *Piceetum* at Mirwart, Belgium (*Ibid*), productivity = 14.6 t/ha/yr.

Open areas: Retained in annual wood and bark increment, aerial and subterranean (except in B, aereal only); vertical lines: released by tree litter; oblique lines: released by dead parts of herbaceous layer; stippled: released by throughfall and stemflow.

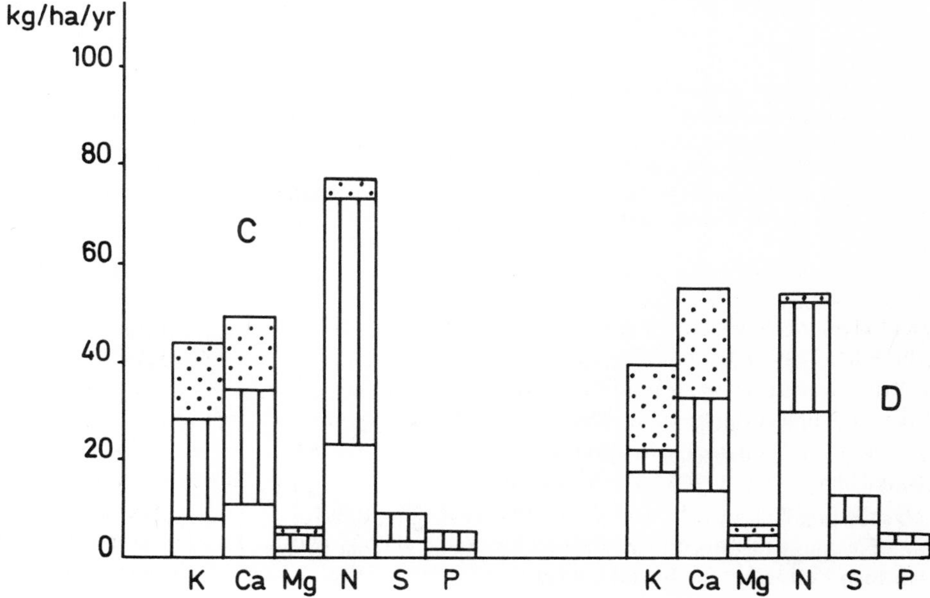

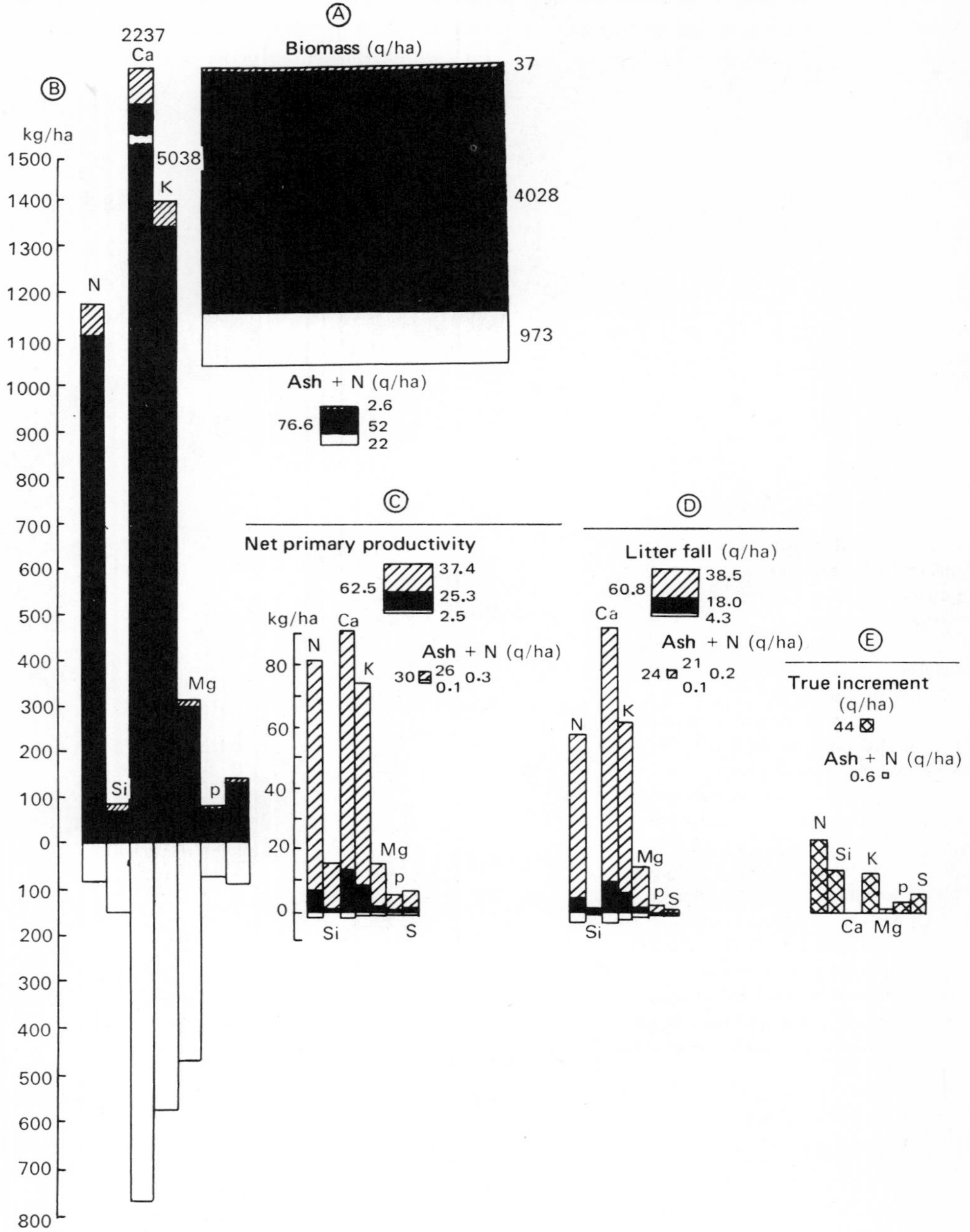

FIGURE 4 Biomass, productivity, and annual biological cycle of nutrients in a 222-year-old oak forest (*Querceto Aegopodietum* "dubrava") Voronesz, U.S.S.R. (Rodin and Bazilevich, 1967). Oblique lines: green parts; black: aerial perennial parts; open: roots; brace: true increment of aerial and subterranean parts. A = biomass; B = mineralomass; C = annual absorption in net primary productivity (green leaves + wood and bark increment); D = released by tree litter; E = retained in true increment (E = C – D).

NPP ("carrying capacity") but do not separate the mineral content of green leaves; consequently there is no possibility to estimate the annual retention in wood and bark increment only. On the other hand, they calculate the mineral content in what they call $\triangle$B (wood and bark NPP minus dead wood), which is very nearly the value of retention *sensu* Ovington, because in the ecosystems studied by these authors dead wood is not very important (±10 percent of the NPP biomass). It is thus possible to make comparison between values obtained following European and Soviet-Scandinavian concepts (Table 1).

The restitution of mineral nutrients is closely related to the turnover of organic matter which involves a complex range of processes from which the main classical stages are as follows:

– litterfall and accumulation
– litter breakdown and transfer by pedofauna
– chemical decomposition of the litter, immobilization or remineralization by soil microflora
– humification and dehumification
– reabsorption by green plants

TABLE 1 Mineral Cycling in a Beech Forest 90 Years Old and in an Adjacent Spruce Forest 55 Years Old; at Kungalund (Sweden), after Nihlgard (1972). Comparison Between Annual Absorption Calculated Following Two Different Concepts (a and b)

	kg/ha/yr					
	K	Ca	Mg	N	S	P
A. Beech Forest						
– Retained in stem and branches increment (minus dead wood) (10.1 t/ha/yr)	16.6	24.8	3.4	83	2.5	3.5
– Returned by						
• litterfall (5.7 t/ha/yr)	14.4	31.7	4.3	69	6.4	5.0
• throughfall + stemflow	11.2	6.6	2.5	1	10.6[a]	0.1
– Absorbed						
a) sensu Ovington (retained + released)	42.2	63.1	10.2	152	19.5	8.5
b) sensu Rodin (absorbed in net primary productivity) (15.1 t/ha/yr)	43.3	49.3	10.9	204	7.4	10.5
B. Spruce Forest						
– Retained in stem and branches increment (minus dead wood) (10.1 t/ha/yr)	8.5	12.0	1.9	14	2.9	1.6
– Returned by						
• litterfall (5.6 t/ha/yr)	10.7	19.8	3.1	58	5.4	4.8
• throughfall and stemflow	25.2	13.9	5.2	16	46.3	0.4
– Absorbed						
a) sensu Ovington (retained + released)	44.4	45.7	10.2	88	54.6	6.8
b) sensu Rodin (absorbed in net primary productivity) (13.8 t/ha/yr)	43.0	12.7	4.8	67	6.0	9.1

[a]40.8 kg/ha/yr beneath *Fagetum* Canopy in Solling (West Germany) (Ulrich *et al.*, 1971).

BIOLOGICAL CYCLES OF MACRONUTRIENTS IN DIFFERENT ECOSYSTEMS

In Western Europe a long time after Albert (see Dengler, 1930), Ovington and Madgwick (1959) again brought into vogue the study of biological cycles in forest ecosystems. In the USSR, Rodin and his co-workers considered the problem on a world scale. A first synthesis of mineral cycling in forest ecosystems was attempted by Duvigneaud and Denaeyer in 1964. An intensive research program on this subject was initiated in Belgium (Virelles) and later due to the IBP similar projects developed in different countries of Europe (West Germany, Solling; Sweden, Kongalund; Denmark, Hestehaven; Great Britian, Meathop; France, Montpellier; Czechoslovakia, Bab; Belgium, Mirwart) as well as on a very large scale in USSR, USA and Japan. The main results of these investigations are summarized in Duvigneaud (1971) and Duvigneaud *et al.* (1971).

In a given ecosystem the biological cycle follows the curve of productivity; for example, in temperate forests, the maximum value is reached between 25 and 40 years (Ovington and Madgwick, 1959; Remezov, 1963). Therefore, comparisons between different ecosystems are most valuable when examining ecosystems which have reached their climax or have the same biomass and belong to a given biome (forests, grasslands, tundras, etc.).

Comparison of Mineralomasses

A comparison made by Tsutsumi (1971) between different ecosystems having nearly the same biomass (±100 t/ha) from the subarctic *Abies* forest to the tropical forest of Thailand has shown that mineralomasses depend on both the forest type and on bedrock chemical composition.

Comparison of Mineral Cycling

A simple graphic method (Figure 5) allows a general comparison of annual absorption, which gives a good idea of mineral cycling in ecosystems. This method is based on a polygonal representation: six axes, starting from a 0 point, leaving between them an angle of 60° and graduated in kg/ha/yr with each axis corresponding to a macronutrient. Connecting the six points corresponding to the annual absorption of each of the six macronutrients yields a polygonal figure whose form and area are characteristic of the ecosystems.

For the Temperate Deciduous Forest biome, the surface of the polygons corresponding to oakwood ecosystems (*Quercetum*) is much larger than the surface of the polygons for beechwood ecosystems (*Fagetum*); this means that the total quantities of absorbed nutrients are higher in the former than in the latter. The form of the polygon depends on the chemical plant/soil relation. Annual Ca absorption is more important and K absorption is lower in calcareous ecosystems (Figure 5b) (K/Ca antagonism). On the other hand, Ca absorption is reduced on very poor soils (Figure 5d); Mg absorption is high on magnesiferous soils (Figure 5c); N absorption is high when mineralization conditions are very favorable (Figure 5i).

In the Temperate Evergreen biome, ecosystems seem to be characterized by a lower annual absorption, especially at the N level (in the two considered ecosystems, Ca absorption is high because of luxury consumption on calcareous soil).

In the Equatorial Forest biome, the annual absorption is much higher (ranging about several hundreds of kg/ha) than in any other ecosystem. This is mainly because of the very high productivity (20 t/ha/year), important rainleaching, and the fact that there are several litterfalls per year.

In the Coniferous Forest biome, the absorption of macronutrients is always low, especially in pineforest ecosystems (*Pinetum*). However, there seems to be an exception for S in a spruce forest of Sweden (Figure 5, 1); the very high absorption of this element results from an abnormal restitution by rainfall as a consequence of atmospheric pollution.

In the Tundra biome, annual absorption varies widely: nearly the same as in Coniferous forest in the *Vaccinium* tundra, but extremely reduced in the spotted (not represented in Figure 5, because it should be only a very little point).

In the same way, the annual absorption in the "Steppe" biome depends on the kind of the considered ecosystem: annual absorption in a grassland ecosystem is comparable to a high productive deciduous forest, but is much more reduced in semidesertic *Artemisia* steppes and in salt steppes. In the latter ecosystems, the very small area of the polygons does not give a real idea of mineral cycling because other elements (mainly Na and Cl) become much more important than the true macronutrients.

For cultivated ecosystems, the areas of polygons may vary widely following crop nature and reflect very well the different degrees of nutrient consumption; but, they are all characterized by a general more elongated form, resulting from K domination and the P high absorption.

BIOLOGICAL CYCLES OF MICRONUTRIENTS

Until recently, biogeochemical cycles of only Mn, Fe and sometimes Al were studied (Klausing, 1956; Ovington and Madgwick, 1959; Rodin and Bazilevich, 1967; Kolli and Reintam, 1970; Ulrich *et al.*, 1971; Nihlgard, 1972). Today owing to the spectrophotometric analysis by atomic absorption, investigations have been extended to other micronutrients. The first results obtained concern however more the individuals (Young and Guinn, 1966; Nilsson, 1972) than the ecosystems. Research programs concerning the distribution and the biological cycle of micronutrients in a beach and adjacent spruce forest are in progress in Sweden (IBP program at Kongalund), in Denmark (Hestehaven IBP program) and in Belgium in the same forest types (IBP program at Mirwart). Figures 6A and 6B show the first results.

INPUT–OUTPUT BUDGET and BIOGEOCHEMICAL CYCLES

Input and Output of Nutrients

In terrestrial ecosystems, input of nutrients occurs by:

–precipitation and dry fallout. Under certain circumstances (semidesert regions, slightly fixed soils), wind-blown dust may constitute an important input of mineral elements. In Guinea, for example, the rainforest is continuously enriched in nutrients by dust removed from Sahara by a very strong North wind. In ecosystems submitted to pollutant emissions as a consequence of human activities (industrial plants, roadsides, etc.), rainfall plays an important part, bringing to the soil many kinds of pollutants (H_2SO_4, Pb, etc.). Vegetation may play a beneficial role, acting as a screen and trapping many airborne particles (forest ecosystems are especially efficient).

– microbial atmospheric N fixation (average of 25 kg/ha/yr by *Azotobacter* or *Clostridium* and 150–160 kg/ha/yr by *Rhizobium* in leguminaceous cultures).

– rock weathering by deep roots (mainly tree roots) which absorb the most soluble elements and restitution of these elements by litterfall (Hartmann's law).

– microbial action on the soil skeleton and on bedrock; this action is performed directly by contact or indirectly by formation of CO_2, oxidizing humic acids or alkaline compounds (Witkamp, 1971).

Output of nutrients is carried out by human activities (harvesting, forest exploitation) (Rennie, 1955), soil water runoff (dissolved and particulate elements). The estimation of output rate is difficult and needs complicated and expensive techniques, (e.g., tension plate lysimeters, evapotranspiration measurement by tritiated water, small watershed systems).

The input–output budget depends above all on the hydrological cycle. Geochemical fluxes and biological cycles complete themselves by lateral exchanges between adjacent ecosystems, as for example, forest and stream ecosystems; research projects in this area have been initiated in the USA (Curlin, 1970) and in Belgium (IBP program at Mirwart).

Comparison of Input-Output Budgets

Comparison of the results obtained from several studies (Table 2) shows that nutrient loss from ecosystems generally

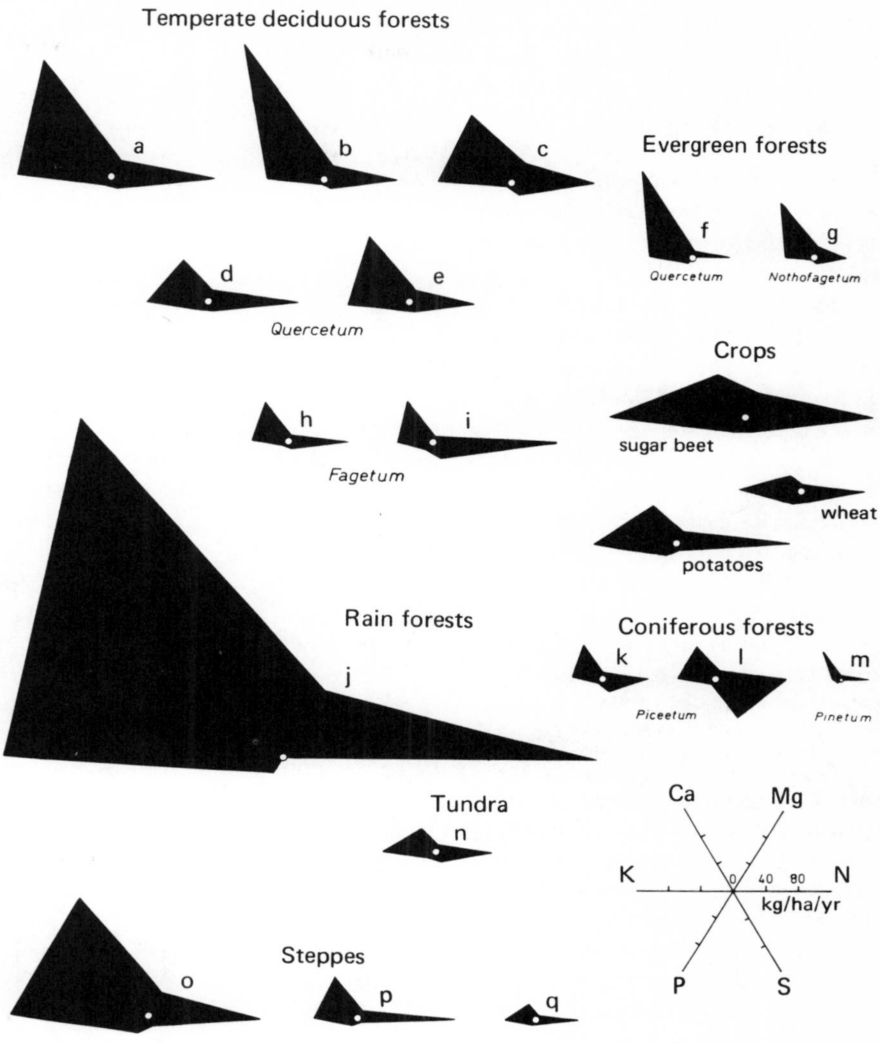

FIGURE 5 Graphical representation and comparison of the annual biological cycle of nutrient elements in different types of terrestrial ecosystems.

a. *Querceto-coryletum,* 117 years old at Ferage-Houyet (High Belgium), forest brown soil rich in exchangeable bases. Aerial increment: 7.9 t/ha/yr; litter: 6.8 t/ha/yr.

b. *Quercetum mixtum,* 35–75 years old at Virelles (High Belgium), dark brown rendzinoidic soil Ca saturated. Aerial increment: 6.3 t/ha/yr; litter: 6 t/ha/yr.

c. *Quercetum,* 90 years old at Villers/Lesse (High Belgium), pseudogley soil rich in Mg. Aerial increment: 5.1 t/ha/yr; litter: 6.6 t/ha/yr.

d. *Quercetum,* 135 years old at Vonèche (High Belgium), pseudogley podzol very poor in exchangeable bases. Aerial increment: 3.1 t/ha/yr; litter: 7.8 t/ha/yr. (a, b, c, and d in Duvigneaud *et al.,* 1971 and Duvigneaud and Denaeyer, 1971).

e. *Querceto-aegopodioso-caricosum,* 222 years old in the province of Voronesz (U.S.S.R.). Aerial increment: 2.5 t/ha/yr; litter: 5.6 t/ha/yr (after Mina, in Rodin and Bazilevich, 1967).

f. *Quercetum ilicis,* about 150 years old at Montpellier (mediterranean France), on red soil. Aerial increment: 2.6 t/ha/yr (provisional results); litter: 5 to 9 t/ha/yr (only leaves). The S cycle has not been established (after Rapp and Lossaint, 1971).

g. *Nothofagetum* about 100 years old, at Silverstream, near Wellington (New Zealand). Strongly leached ochre-brown soil. Litter: 6 t/ha/yr (after Miller, 1963).

h. *Fagetum,* 130 years old, at Mirwart (High Belgium), acid limono-stony brown soil, poor in exchangeable bases. Aerial increment: 6.2 t/ha/yr; litter: 5.2 t/ha/yr (after Denaeyer and Duvigneaud, 1972).

i. *Fagetum,* 90 years old, at Kongalund (Sweden), acid brown soil on moraine. Aerial increment: 11.3 t/ha/yr; litter: 5.7 t/ha/yr (after Nihlgard, 1972).

j. Dense ombrophyte forest, 50 years old (secondary forest) at Ghana. Aerial increment: 21.5 t/ha/yr; litter: 21.7 t/ha/yr. The S cycle has not been established. (After Greenland and Kowal, 1960; Nye, 1961).

k. *Piceetum,* 55 years old, at Mirwart (High Belgium) on acid limono-stony brown soil, poor in exchangeable bases. Aerial increment: 10.5 t/ha/yr; litter: 2.7 t/ha/yr (after Denaeyer and Duvigneaud, 1972).

l. *Piceetum,* 55 years old, at Kongalund (Sweden), on acid brown soil on moraine. Aerial increment: 11.1 t/ha/yr; litter: 5.6 t/ha/yr (after Nihlgard, 1972).

m. *Pinetum,* mean values obtained for different forests in Germany (RDA). Aerial increment: 4 t/ha/yr; litter: 3.5 t/ha/yr (after Ehwald, 1957). The S cycle has not been studied.

n. Tundra with *Vaccinium myrtillus* (USSR). Aerial productivity: 2.2 t/ha/yr (after Chepurko, 1972).

o. Meadow steppe with *Grammaceous* sp. and *Filipendula*/Chernozem (occidental Siberia); litter: 10.3 t/ha/yr.

p. Arid steppe with *Grammaceous* sp. and *Artemisia* (plains of USSR); litter: 8.7 t/ha/yr.

q. Steppe on Solonetz at Khazakstan (USSR); annual litter: 3.5 t/ha/yr. (o, p, and q after Rodin and Bazilevich, 1967.) (Crops, after Mengel, 1968.)

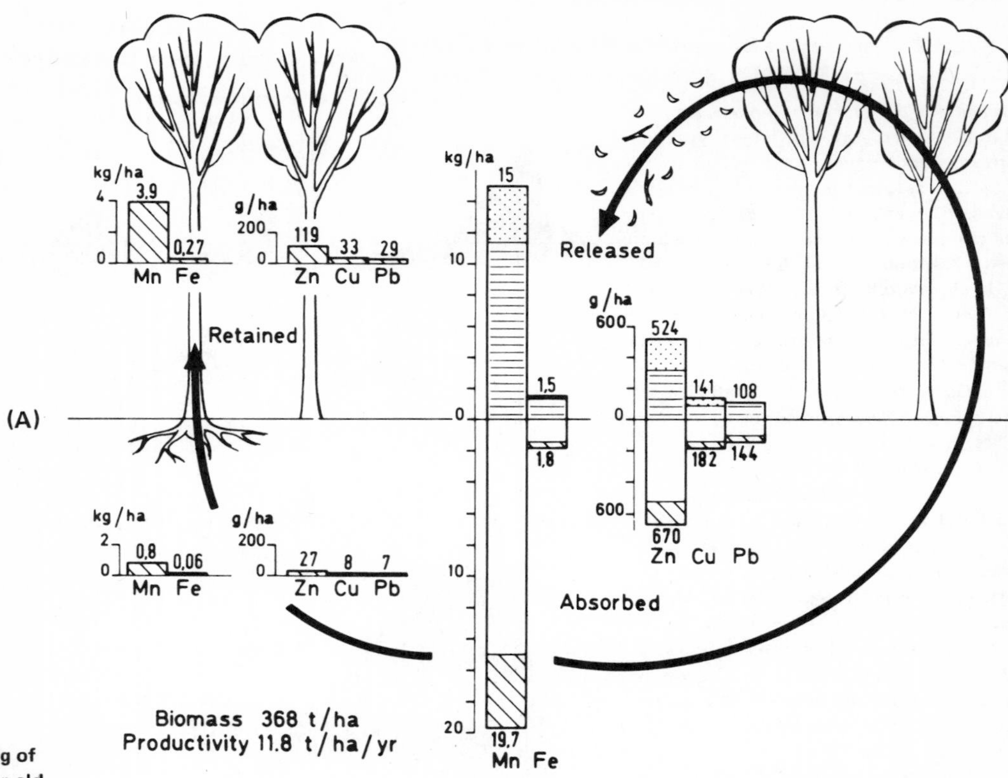

FIGURE 6 Annual mineral cycling of several micronutrients in a 130-year-old beech forest (A) and in an adjacent 55-year-old spruce plantation (B) (IBP Mirwart project, Belgium); oblique lines = retained in increment; horizontal lines = released by litter; dotted lines = released by rainleaching.

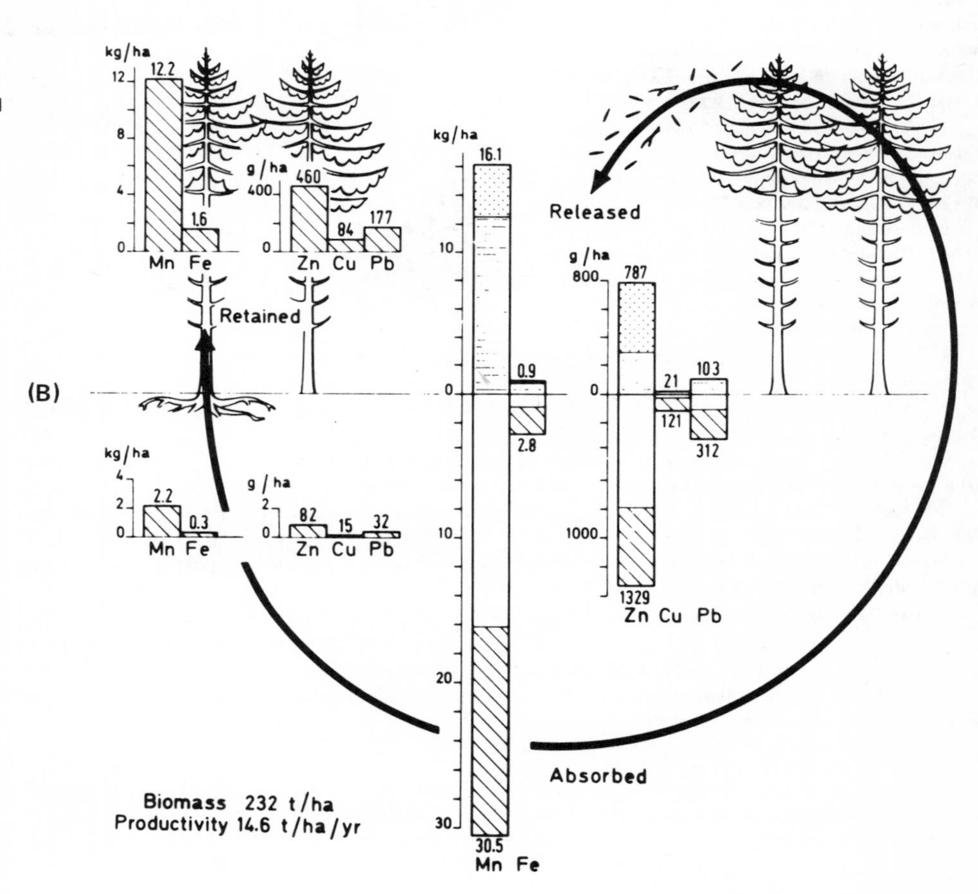

TABLE 2 Comparison Between the Geochemical Cycle of Several Chemical Elements in Different Types of Forest Ecosystems

		kg/ha/yr		
		Input	Output	Net Gain or Loss
K	Coweeta (USA)[a]	3.16	5.17	- 2.02
	Walker Branch (USA)[b]	4.8	5.6	- 0.8
	Hubbard Brook (USA)[c]	1.4	1.5	- 0.1
	Brookhaven (USA)[d]	2.4	3.3	- 0.9
	Cedar River (USA)[e]	0.8	1.0	- 0.2
	Solling (WG)[f]	2.0	1.6	+ 0.4
Ca	Coweeta (USA)[a]	6.16	6.92	- 0.76
	Walker Branch (USA)[b]	28.6	138.4	-105.8
	Hubbard Brook (USA)[c]	2.6	10.6	- 8.0
	Brookhaven (USA)[d]	3.3	8.0	- 4.7
	Cedar River (USA)[e]	2.8	4.5	- 1.7
	Solling (WG)[f]	12.4	14.1	- 1.7
Mg	Coweeta (USA)[a]	1.26	3.0	- 1.82
	Walker Branch (USA)[b]	3.2	69.6	- 66.4
	Hubbard Brook (USA)[c]	0.7	2.5	- 1.8
	Brookhaven (USA)[d]	2.1	6.1	- 4.0
	Solling (WG)[f]	1.8	2.4	- 0.6
N	Hubbard Brook (USA)[c]	5.6	2.2	+ 3.4
	Cedar River (USA)[e]	1.1	0.6	+ 0.5
	Solling (WG)[f]	23.9	6.2	+ 17.7
S	Hubbard Brook (USA)[c]	12.8	16.2	- 3.4
	Solling (WG)[f]	24.8	24.7	+ 0.1
P	Cedar River (USA)[e]	tr	0.02	- 0.02
	Solling (WG)[f]	0.48	0.01	+ 0.47
Na	Coweeta (USA)[a]	5.4	9.74	- 4.3
	Walker Branch (USA)[b]	9.2	5.3	+ 3.9
	Hubbard Brook (USA)[c]	1.5	6.1	- 4.6
	Brookhaven (USA)[d]	17	19.4	- 2.4
	Solling (WG)[f]	7.3	8.8	- 1.5
Cl	Hubbard Brook (USA)[c]	5.2	4.9	+ 0.3
	Solling (WG)[f]	17.8	17.8	0
Al	Hubbard Brook (USA)[c]	–	1.8	- 1.8
	Solling (WG)[f]	3.1	10.3	- 7.2
Fe	Solling (WG)[f]	1.2	0.07	+ 1.13
Mn	Solling (WG)[f]	0.22	4.3	- 4.1
SiO_2	Hubbard Brook (USA)[c]	–	35.1	- 35.1

[a]Coweeta, North Carolina, oak-hickory (*Quercus, Carya*) mature forest, precambrian gneiss (including granite, diorite, mica gneiss and mica schist) (Johnson and Swank, 1973) (small watersheds method).
[b]Walker Branch, eastern Tennessee, oak-hickory (*Quercus, Carya*) mature forest, dolomite (Johnson and Swank, 1973) (small watersheds method).
[c]Hubbard Brook, New Hampshire, mixed forest (*Acer saccharum, Fagus grandiflora, Betula alleghaniensis, Picea rubeus*), bedrock: quartz, plagioclase, biotite (Likens *et al.*, 1971) (small watersheds method).
[d]Brookhaven, Long Island, N.Y., New Hampshire, late successional oak-pine forest (*Pinus rigida, Quercus alba, Q. coccinea, Q. velutina*), (lysimetric method), glacial outwash sands (Woodwell and Whittaker, 1968).
[e]Cedar River research area, western Washington, 36-year-old Douglas fir (*Pseudotsuga menziesii*) plantation on a glacial outwork soil (lysimetric method) (Cole *et al.*, 1967).
[f]Solling, near Göttingen (Western Germany), 125-year-old beech (*Fagus sylvatica*) forest, bedrock: triassic sandstone (Ulrich *et al.*, 1971) (lysimetric method).

exceeds rainfall input except for N; for other elements, it is not yet possible to give general rules because of insufficient data. Table 2 also shows that loss or gain of an element is always low, except for ecosystems established on rich and relatively soluble bedrock (dolomite at Walker Branch) or for ecosystems submitted to emission of pollutants (especially N oxides and S compounds).

A comparison between different ecosystems of the same little watersheds system and established on the same bedrock (Table 3) show that the cations input–output budget may vary widely following the type of phytocenosis; it seems not yet possible to give general conclusions on this subject.

BIOGEOCHEMICAL CYCLES

Cole *et al.* (1967) were the first to establish the relation between biological cycles and geochemical fluxes. They found (Figure 7) in a second growth *Pseudotsuga Menziesii* forest a low output of N, P, K and Ca but a very high accumulation rate of these elements in the ecosystem; they calculated that such an accumulation would result in a depletion of soil nutrients after 12 years for K (exchangeable), 64 years for Ca (exchangeable), 125 years for N (total) and 582 years for P (total).

Recently, Ulrich *et al.*, (1971) calculated what they call the "biogeochemical flux" for a 130-year-old beachwood (IBP Solling project) corresponding to the difference be-

TABLE 3 Average Annual Cation Budgets for One Undisturbed and Three Manipulated Watersheds during Two Water Years (June–May, 1969–71, Coweeta, U.S.A.) (After Johnson and Swank, 1973)

	Forest Type			
	Field-to-Forest Succession	Coppice	White Pine	Mature Hardwood
K^+				
Input	3.02	3.25	3.32	3.16
Output	5.98	4.62	3.56	5.17
Net loss or gain	- 2.96	-1.38	-0.24	-2.02
Ca^{++}				
Input	5.73	5.76	6.51	6.16
Output	10.40	5.01	4.10	6.92
Net loss or gain	- 4.68	+0.75	+2.42	-0.76
Mg^{++}				
Input	1.20	1.34	1.34	1.26
Output	6.26	2.68	1.69	3.09
Net loss or gain	- 5.06	-1.34	-0.35	-1.82
Na^+				
Input	5.11	5.40	5.70	5.40
Output	10.86	6.82	6.06	9.74
Net loss or gain	- 5.75	-1.42	-0.36	-4.34

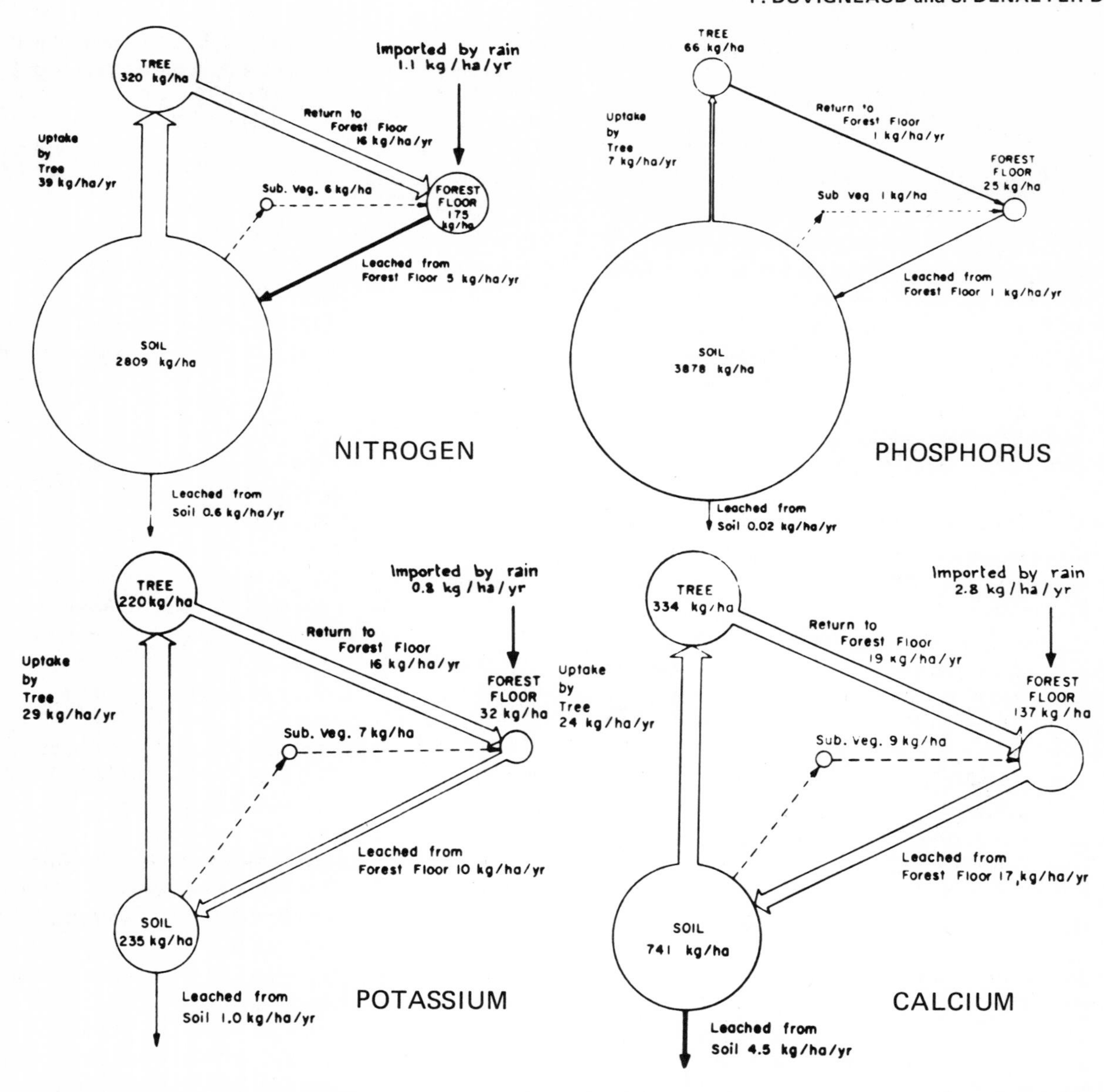

FIGURE 7 Distribution and annual biogeochemical cycle of N, P. K, and Ca in a second-growth Douglas fir ecosystem (after Cole *et al.*, 1967).

tween "soil input" (= nutrient input by rainfall and litter-fall) and "soil output" (= nutrient losses by percolation water, measured 1 m beneath the soil surface). It is interesting to point out that the values obtained in this way are nearly the same as those obtained for a very similar forest ecosystem in Belgium (IBP Mirwart project) but by a quite different method (direct measurement of annual nutrient retention and restitution).

Intensity of the Biological Cycle

The intensity of the biological cycle depends obviously on the rate of absorption and restitution of elements. In ecosystems established on normal soils absorption depends on primary productivity (= biomass). We know also that it may depend (at least in forests) on the rate of root mycorrhization; the root-microbe association may result in 3.5 times greater uptake of P, 75 percent more K and 85 percent more N by trees (Bjorkman, 1970). The greatest effect of mycorrhizae on the rate of mineral cycling may be found in the almost closed cycle of minerals from litter to microflora (endotrophic mycorrhizae) to tree roots in equatorial forests (direct mineral cycling: Went and Stark, 1968); the same phenomenon appears to happen in the modern flora of temperate European forests. Restitution depends mainly on the turnover rate and for some nutrients (especially for K) on the importance of precipitation (canopy leaching and stem-flow).

It woud not be possible to review here the very large number of publications on the action of the pedofauna and the pedoflora on the rapidity of decomposition of litter, which is much higher on mull soil (high density of lumbrics and centipedes) than on moder or mor soil. Nutrients are not completely liberated by soil-organisms before their metabolism is depressed due to dietary impoverishment. Disappearance of the basic hydrocarbon matrix leads to progressive N and P enrichment, with the C/N ratio decreasing during decomposition. At the same time excess nutrients are liberated and progressive, and buffered remineralization occurs. It is thus conceivable that the rate of recycling may be measured by the cellulolitic activity of microorganisms (Witkamp, 1971) and in fact depends on the turnover of organic matter.

The litter → pedofauna ⇋ microflora → soil → root pathway is promoted by high temperature and precipitations (Witkamp and Vanderdrift, 1961) (e.g., the very high rate of mineral cycling in equatorial forests). However, before complete remineralization of nutrients occurs, nutrients become immobilized in the soil organisms, especially in microflora. Such immobilization is very high and may lead to concentration, e.g., bacterial content 15 percent N (C/N between 4.5 and 6.5), fungal content 1.5 to 10 percent N (C/N between 8.5 and 13). Bacteria may also have a luxury consumption of Ca and K. Soil microorganisms act as a "biobuffer" for P, preventing abiotic and hardly reversible fixation of this element.

Nutrient immobilization may lead to a momentary deficiency of some nutrients (e.g., Fe and P) in the phytocenosis. The N cycle is much more intensive in tropical regions because N uptake and rate of turnover are higher. On the other hand, the tropical N cycle is based on NO_3^- uptake, whereas in boreal regions it is mainly based on NH_4^+ uptake (Ellenberg, 1971). The intensity of the biological cycle may be estimated in a simple manner (Rodin and Bazilevich, 1967) by determination of the decomposition rate of litter,

$$\frac{L}{FL}$$ (L = total litter accumulated on soil surface, FL = fresh annual litter fall).

Table 4 gives some data obtained by these authors and shows that the intensity of the biological cycle decreases with increasing latitude. Using system analysis, Jordan and Kline (1972) come to the same conclusion. Some authors (Manil *et al.*, 1963) apply the more complicated "Jenny" coefficient for the mineral cycle rate.

Stability and Evolution of the Cycles

Jordan and Kline (1972) define the relative stability of an ecosystem by the time required to recover the initial state after a perturbation—the shorter the time, the more stable the ecosystem. They show that the cycle of essential nutrients (such as Ca, K, Mg, Fe and Cu) are stable; nonessential elements, such as Na imported by sea spray or ^{137}Cs injected in tree stems, are not recycled in the living compartments of the ecosystem.

During the evolution of ecosystems from the early development stages until the mature climax stages, important modifications affect the cycles (Odum, 1969):

– extrabiotic nutrients become more and more intrabiotic
– cycles, initially open, close progressively
– the rate of nutrient exchange between organisms and their environment, initially quick, becomes slow
– the role played by detritus in the remineralization becomes more and more important.

During a succession, as a rule, one ecosystem promotes the conditions of soil nutrients for the following one (improvement of sylvicultural properties of the soil by the plants themselves) (Remezov and Pogrebniyak, 1969). This phenomenon is aptly illustrated by the manipulated watershed studies of the US/IBP (Table 3).

TABLE 4 Intensity of Mineral Cycling in Some Woodland Ecosystems (After Rodin and Bazilevich, 1967)

	Soil Litter (L) (t/ha)	L / Annual Litter Fall	Cycle
Sphagnum forest	80	50	stagnant
Taiga	30 to 45	10 to 17	very retarded
Deciduous forest	*a*	3 to 4	retarded
Steppe	4 to 6	1 to 1.5	intensive
Subtropical forest	10	0.7	intensive
Savanna	1 to 2	0.2	very intensive
Equatorial forest	–	0.1	very intensive

[a]Changing with time of the year and type of humus.

Mineral Cycling and System Analysis

Recently, system analysis has been applied to mineral cycling. The studied ecosystem is divided into compartments; the nutrient content of each compartment and nutrient transfer from one of them to the other are established. For each compartment, a differential equation gives the change of content, which equals the difference between what enters and leaves that compartment. These equations are solved by computer and constitute the mathematical model of the mineral cycling of the ecosystem. If the transfer coefficients for the most important compartments are known, it becomes possible to predict the dynamics of nutrient concentrations in various ecosystem components (at least if the ecosystem is still functioning in the same way as when those transfer coefficients were established).

An interesting experience consists in modifying (with computer) the content of one compartment or the value of a flux of the model and to observe the effects on the whole ecosystem. In this manner, it is possible to point out the most sensitive parts of the model ecosystem and to develop appropriate research programs. Such mathematical models are of particular utility in studying ecosystems submitted to toxic element pollution.

MINERAL CYCLING AND ENVIRONMENTAL POLLUTION

Research on mineral cycling is of great interest when toxic elements are involved, because the knowledge of their distribution and transfer pathways through the different compartments of the ecosystems allows one to detect their possible accumulation along given food chains. In Sweden, for example, the study of S cycling in forest ecosystems (Table 1) has shown that rainfall carries into the soil very high quantities of sulfur (mainly H_2SO_4) which acidify the upper layer of the soil and are responsible for an increased nutrient leaching which will probably affect the primary productivity of the ecosystems. The same phenomenon also has been observed in Germany and the U.S.

Another example is the input of heavy metals (Zn, Pb, Cd, Cr, Ni) originating from industrial activities and roadside emissions which contaminate the aerial parts of vegetation, pollute rainwater and accumulate in dead organic matter, as shown by Tyler (1972), for several types of terrestrial and seashore meadow ecosystems in Sweden. A detailed study of Cd distribution in a contaminated spruce forest (Table 5) shows that Cd accumulation is highest in the most decomposed organic matter (humus layer) from where it may be absorbed by shallow-rooted plant species. Plant to animal transfers already demonstrated for Pb by Schwickerath (1931), have been observed recently for Cd in Great Britain [death of horses having consumed contaminated feed (Goodman and Roberts, 1971)].

Another interesting aspect of mineral cycling is the study of elements chemically analogous to toxic elements and occurring in undisturbed ecosystems, as for example, stable Sr whose behavior is probably very similar to that of radioactive ^{90}Sr contaminating several ecosystems as a consequence of nuclear activities (Alexahin and Ravikovich, 1966).

BIOGEOCHEMICAL LANDSCAPES

Starting from Dokuchaev's conceptions (1883, 1886, 1889) on the zonality of the soils and corresponding vegetation, the

TABLE 5 Cadmium Content of a Spruce Forest in Central Sweden Polluted from a Local Industrial Source (After Tyler, 1972)

Forest Component	mg Cd/Dry Matter
Picea abies	
roots <5 mm diameter	2.7
roots >5 mm diameter	1.5
stem wood	<0.1
stem bark	2.5
Twigs	
1st year	5.4
2nd year	4.6
3rd year	4.2
4th year	3.3
5th–7th year	2.7
Needles	
1st year	0.6
2nd year	0.4
3rd year	0.5
4th year	0.5
5th–7th year	1.0
Vaccinium vitis idaea	
aboveground biomass	3.2
Vaccinium myrtillus	
aboveground biomass	4.4
Deschampsia flexuosa	
leaves	7.6
leaf litter	12
roots	11
Parmelia physodes	
whole plant	12
Hypnum cupressiforme	
whole plant	30
Needle litter	24
Raw humus	44

"natural landscape" has been defined as an area where relief, climate, vegetation and soil characters are considered as forming a harmonious whole which repeats itself along a given terrestrial zone. In the classical study on "Landscape-Geographical zones of the USSR," landscapes were classified as follows: facies → landscape → zone. Facies correspond nearly to the biogeocenosis of Sukachev (1947) and to the ecosystem of Lindeman (1942).

In one and the same landscape, the lateral migration of chemical elements links the soils of the elevations, slopes and depressions, which form a "*Catena*" (Milne, 1936), characterized by an eluvial (top), colluvial (slope) and illuvial complex (depression) to which correspond ecosociologic groups of plant species (Duvigneaud, 1955). Water movements, which are of prime importance for this topographic ecology were studied in 1928 by Vysotskii. At a more "mineral" level, this concept of chemical unit of landscape has been used by geologists (Perelman, 1964; Glazovskaya, 1968) who initiated studies on the landscape biogeochemistry.

The type of rockweathering and of transformation of the dead organic matter in the eluvial complex are important:

from this is determined the composition of water running on the slope (transalluvial or transaccumulative complexes) and accumulating in the depressions (accumulative, often "superacquous" complexes) where it forms very shallow tables. The difference in mobility of the chemical compounds dissolved in the eluvial complex forms *geochemical sequences* very typical in deserts. The classical sequence observed in the cold desert is:

$$Fe_2O_3 < MnO < CaCO_3 < Na_2SO_4 < CaCl_2, MgCl_2, (NaCl).$$

The same occurs in hot deserts, plus an intercalation of $CaSO_4$ and nitrates.

In more humid landscapes, solifluction complicates the situation and soil depth increases at the bottom of the slopes. Glazovskaya (1968) shows that in the subarctic catena in Scotland, Norway and USSR the high mobility of Al versus Fe depends on the mobility of the complexes of those elements with humic matter:

$$Fe^{\#} \text{ humate} < Fe^{\#} \text{ fulvate} < Al^{\#} \text{ fulvate} < SiO_2 \text{ exeunt}, Ca(HCO_3)_2.$$

The more contrasting geochemical landscapes (linked with arid climate) are obtained when the eluvial complex develops a strongly acid environment, which is transformed in a neutral or alkaline one in the accumulative complex whose evaporation is very high. The succession is as follows: top, acid and ferruginous soils; slope, ferruginous and neutral soils, with Si precipitation (opal); transaccumulative soils in the lower part of the slope (carbonated chernozem); accumulative bottom soils (solonchak) and salty lakes (Na) studies in Siberia and Far-East (Kovda, 1946). The sequence is as follows:

$$Fe_2O_3 < SiO_2 < CaCO_3 < NaHCO_3 < NaCl.$$

In less extreme landscapes, with steppe vegetation, we have:

$$CaCO_3 < CaSO_4 < Na_2SO_4 < NaCl.$$

So it appears that the future of ecology consists of an extension of the ecosystem concept to a basin, a catena or a geochemical landscape, where ecological, geological and hydrological problems are strongly linked.

WORLDWIDE MINERAL CYCLING—BIOGEOCHEMICAL TERRITORIES

Classification of the Biological Cycles

We may not insist here on the tightness of the relationships between biogeochemistry and biosphere, well-defined following the conceptions and investigations of Dokuchaev (1889), Vernadskii (1926) and Grigorieff (1965).

Since 1965, Rodin and Bazilevich (1967) have ameliorated continuously a classification of the world biomes in terms of their type of biological cycle. On the basis of the "Geography of Productivity"—a main objective of the IBP—they established provisionally the patterns of the circulation of nutrients (ash elements and N) in the main vegetation types of the world, following a zonality of the Russian schools of pedology and phytosociology. They first proposed a classification of the types of nutrient circulation based on litter and litter turnover. A 10-point scale was used consisting of the following parameters: biomass of the climax (B), productivity (P), annual litterfall (L), decay (D = litter/litterfall), and the mean nutrient content of fresh litter (Table 6).

Bazilevich and Rodin (1971) published a world map based on a 10-point scale of productivity and the dominant nutrients of the carrying capacity (Figure 8); the basic network is the map of the vegetal formations from the "Atlas

TABLE 6 World Classification of Mineral Cycling (After Rodin and Bazilevich, 1967)

Class	Group	Ash Content	Productivity	Cycle	Vegetation
Equatorial silicic	N-SiO_2	mean	very high	very intensive	Equatorial forest Si > N (Al-Fe-Mn-S) B10, P9, L9, D9, A5
Subboreal calcic	N-Ca	mean	mean	retarded	Temperate deciduous forest Ca ≈ N B8, P7, L6, D6, A4
Boreal nitric	N	low	low	stagnant	Tundra N ≫ Ca (K, Mn) B2-3, P2, L2, D1-2, A2
Semidesertic chlorinated	Na-Cl	very high	very low	very intensive	Semideserts on Solonchak Cl > Na > N (mg-S-Ca) B1-2, P1, L1-2, D10, A10
Continental silicic	SiO_2-N-K	mean	mean	intensive	Steppes Si > N (K aerial-Ca subt.) B3, P5, L6, A3-4

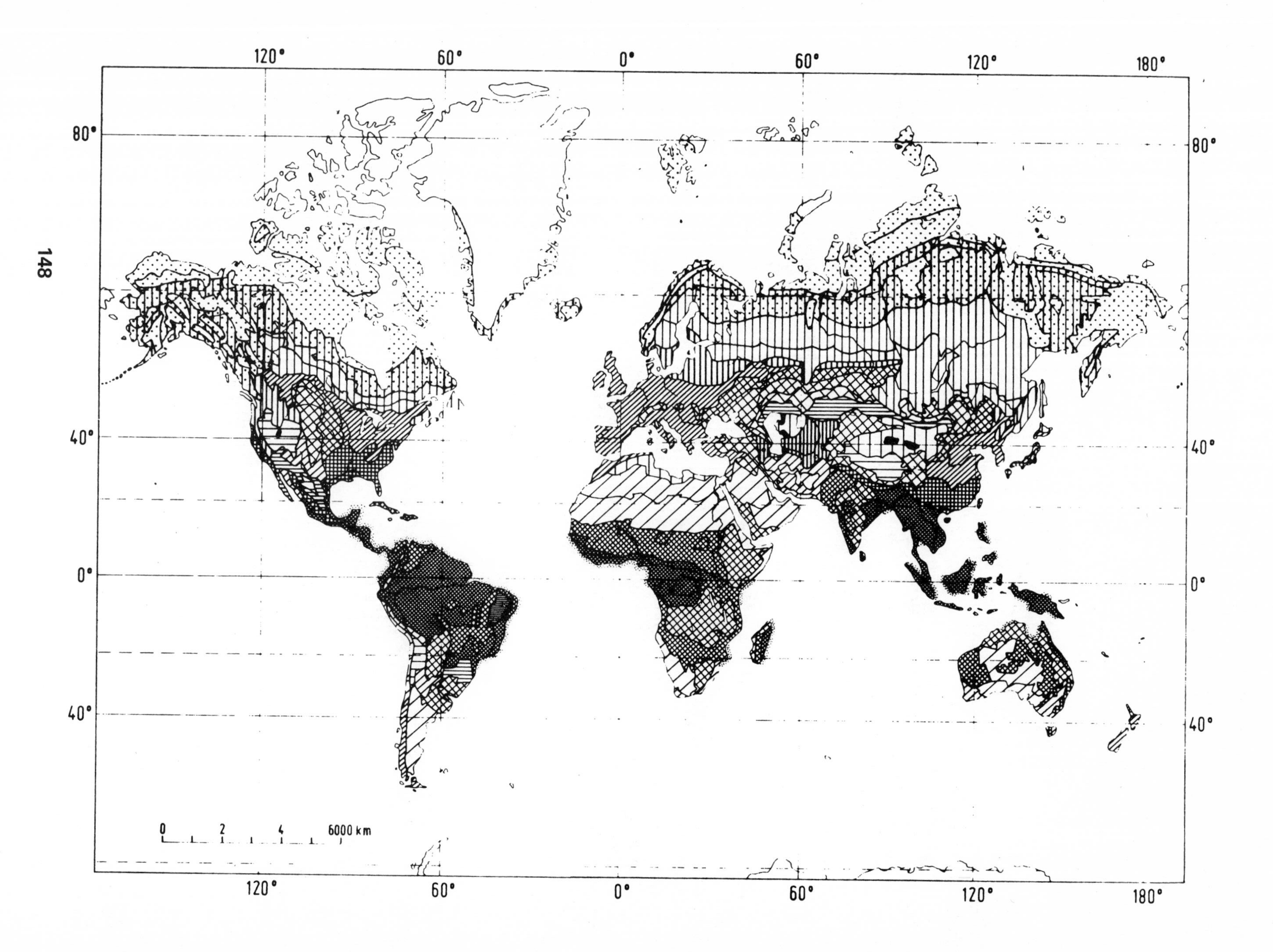
120°
60°
0°
60°
120°
180°
80°
40°
0°
40°
0
2
4
6000 km

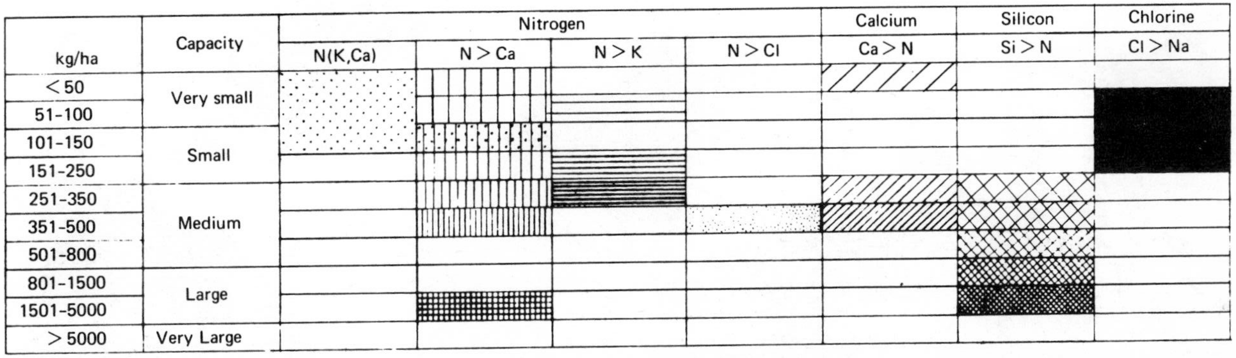

kg/ha	Capacity	Nitrogen				Calcium	Silicon	Chlorine
		N(K,Ca)	N > Ca	N > K	N > Cl	Ca > N	Si > N	Cl > Na
< 50	Very small							
51-100								
101-150	Small							
151-250								
251-350	Medium							
351-500								
501-800								
801-1500	Large							
1501-5000								
> 5000	Very Large							

FIGURE 8 Worldscale classification of the carrying capacity of the biological cycle of elements in the main vegetation types. Top of columns: the two dominant elements. After Bazilevich and Rodin, 1971.

of Physical Geography of the World" (Senderova, 1964). Following parameters were used: phytomass, NPP, mean content of the primary productivity (leaves, stems, roots) and "carrying capacity" of the biological cycle (see p. 135), which depends on the quantity of organic matter produced and on the structure of the phytocenosis and thus on the dominant life form. The maps of the chemical composition and types of biological cycles show the importance of global zonality (chemical and geographical) on biogeochemical cycles:

- N dominance in peat, bogs, tundra forests, coniferous and small leaved forests, humid subtropical forests,
- Si dominance in steppe formations,
- Ca dominance in broadleaved forests and steppes,
- Cl dominance in salt deserts.

Several regularities appear—maps of annual absorption of Ca and N may be superimposed on that of carrying capacity (N absorption depends directly on NPP and P absorption is very low in steppe and desertic formations). Phosphorus appears to be insufficient in the broadleaved forests of subtropical and humid regions (perhaps because of retention of P as P_2O_3 which is very abundant in soil). Potassium absorption does not seem to be sufficient in steppes and broadleaved forests, at least with regard to the other nutrients.

A detailed study of the biological productivity and mineral cycling of terrestrial communities of USSR has been published recently by several Soviet IBP workers (Bazilevich and Rodin, 1971). Several maps have been established following the same principles used in the world maps discussed earlier. The following data (Table 7) show how, in some ecosystems, the accumulation of dead organic matter may retard considerably the biological cycles of elements.

TABLE 7 Comparison of the Rapidity of the Biological Cycling of Elements with the Amount of Accumulated Litter in the Ecosystem (After Bazilevich and Rodin, 1971)

Class[a]	Formation	Accumulated Litter
2	desert	0.5 t/ha
3	subdesert	0.5–2.5 t/ha
4	subtropical	2.5–7.5 t/ha
7	tundra forests	25–40 t/ha
8	coniferous forests	40–60 t/ha
9	forests and swamp	60–100 t/ha
10	forests and swampy soils with mosses	>100 t/ha

[a]Indicates relative rapidity of the biological cycle of elements, from rapid (low numbers) to slow (high numbers) ecosystems.

Biogeochemical Territories

The concept of biogeochemical landscapes may be extended to entire regions characterized by either the abundance or dominance of given geological conditions. This concept was mainly developed by Soviet scientists influenced by the soil zonality of Dokuchaev. Following Kovalsky (1963), the USSR may be divided into biogeochemical territories corresponding with the large phytogeographic zones: taiga or black earth (excess of Co, Ca, Cu and I), woody steppes and steppes on black earth (sufficiency of Co, Ca, Cu and I), dry steppes, semideserts and deserts (excess of Na, Ca, Cl, SO_4 and B), mountains (I and sometimes Cu and Co deficiency). One of the most interesting aspects of this special geography is the fact that it corresponds to endemic diseases: goiter by I deficiency; endemic gout when Mo is more abundant than Cu; endemic dwarfness when the Si/Ca ratio is too high; and enterite by B excess. Recently, Bazilevich and Rodin (1971) have published a map of the continents, showing that, as well as the flora, biogeochemistry characterizes the main biogeochemical territories.

More recently, Cannon (1969, 1970) demonstrated that the USA may be divided into several biogeochemical provincies, e.g., limestone areas of the Eastern USA (possible excess of P, K, and possible deficiency of Mo, Sr, Ca and Mg), coastal plain sands (possible deficiency of Fe, Mr, Cu, Co and Bo), the shales of the northcentral plains (excess of Mo and Se and possible deficiency of P), the evaporative basins (possible excess of Si and deficiency of Cr inducing endemic diabetes by Indian peoples).

Biogeochemical Epidemiology (Diseases induced by biogeochemical anomalies)

Pedological anomalies may modify the chemical composition of plants serving as a food source for consumers.* In some regions of the world (USA and Central Asia), bedrock contains Se in amounts a little higher than typical soils; by shedding of the aerial organs of some "concentrator" plant species Se is brought to the soil surface and then absorbed by meadow or cultured plants which transform Se in toxic forms for consumption by animals and men (several kinds of seleniosis). Similar phenomena are induced by the deficiency of some elements essential for animals and men, e.g., cattle rachitis induced by Co (constituent of vitamin B_{12}) deficiency of grass.

The most spectacular case of element deficiency is the goiter, which frequently is induced by an I deficiency (regions far from the sea). Before the administration of artifically I enriched food, goiters killed 10^6 pigs/year in Montana (USA goiter belt). Observations in Great Britain and in the Netherlands show that goiters also depend on the nature of the bedrock: carboniferous and dolomitic limestone seem to favor goiters, whereas chalky and eruptive rocks

* The simplest example is given by the accumulation of a toxic element along a trophic chain.

do not. Surveys made in England, Wales, Normandy and Netherlands show a positive correlation between mortality by gastric cancer and the organic matter content of cultivated soils (Legon, 1952); organic matter content is especially high in boggy regions and in valleys with clay soils ("cancer valleys"). Low organic soils suffer from a deficiency of available Cu and from an excess of Zn, Co and Cr. Some people consider this dietary imbalance to be responsible for gastric cancer. The issue remains scientifically unresolved.

Certain forms of esophageal cancer are linked with given regions of the world, for example, in western China, esophageal cancer is linked with the cotton agriculture. The cancer rate in most regions with highest incidence is 100–200 times higher than in other regions. Such differences may be observed over a very short distance. In certain regions, this increase in esophageal cancer has occurred recently. Esophageal cancer seems to be a "guinea pig" for studying the relations between certain factors of our environment (food is probably the most important) and human diseases. It is also an important way to study cancer origin. This is the reason why many studies are in progress now in the esophageal cancer belts of Africa and Asia. On the Caspian Sea shore, a very high esophageal cancer rate (1.1 percent in men, 1.7 percent in women) affects turcoman population of the northeast; this region is semidesertic with a predominance of salty soils (*Artemisia* and *Astragalus* steppes) with a very high pH which reduces the availability of Fe, B, Cu and Zn. Human food in this region is lacking in animal proteins, vitamin A, riboflavin and vitamin C. The regions with the lower cancer rate have a very high rainfall and associated soil leaching and, as a consequence, a type of agriculture with higher Fe, B, Ca and Zn content of crops (Kmet and Mahboudi, 1972). These observations may be related to those made in Transkei (South Africa) where severe esophageal cancer affects the Bantu population in a region with high soil erosion and a more vegetarian diet. It seems that the juice of *Solanum incanum* used to curdle milk is responsible.

In Iceland, Armstrong (1964a, 1964b and 1967) could not find a relationship between stomach cancer and the content of 22 chemical elements in vegetables, milk, water and herbage of grasslands. In Kenya, Robinson and Clifford (1968) did not find significant differences in the content of 8 chemical elements in *Zea mays*, growing near farmhouses and the nose-pharynx cancer rate. In the USA several meetings have been organized: "Trace substances in environmental health" (Cannon, 1969-70) and "Geochemical environment in relation to health and disease" (Hopps and Cannon, 1972). Reports overviewed the inhibitional action of trace elements such as As, Cu, Pt, Se and Zn, on neoplastic growth (Pories *et al.*, 1972), the antidiabetic action of Cr, the relation between Cd and high blood pressure (Perry, 1972), the inverse relation between water hardness and the entire group of cardiovascular diseases (Correa and Strong, 1972). Feeding patterns are very complicated, even in the most primitive societies and it is necessary to take into account the food allocation along food chains, the cooking and food preparation methods, dietary preferences, etc. Collaboration is necessary between geologists, ecologists, agronomists, and biochemists having a common interest in biogeochemistry and the consequences on human health.

This modern aspect of ecology—biogeochemical epidemiology—also concerns health problems in relation to technology. For example, the addition of an excess of nitrate to a soil lacking Mo may induce accumulation of carcinogenic nitrosamine, because of the poor N metabolism of the plants.

CONCLUSIONS

Mineral cycling is one of the most important parameters in the analysis of ecosystems, because mineral elements, as much as energy and water, are essential to maintain the continuity and stability of these systems. Although the processes of absorption, retention and restitution of mineral elements are not yet well known at the ecosystem level, comparison of initial data obtained from several terrestrial ecosystem types show some general regularities that begin to emerge. Mineral cycling appears to be one of the best and easiest ways to characterize the general metabolism and functioning of ecosystems, but many years are still required (several IBP or SCOPE programs) to obtain sufficient data to make valuable models and predictions.

Knowledge of mineral cycling is of greatest practical interest in agronomy and forestry, because it will allow us to provide the highest utilization of the solar energy in the productivity of plant communities. It will optimize the utilization rate of fertilizers by selection of the best ecosystem types in relation to environmental conditions ("biological agriculture" leading to a maximum of productivity and quality). Mineral cycling studies should also give a good idea of the role played by bedrock in the nutrition of terrestrial ecosystems.

The accurate knowledge of mineral cycling will allow ecologists to make practical recommendations for better quantitative and qualitative productivity. The cycles of mineral elements at the ecosystem level combine to form the overall mineral cycle at the biosphere level, upon which depends man's future. It is only with a sufficient knowledge of these large cycles that it will be possible to make quantitative predictions and answer some of the pressing environmental questions. Is the earth in process of eutrophication by an increasing nitrogen enrichment, or in the process of dystrophication by a blocking up of the phosphorous cycle in the deep oceans, or in the process of acidification by sulfuric and hydrochloric acid? What are the global effects of biotic pollution by lead, cadmium, mercury? So many questions which it will be possible to answer only if we know accurately the cycles of elements on an ecosystem and biospheric level.

REFERENCES

Alexahin, R. M., and M. M. Ravikovich. 1966. Radioecological concentration processes, p. 443. *In* B. Aberg and F. P. Hungate (eds.) Proceedings of an international symposium held in Stockholm, 25-29 April. Pergamon Press.

Armstrong, R. W. 1964a. Spectographic analysis of pasture grass and drinking water in relation to stomach cancer mortality in Iceland. Acta. Agr. Scand. 14:65-76.

Armstrong, R. W. 1964b. Environmental factors involved in studying the relationship between soil elements and disease. Am. J. Public Health 54:1536-1544.

Armstrong, R. W. 1967. Milk and stomach cancer in Iceland. Acta. Agr. Scand. 17:30-32.

Bazilevich, N. I., and L. E. Rodin. 1971. Geographical regularities in productivity and the circulation of chemical elements in the earth's main vegetation types. Am. Geogr. Soc. New York.

Bjorkman, E. 1970. Forest tree mycorrhiza. The conditions for its formation and the significance for tree growth and afforestation. Plant and Soil 32:589-610.

Boriskina, Titova, and P. S. Pogrebniyak. 1955. Fundamentals of forest typology. Kiev. Acad. Sci. Ukr. SSR.

Boyko, H. (ed.) 1966. Salinity and aridity. Junk. Mon. Bid. 16. The Hague, Netherlands. 408 p.

Cannon, H. L. 1969-1970. Trace element excesses and deficiencies in some geochemical provinces of the U.S. Proceedings Univ. Missouri Fourth Ann. Conference on Trace Substances. Environmental Health 21-43. Univ. Missouri, Columbia.

Carlisle, A., A. H. F. Brown, and E. J. White. 1967. The nutrient content of tree stem flow and ground flora litter and leaches in a sessile oak (*Quercus petraea*) woodland. J. Ecol. 55(3): 615-627.

Chepurko, N. L. 1972. The biological productivity and the cycle of nitrogen and ash elements in the dwarf shrub tundra ecosystems of the Kibini mountains (Kola Peninsula), p. 236-320. *In* F. E. Wielgolaski and Th. Rosswall (eds.) On the biological productivity of tundra. Proceedings, IV International Tundra Biome meeting. Swedish IBP Committee, Stockholm.

Cole, D. W., S. P. Gessel, and S. F. Dice. 1967. Distribution and cycling of nitrogen, phosphorus, potassium and calcium in a second-growth Douglas-fir ecosystem, p. 197-232. *In* H. E. Young (ed.) Symposium on Primary Productivity and Mineral Cycling in Natural Ecosystems. Univ. Maine Press, Orono.

Correa, P., and J. P. Strong. 1972. Atherosclerosis and the geochemical environment: a critical review, p. 217-228. *In* H. C. Hopps and H. L. Cannon (eds.) Geochemical environment in relation to health disease. Ann. N.Y. Acad. Sci., IGY.

Curlin, J. W. 1970. Models of the hydrologic cycle, p. 268-285. *In* D. E. Reichle (ed.) Analysis of temperate forest ecosystems. Ecological studies 1. Springer-Verlag, New York, Heidelberg, Berlin.

Davis, G. K. 1972. Competition among mineral elements relating to absorption by animals, p. 62-69. *In* H. C. Hopps and H. L. Cannon (eds.) Geochemical environment in relation to health and disease. Ann. N.Y. Acad. Sci., 199.

Denaeyer-De Smet, S. 1966. Bilan annuel des apports d'éléments minéraux par les eaux de précipitation sous couvert forestier dans la forêt caducifoliée de Blaimont. Bull. Soc. Roy. Bot. de Belgique 99:345-375.

Denaeyer-De Smet, S. 1969. Recherches sur l'écosystème forêt. La Chenaie mélangée calcicocole de Virelles-Blaimont. Apports d'éléments minéraux par les eaux de précipitations, d'égouttement sous couvert forestier et d'écoulement le long des troncs (65, 66, 67). Bull. Soc. Roy. Bot. de Belgique 102:355-372.

Denaeyer-De Smet, S., and P. Duvigneaud. 1972. Comparison du cycle des polyéléments biogènes dans une hêtraie (Fagetum) et une pessière (Piceetum) établies sur même roche-mère à Mirwart (Ardenne Luxembourgeoise). Bull. Soc. Roy. Bot. de Belgique 105:197-205.

Denaeyer, S., J. Lejoly, and P. Duvigneaud. 1968. Note sur la spécificité biogéochimique des halophytes du littoral belge. Bull. Soc. Roy. Bot. Belgique 101:293-301.

Dengler, A. 1930. Waldbau auf ökologischer Grundlage. Springer-Verlag, Berlin.

Dokuchaev, V. V. 1883. Report to the provincial *zemstvo* (local authority) of Nizhnii-Novgorod (i.e., now Gor'kii) fasc. I: main phases in the history of land assessment in European Russia, with classification of Russian soils. Sochineniya (collected works), Vol. IV, izd. Acad. Sci. USSR, Moscow-Leningrad.

Dokuchaev, V. V. 1886. Place and role of present-day pedology in science and life. Sochineniya (collected writings), Vol. IV, izd. Acad. Sci. USSR, Moscow-Leningrad.

Dokuchaev, V. V. 1889. The doctrine of natural zones. Sochineniya, Vol. VI. Acad. Sci., USSR, Moscow-Leningrad.

Duchaufour, Ph. 1957. Pédologie. Tableaux descriptifs et analytiques des sols. Ecole Nat. Eaux et Forêts, Nancy. 87 p.

Duvigneaud, P. 1946. La variabilité des associations végétales. Bull. Soc. Roy. Bot. de Belgique 78:107-134.

Duvigneaud, P. 1949. Les savanes du Bas-Congo. Essai de phytosociologie topographique. Lejeunia Mem, 10. Brussels, Belgium. 192 p.

Duvigneaud, P. 1955. Études écologiques de la végétation de l'Afrique tropicale. Ann. Biol. 31(5-6):375-392.

Duvigneaud, P. 1968. Recherches sur l'écosystème forêt. La Chênaie-Frênaie à Courdrier du Bois de Wève. Apercu sur la biomasse, la productivité et le cycle des élements biògenes. Bull. Soc. Roy. Bot. de Belgique. 101:111-127.

Duvigneaud, P. (ed.) 1971. Productivity of forest ecosystems. Proceedings of the Brussels symposium, October 1969. UNESCO, Paris. 707 p.

Duvigneaud, P., and S. Denaeyer-De Smet. 1964. Le cycle des éléments biogènes dans l'écosystème forêt. Lejeunia, N.S. 28:1-143.

Duvigneaud, P., and S. Denaeyer-De Smet. 1966. Accumulation du soufre dans quelques espèces gypsophiles d'Espagne. Bull. Soc. Roy. Bot. Belg. 9:263-269.

Duvigneaud, P., and S. Denaeyer-De Smet. 1968. Essai de classification chimique (éléments minéraux) des plantes gypsicoles du bassin de l'Ebre. Bull. Soc. Roy. Bot. Belg. 101:279-391.

Duvigneaud, P., and S. Denaeyer-De Smet. 1970a. Phytogéochimie des groupes écosociologiques forestiers de Haute-Belgique. Oecol. Plant 5(1):1-32.

Duvigneaud, P., and S. Denaeyer-De Smet. 1970b. Biological cycling of minerals in temperate deciduous forests, p. 199-225. *In* D. E. Reichle (ed.) Analysis of temperate forest ecosystems. Ecological studies 1. Springer-Verlag, New York, Heidelberg, Berlin.

Duvigneaud, P., and S. Denaeyer-De Smet. 1971. Cycle of the biogenic elements in the forest ecosystems of Europe (chiefly deciduous forests), p. 527-542. *In* P. Duvigneaud (ed.) Productivity of forest ecosystems. Proceedings of the Brussels Symposium, October, 1969. UNESCO, Paris.

Duvigneaud, P., and S. Denaeyer-De Smet. 1973. Considerations sur l'écologie de la nutrition minérale des tapis végétaux naturels. Oecol. Plant 8(3):219-246.

Duvigneaud, P., S. Denaeyer-De Smet, P. Ambroes, and J. Timperman. 1971. Recherches sur l'écosystème forêt. Biomasse, Productivité et cycle des polyéléménts biogènes dans l'écosystème "Chênaie caducifoliée." Essai de phytogéochimie forestière. Inst. Roy. des Sci. Nat. de Belgique. Memoire 164. 101 p.

Edwards, C. A., D. E. Reichle, and D. A. Crossley, Jr. 1970. The role of soil invertebrates in turnover of organic matter and nutrients, p. 147-172. *In* D. E. Reichle (ed.) Analysis of temperate forest ecosystems. Ecological studies 1. Springer-Verlag. New York, Heidelberg, Berlin.

Ehwald, E. 1957. Über den Nährstoffkreislauf des Waldes. Deut. Akad. Landw. Wiss. Sitz. 6:1-56.

Ellenberg, H. 1954. Landwirtschaftliche pflanzensoziologie. I. Unkrautgemeinschaften als Zeiger für Klima und Boden. Stuttgart, Ulmer. 141 p.

Ellenberg, H. 1966. Vegetationskunde (soziologische Geobotanik). Fortschr. der Botanik. 28:289-296.

Ellenberg, H. 1971. Nitrogen content, mineralization and cycling, p. 509-514. *In* P. Duvigneaud (ed.) Productivity of forest ecosystems. Proceedings of Brussels Symposium, October 1969. UNESCO, Paris.

Epstein, E. 1972. A blind spot in biology. Science 176(4032):235.

Glazovskaya, M. A. 1968. Geochemical landscapes and types of geochemical soil sequences. 9th Int. Cong. Soil Sci. Transact. Adelaide, Australia 4.

Goodman, G. T., and T. M. Roberts. 1971. Plant and soils as indicators of metals in the air. Nature 231:287-292.

Greenland, D. J., and J. M. L. Kowal. 1960. Nutrient content of the moist tropical forest of Ghana. Plant and Soil 12:154-174.

Grigorieff, A. A. 1965. The relationship between heat and moisture budgets and the intensity of geographical processes. Doklady. USSR, 162(1):151-154.

Hemphill, D. D. 1972. Availability of trace elements to plants with respect to soil-plant interaction, p. 46-61. *In* H. C. Hopps and H. L. Cannon (eds.) Geochemical environment in relation to health and disease. Ann. N.Y. Acad. Sci., 199.

Hohne, H. 1962. Vergleichende Untersuchungen über Mineralstoff– und Sticksoffgehalt sowie Trockensunstanzproduktion von Waldbodenpflanzen. Arch. für Forstw. 11(10):1085-1141.

Hopps, H. C., and H. L. Cannon (eds.) 1972. Geochemical environment in relation to health and disease. Ann. N.Y. Acad. Sci., 199. 325 p.

Horvath, D. J. 1972. An overview of soil/plant/animal relationships with respect to utilization of trace elements, p. 82-94. *In* H. C. Hopps and H. L. Cannon (eds.) Geochemical environment in relation to health and disease. Ann. N.Y. Acad. Sci., 199.

Johnson, P. L., and W. T. Swank. 1973. Studies of cation budgets in the southern Appalachians on four experimental watersheds with contrasting vegetation. Ecology 54(1):70-80.

Jordan, C. F., and J. R. Kline. 1972. Mineral cycling: some basic concepts and their application in a tropical rain forest. Adv. in Ecol. and Syst. 33-50.

Klausing, O. 1956. Untersuchungen über den Mineralumsatz in Buchenwäldern auf Granit und Diorit. Forstwiss Cbl. 75:18-32.

Kmet, J., and E. Mahboudi. 1972. Esophageal cancer in the Caspian littoral of Iran. Initial studies. Sciences 175:846-853.

Kolli, R., and L. Reintam. 1970. Content of nitrogen and ash elements in the phytomass of spruce stands on brown forest soils. Acad. Sci. of the Estonian USSR Committee for the IBP. 10-48.

Kovda, V. A. 1946. Origin and regime of saline soils. USSR Acad. Sci., Moscow-Leningrad.

Kovalsky, V. V. 1963. Geochemical ecology and its evolutionary trends. Izv. Akad. Nauk. SSSR.

Kovalsky, V. V. 1974. Geochemical Ecology. Akad. Nauk. SSR. Moscow. 297 p. (in Russian)

Kubota, J., and W. H. Allaway. 1972. Geographic distribution of trace element problems, p. 525-554. *In* Micronutrients in agriculture. Proceedings of the symposium held at Muscle Shoals, Alabama, April 20-22, 1971. Soil Science Society of America, Inc. Madison, Wisconsin.

Legon, C. D. 1952. The aetiological significance of geographical variations in cancer mortality. Brit. Red. July-Dec.:700-702.

Likens, G. E., F. H. Bormann, R. S. Pierce, and D. W. Fisher. 1971. Nutrient-hydrologic cycle interaction in small forested watershed-ecosystems, p. 553-563. *In* P. Duvigneaud (ed.) Productivity of forest ecosystems. Proceedings, Brussels Symposium. October 1969. UNESCO, Paris.

Lindeman, R. L. 1942. The trophic-dynamic aspect of ecology. Ecology 23:399-418.

Lossaint, P., and M. Rapp. 1971. Distribution of organic matter, productivity and mineral cycles in ecosystems in Mediterranean climate, p. 597-617. *In* P. Duvigneaud (ed.) Productivity of forest ecosystems. Proceedings, Brussels Symposium, October 1969. UNESCO, Paris.

Manil, G., F. Delecour, G. Forget, and A. El Attar. 1963. L'humus, facteur de station dans les hêtraies acidophiles de Belgique. Bull. Inst. Agron. Stat. Rech. Gembloux 31:28-102; 183-222.

Mengel, K. 1968. Ernährung und Stoffwechel der Pflanze. Stuttgart. Fischer Verlag. 425 p.

Miller, R. B. 1963. Plant nutrients in hard beech. III. The cycle of nutrients. N.Z. J. Sci. 6:388-413.

Milme, G. 1936. A provisional soil map of east Africa. Amour Memoirs.

Nihlgard, B. 1972. Plant biomass, primary production and distribution of chemical elements in a beech and a planted spruce forest in South Sweden. Oikos 23:68-81.

Nilsson, I. 1972. Accumulation of metals in spruce needles and needle litter. Oikos 23:132-136.

Nye, P. H. 1961. Organic matter and nutrient cycles under moist tropical forest. Plant and Soil 13:333-346.

Odum, E. P. 1969. The strategy of ecosystem development. Science 144:262-270.

Ovington, J. D. 1968. Some factors affecting nutrient distribution within ecosystems, p. 95-105. *In* Proceedings of the Copenhagen Symposium. Natural Resources Res. 5, UNESCO, Paris.

Ovington, J. D., and H. A. I. Madgwick. 1959. The growth and composition of natural stands of beech. II. The uptake of mineral nutrients. Plant and Soil 10:389-400.

Perelman, A. I. 1964. Geochemistry of landscapes. Higher School Pretl. House, Moscow.

Perry, H. M. 1972. Hypertension and the geochemical environment, p. 202-216. *In* H. C. Hopps and H. L. Cannon (eds.) Geochemical environment in relation to health and disease. Ann. N.Y. Acad. Sci., 199.

Pories, W. J., E. G. Mansour, and W. H. Strain. 1972. Trace elements that act to inhibit neoplastic growth, p. 265-273. *In* H. C. Hopps and H. L. Cannon (eds.) Geochemical environment in relation to health and disease. Ann. N.Y. Acad. Sci., 199.

Reichle, D. E. 1971. Energy and nutrient metabolism of soil and litter invertebrates, p. 465-477. *In* P. Duvigneaud (ed.) Productivity of forest ecosystems. Proceedings of the Brussels Symposium, October 1969. UNESCO, Paris.

Remezov, N. P. 1963. Über den biologischen Stoffkrelauf in der Wäldern des europäisches Teils der Sowjetunion. Arch. Forstwes. 12:1-43.

Remezov, N. P., and P. S. Pogrebriyak. 1969. Forest soil science. Israel Progr. for. Sc. tr. Jerusalem. 261 p.

Rennie, P. J. 1955. The uptake of nutrients by mature forest growth. Plant and Soil 7:49-95.

Robinson, J. B. D., and P. Clifford. 1968. Trace element levels and carcinoma of the nasopharynx in Kenya. E. African Med. J. 45(2):694-700.

Rodin, L. E., and N. I. Bazilevich. 1967. Production and mineral cycling in terrestrial vegetation. (English translation by G. E. Fogg.) Oliver and Boyd, London 288 p.

Schlenker, G. 1939. Die natürlichen Waldgesellschaften im Laubwaldgebiet des Würtemberges Unterlandes. Veröff. d. Württ. Landesst. f. Naturschutz 15:103–150.

Schlenker, G. 1960. Zum Problem der Einordnung klimatischer Unterschiede in das System der Waldstandorte Boden-Württembergs. Mitt. Ver. Forstl. Stand. und Forstpflanzenzüchtung. 9:3–15.

Schwickerath, M. 1931. Das Violetum calaminariae der Zinkböden in der Umgebung Aachens. Beitrage zur Naturdenkmalpflege 14:463–503.

Senderova, G. M. (ed.) 1964. Physical-Geographical Atlas of the World. USSR Akad. Sci. and Main Admin. of Geodesy and Cartography.

Sukachev, V. N. 1947. The theory of bio-geo-coenology. *In* Collection of the Acad. of Sci. USSR, in commemoration of the 30th anniversary of the Revolution, part II. Moscow–Leningrad, USSR.

Trelease, S. F., and H. M. Trelease. 1938. Selenium as a stimulating and possibly essential element for indicator plants. Amer. J. Bot. 25:372–380.

Tsutsumi, T. 1971. Accumulation and circulation of nutrient elements in forest ecosystems, p. 543–552. *In* P. Duvigneaud (ed.) Productivity of forest ecosystems. Proceedings of Brussels Symposium, October 1969. UNESCO, Paris.

Tyler, G. 1972. Heavy metals pollute nature, may reduce productivity. Ambio 1(2):52–59.

Ulrich, B., R. Mayer, and M. Pavlov. 1971. Investigations on bioelement stores and bioelement cycling in beech and spruce stands including input-output analysis, p. 87–113. *In* Th. Rosswall (ed.) Systems analysis in northern coniferous forests. Ecological Research Committee Bulletin 14, Swedish National Committee for IBP, Stockholm.

Vernadskii, V. I. 1926. The biosphere. Biosfera Publ. House for Chemical Technical Sciences, Leningrad, USSR.

Vysotskii, G. N. 1928. Ombro-evaporometrical correlations. Pulsivity and dispulsivity of infratables and ground water. Journ. "Pochvovedenie," No. 3–4.

Wacquant, J. P. 1968. Capacité d'échange cationique racinaire chez quelques populations d'Anagallis arvensis L. sensu lato. C. R. Acad. Sci., Paris. Série D. 266:1580–1582.

Wacquant, J. P. 1969. Adsorption cationique préférentielle et écologie végétale. Soc. franc de Physiologie vég. Colloque du 15 Mar. 1969 sur les Mécanismes de l'Absorption minérale.

Went, F. W., and N. Stark. 1968. Mycorrhiza. Bioscience 18(11): 1035–1039.

Witkamp, M. 1971. Soils as components of ecosystems. Ann. Rev. Ecol. and Syst. 1:85–111.

Woodwell, G. M. 1970. Effects of pollution on the structure and physiology of ecosystems. Science 168(3930):429–432.

Woodwell, G. M., and R. H. Whittaker. 1968. Primary production and the cation budget of the Brookhaven forest, p. 151–166. *In* H. E. Young (ed.) Symposium on primary productivity and mineral cycling in natural ecosystems. Univ. Maine Press, Orono.

Young, H. E., and V. P. Guinn. 1966. Chemical elements in complete mature trees of seven species in Maine. Tappi 49:190–197.

HYDROLOGIC TRANSPORT MODELS

D. D. HUFF

INTRODUCTION

There is little question that water is a key factor in virtually all ecosystems. Yet, assessing the importance of water in an ecosystem in precise quantitative terms is an elusive goal. Two factors have arisen over the past decade, however, to bring that goal closer to a reality than ever before. The first factor is the recent burst of progress in computer hardware and software technology. In particular, the storage capabilities and speed of computational operations have finally made it possible to deal in detail with problems on the scale of natural environmental systems.

The second factor is an emerging attitude toward cooperation between scientists from diverse disciplines. Namely, the growing willingness on the part of many to sacrifice the time and effort necessary to establish a meaningful communication link with those in other disciplines. That dialogue can and has provided a synergistic richness evident in the results of many of the studies supported through the International Biological Program (IBP). In the spirit of enhancing that dialogue, the following material is presented. It is an attempt to selectively highlight aspects of hydrologic transport modeling of primary importance to those working within the IBP. The review is not exhaustive, however; it is an attempt to place the role of hydrologic transport modeling in a proper perspective relative to analysis of ecosystems. It is further hoped that the examples cited will serve to stimulate the exchange of information and goals between disciplines.

CLASSES OF HYDROLOGIC TRANSPORT MODELS

When hydrologic transport models are discussed with reference to analysis of ecosystems, it is useful to classify types of models on the basis of the physical system represented and the general analytic tools used in model implementation. Figure 1 represents a simple classification scheme which may be used to illustrate both the range of possible model types and more closely identify a category of hydrologic transport models that have received the most attention in ecosystem analysis research.

TERRESTRIAL AND AQUATIC SYSTEM MODELS

It is most convenient to separate ecosystems into either a terrestrial or an aquatic classification. It is well known that this separation is artificial since the two are intimately connected, yet, from a practical point of view, implementing a detailed model for the combined system is not operationally feasible at present. There are examples of very good hydrologic transport models for both types of systems, but it is a fact that the aquatic class of hydrologic transport models has reached a level of development well ahead of terrestrial counterparts. To a large degree, this is a result of emphasis on developing water quality models for managing streams, lakes, and estuaries. Although the motivations for such model development were different from those stimulating the IBP, there is a good deal of information available

ORNL-DWG 74-4171

HYDROLOGIC TRANSPORT MODELS

ECOSYSTEM

TERRESTRIAL AQUATIC

MATERIALS TRANSPORTED

POINT SOURCES DIFFUSE SOURCES

MODEL CLASSIFICATION

MECHANISTIC PARAMETRIC STOCHASTIC

FIGURE 1 A classification scheme for hydrologic transport models.

through studying such models. Fortunately, these models have been reported widely in the water quality literature, and those interested in pursuing the available information may be referred to an excellent review of modeling hydrologic transport and corresponding biological response in aquatic systems (Orlob, 1972). An article by Simons (1972) will serve as a good starting point for those interested in the modeling of lake circulation and associated solute transport. Woolhiser (1973) has reviewed watershed and associated water quality component models and presents a very useful list of references of hydrologic transport models for both terrestrial and aquatic ecosystems. His review clearly makes the point that relatively little has been accomplished in modeling hydrologic transport in terrestrial ecosystems, and that such work must involve several disciplines because of the scope of the problem. For this reason, much of the following material deals with terrestrial hydrologic transport modeling.

DIFFUSE AND POINT SOURCES

A useful distinction may be made between point and diffuse sources of materials when dealing with hydrologic transport and interaction of materials with ecosystems. Clearly, many point sources of materials such as outfalls from waste treatment facilities, irrigation return flow collection systems, or industrial stacks discharging to the atmosphere, ultimately contribute to the pool of diffuse materials in the environment. However, because of the relatively high concentrations in point sources, it is usually most important initially to evaluate their effects independently. A good example is the extensive body of research on the effects of point sources of pollution on water quality variables such as dissolved oxygen. Since point sources have been so extensively studied, emphasis here will be confined to the transport of materials from diffuse sources, such as geologic formations or litterfall in a forested drainage basin.

TYPES OF ANALYTIC MODELS

There are three general types of analytic approaches to constructing hydrologic transport models. The mechanistic approach is based upon an understanding of physical processes. Each process is represented by accurate mathematical expressions, then the processes are linked together to form a mathematical model of the whole system. Linkages are determined by the physical system modeled, and each portion of the model has a physical counterpart which it simulates. The major argument for developing this type of model is that it ultimately offers the potential for predicting system behavior for conditions that have not been observed in the natural system. In fact, it is only this general class of "theoretical" models that offers the ability to evaluate environmental impact of the introduction of new substances or new land or water use policies prior to actual implementation (Woolhiser, 1973). A strong counter argument (or at least a limitation) is that even though a system may be totally deterministic, it is not possible to describe the system in enough detail to predict its behavior mechanistically.

A second type of analytic model formulation is parametric, which is characterized by compartment type models (compartment type models can also be mechanistic) based upon linear regression representations of system and subsystem behavior. At the present time, parametric models are probably the most useful of the three types for ecosystem modeling because of their empirical base and their ability to describe complex systems with simple mathematical expressions. A fine example of a parametric model is the ELM (Ecosystem Level Model) (Anway *et al.*, 1972) developed by modelers in the Grasslands Biome portion of the US/IBP Analysis of Ecosystems (AOE) studies. The major point of difference between a mechanistic and parametric model is that the loss rates for a storage in a mechanistic system will depend on known physical laws and the state of the system. In a parametric model, the transfer rates are determined from analysis of system response without attempting to explicitly describe individual contributing processes.

Finally, stochastic models describe or recreate statistical properties of observed variables of the system and are based upon probabilistic considerations. For example, a stochastic model could be used for generating a synthetic temporal sequence of cation flux from a watershed. The statistical properties of the synthetic and observed records for the basin, such as monthly mean, variance, and serial and cross correlations with other parameters, would be stastically indistinguishable.

HYDROLOGIC TRANSPORT MODELS FOR TERRESTRIAL ECOSYSTEMS

The focus on "process studies" within the IBP most appropriately leads to emphasis on a specific type of hydrologic transport model: terrestrial, diffuse source, and of a predominantly deterministic and mechanistic type. One key concept in the development of such a model is that water is a carrier of materials, thus modeling hydrologic transport of materials may be accomplished by linking chemical and biological transformations of materials within an ecosystem to the presence and movement of water. The logical point for beginning the discussion is with a hydrologic simulation model.

HYDROLOGIC CYCLE

Figure 2 presents a diagram of the most generally important components of the hydrologic cycle in schematic form. It is also a representation of a flow chart for a comprehensive hydrologic simulation model such as that presented by

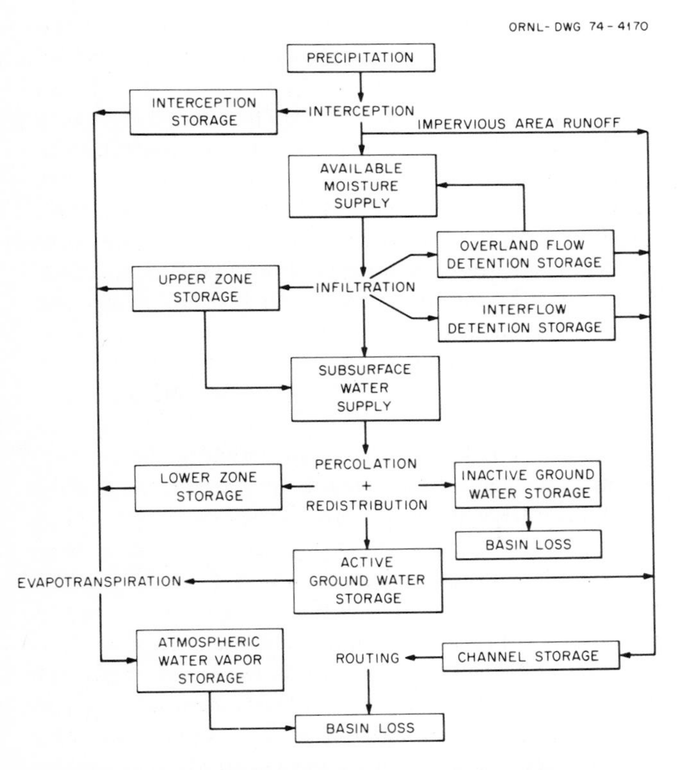

FIGURE 2 A schematic diagram of the most important components of the hydrologic cycle for modeling purposes. Consideration of snow is omitted.

Crawford and Linsley (1966). Precipitation is divided between that held as interception storage on the basin canopy, and the throughfall portion which reaches the basin surface. In climates where snowfall is possible, an additional snowpack storage component is added and the moisture delivered to the basin surface is that which is released from the snowpack. Throughfall is further divided between the portions reaching impervious and pervious surfaces to obtain both impervious area runoff to stream channels (includes water falling directly on channels), and the incremental addition of water to the moisture supply available at the basin surface at any time. The available moisture supply may either infiltrate into the soil profile (subsurface water supply) or be partitioned into any one of three storage components. The upper zone storage represents depression storage on the basin surface, more commonly known as mud puddles. The interflow detention storage represents a fraction of storm runoff with a delayed input to the basin drainage network. Flow from this storage is termed interflow and is assumed to have a rate directly proportional to the quantity of interflow detention storage. It is often stated that interflow represents lateral, shallow subsurface flow to the stream channel, however this concept has recently been challenged (Dunne and Black, 1970).

If infiltration rates are lower than the available moisture supply rate, water may move across the pervious soil surface. The quantity of water moving over the basin surface at any time is termed overland flow detention storage, and the flow rate to the channel may be expressed as a function of this storage from considerations of open channel hydraulics. Note that overland flow detention storage as well as upper zone storage are both available for subsequent infiltration. Of the quantity of water infiltrated into the soil, a portion remains as soil moisture in the upper horizons of the soil, and some may percolate downward into the saturated or groundwater zone in the basin. The active groundwater storage feeds directly into the basin drainage network and is measured as base or "dry weather" flow. The inactive groundwater storage represents groundwater lost from the basin by any of several processes. For example, water pumped for use in a municipal water supply system, then diverted out of the basin through a waste treatment plant would be included in the inactive groundwater category. Evapotranspiration may occur from any of the storages. Inflow to the stream channel, consisting of impervious area runoff, overland flow, interflow, and groundwater flow is introduced into the appropriate section of the channel system. The entire combined flow is simulated section by section from the most remote parts of the channel system to the basin outlet.

In addition to explicitly quantifying four separate components of flow to basin channels, several basin state variables (such as soil water content) are also simulated. These variables have great utility for modeling ecosystem processes such as primary productivity. Thus, the framework for the hydrologic model can serve as a part of the physical description of the state of the basin ecosystem.

The fact that there are four separate sources of water for streamflow becomes important when it is recognized that the chemical quality of the water is highly dependent on its past history. The most significant changes in waterborne materials content probably occur when the water comes in contact with the soil and litter surface. Therefore it is appropriate to begin the discussion of examples of hydrologic transport models with a review of some research on solute transport in soils.

HYDROLOGIC SOLUTE TRANSPORT IN SOILS

A very generalized formulation of a combined model for water and salt flow in soil has been presented by Endelman *et al.* (1972). A schematic diagram of the model and associated equations is shown in Figure 3. In simple terms, the model is a statement of the principle of conservation of mass,

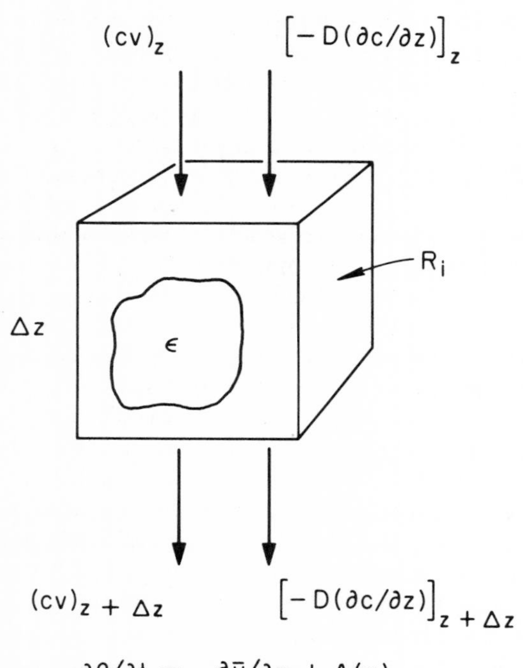

FIGURE 3 A schematic diagram of a soil water and salt flow model, and associated equations (after Endelman *et al.*, 1972).

coupled with the Darcy law for one-dimensional (vertical) flow of water in soil. The figure shows a volume element of unit cross-sectional area and length Δz. A fraction of the volume element (ϵ) is occupied by the carrier, water. Two transport processes are considered. Convective flow transport $[(cv)_z]$ represents transport of material carried along by the water in terms of solute concentration (c) and carrier velocity (v). Dispersive transport $\{[-D(\partial c/\partial z)]_z\}$ represents movement of material relative to the carrier by virtue of a concentration gradient along the flow path. An additional term, the net rate at which the solute is consumed, produced, or transformed (R_i) within the control volume, is also included.

For application of the model, the continuity equation for one-dimensional water flow is solved to determine volumetric water content within the soil profile at some point in time, t. This profile may in turn be used to derive the velocity (v) profile for water flux and then to solve the general equation for salt transport between layers in the soil.

An example of the results produced by a model of this type, as compared to field observations, is shown in Figure 4 (Gupta, 1972). The research was conducted at Hullinger Farm, near Vernal, Utah, in cooperation with the US/IBP Desert Biome program. In Figure 4a, computed and measured soil water content profiles at selected times are shown. The profiles resulted from a wetting and drying cycle. Figure 4b shows computed and observed salt concentrations at the same times. The procedure used was to apply crystalline salt to the soil surface, irrigate, then allow the soil to dry for an extended period.

One rather interesting outcome is the model prediction of an upward water flux in response to evaporative stress (Figure 4). For purposes of modeling salt movement, three basic chemical processes were considered as influencing net rate of consumption, production or exchange within a layer:

a. Dissolution or precipitation
b. Ionization (dissociation)
c. Ion exchange between soil and water solutions.

By using simultaneous equations representing the three processes noted, together with computed water fluxes, the computed concentrations shown in the lower half of Figure 4 were derived. Even though the materials considered were limited to the cations Ca^{++}, Mg^{++}, and Na^{+}, and the Cl^{-} anion, the implications for modeling movement of more complicated chemical species such as N, C and P compounds are encouraging. It must be clearly understood however, that much work yet remains before models of this level of resolution will be available for general simulation of transport of N, C, and P through ecosystems.

The work of Gupta (1972) addressed determination of the chemical quality of irrigation return flows, which should

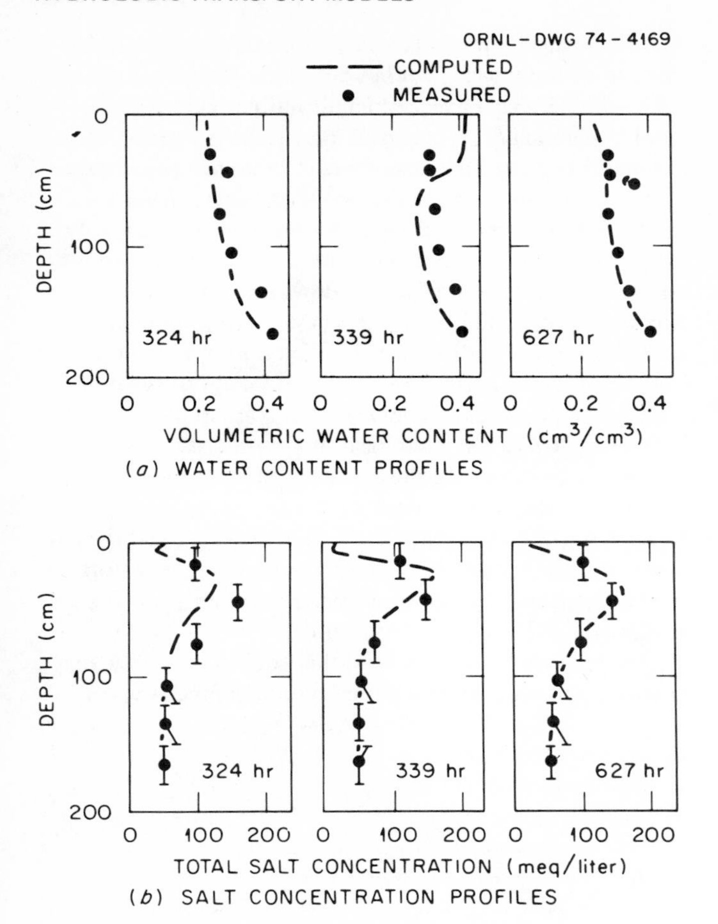

FIGURE 4 Results of a soil water and salt flow model simulation for (a) soil water content and (b) salt concentration compared against field observations (after Gupta, 1972).

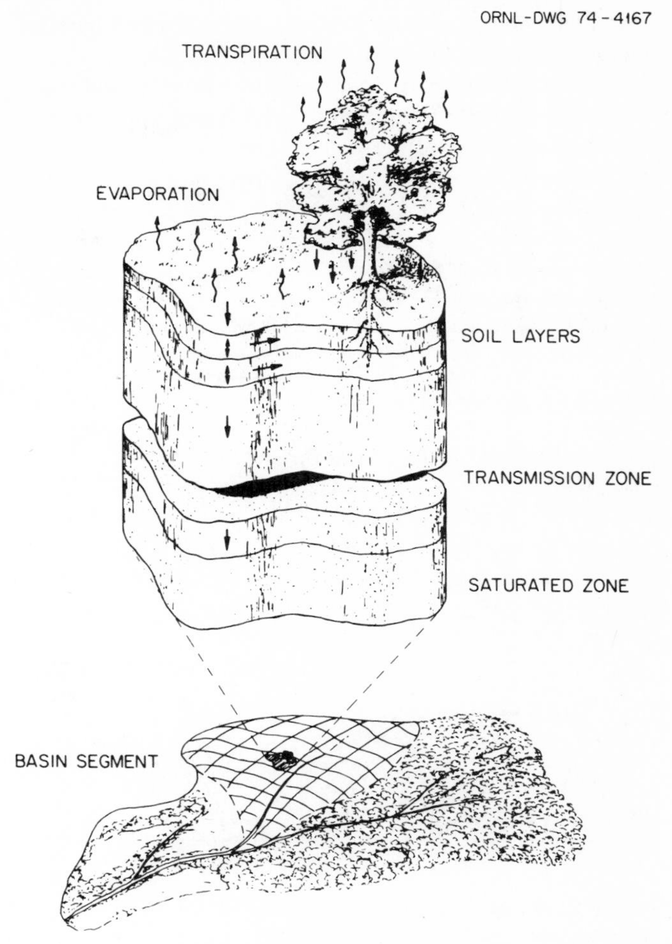

FIGURE 5 A schematic representation of the typical basin segment used in preparing hydrologic simulations.

be considered as point sources of solutes. To illustrate how work of this sort may be used on a basin-wide scale, it is useful to examine a different level of resolution in predictive hydrologic transport models.

BASIN-WIDE TRANSPORT MODELING

For purposes of watershed scale hydrologic transport models, Figure 2 is illustrative of the general structure and hydrologic components which are explicitly considered. Figure 5 shows the physical representation of a typical basin segment. The segment forms the basic element for watershed modeling, since the combined effects of one or more segments may be used to represent the response of any watershed. The usual sequence of simulation progresses from computing the complete hydrologic cycle for each segment to passing combined inflows through the stream channel system to generate a runoff hydrograph (discharge versus time) at the basin outlet. In the process, the hydrologic state variables of the basin are allowed to vary with time, and it is possible to obtain the quantity of any individual component of flow into the channel at any time.

HYDROLOGIC SIMULATION

As an example of basin-wide hydrologic simulation, Figure 6 depicts a total flow hydrograph that has been disaggregated into its simulated components (Shih *et al.*, 1972). The study basin is the H. J. Andrews Experimental Catchment, which is a focus for US/IBP Coniferous Forest Biome research. As a first step in that project, the observed hydrographs were used for comparison with simulation results to calibrate the watershed model. (Note that for clarity, the scale used for presenting surface runoff is only 17% of the other two scales.)

One important point to be made from Figure 6 is that there are four components of the total flow which can be isolated. Clearly, there is a reasonable probability that some compensating errors are present between the flow components shown. In other words, it is possible to get a "correct" total

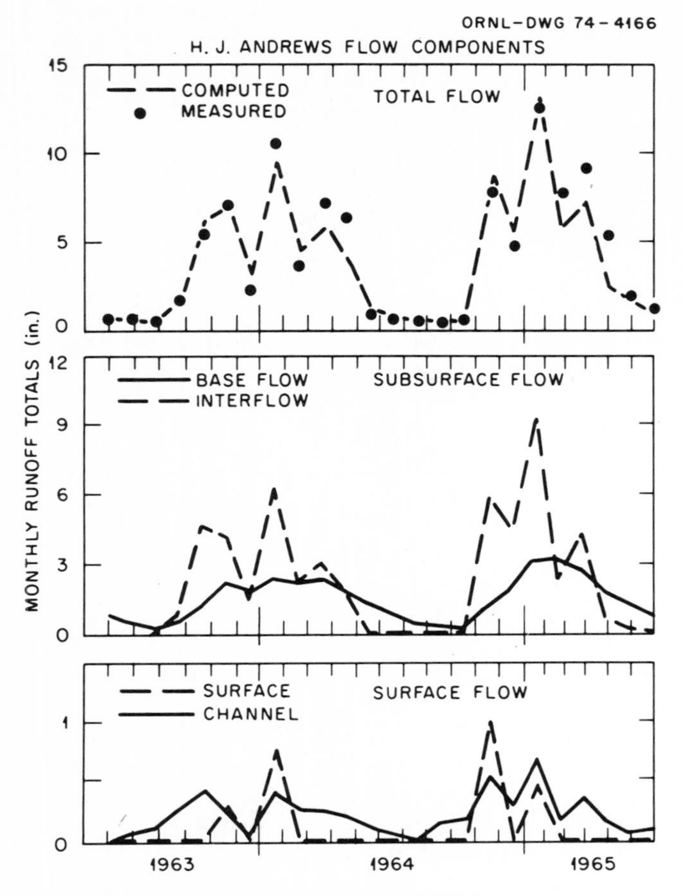

FIGURE 6 Disaggregated total flow hydrograph for the H. J. Andrews Experimental Catchment for 1963–1965 (after Shih *et al.*, 1972).

flow simulation with some errors in individual components of flow. Note, however, that since one generally expects very significant differences in solute concentrations among the various flow components, it is much less likely that compensating errors in simulations can adequately reproduce observed water and solute fluxes at the same time. Thus, the allowable tolerances for error in a detailed hydrologic transport model are even more stringent than those for general hydrologic simulation models. At the same time, with more information available, it is likely that it will be possible to further refine hydrologic simulation models through studies of hydrologic transport processes.

TRANSPORT SIMULATION

A study by Thomas *et al.* (1971) may be used to illustrate the combined use of detailed and basin-wide type simulation models. The overall model contains both parametric and mechanistic components, and gives an indication of how model types can be combined. One objective of the study was to simulate water quality and quantity in the Bear River, Utah, drainage basin, especially as it is influenced by irrigation return flows with high salt concentrations.

The model may be separated into three general subsections. They include hydrologic simulation, estimation of salt content in natural flow, and detailed examination of irrigation and corresponding return flow salt concentrations. A model following the work of Dutt *et al.* (1972) and Gupta (1972), was used to obtain computed values of the quantity and salt concentration of return flows from specified irrigation practices. In addition, a general hydrologic model was used to compute hydrographs of natural flow at several points in the Bear River Basin. The third component of the model used derived correlations between salt content and quantity of natural flows at selected locations in the basin. During simulation tests, natural flows were calculated, then used to derive corresponding natural salt concentrations using regression (parametric) equations. At the same time, the detailed salt flow model was used to compute the magnitude of water and salt quantities in irrigation return flow. Finally, the natural and irrigation return flows were combined to simulate the total water and salt flux for monthly time increments. Figure 7 presents some typical comparisons of

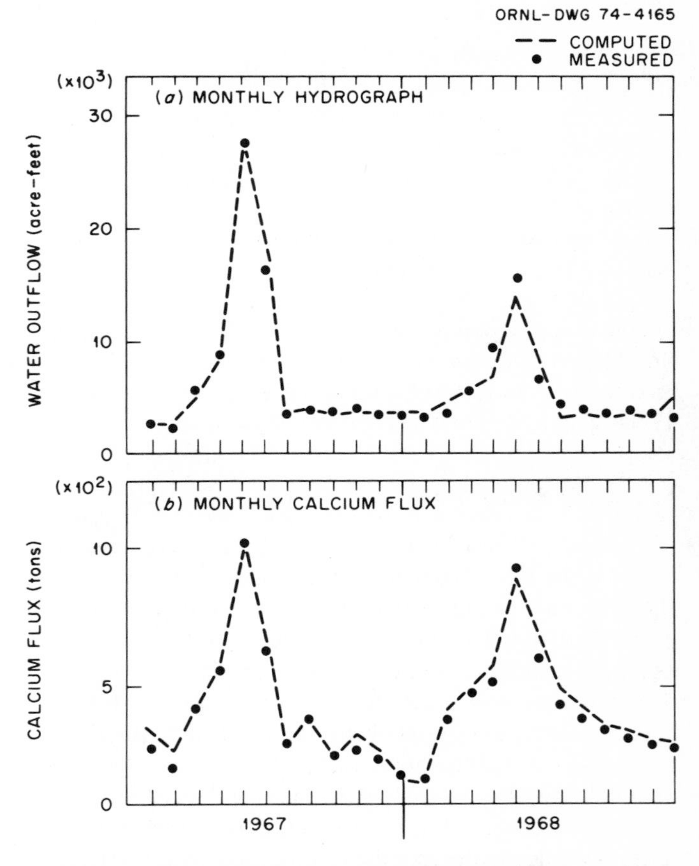

FIGURE 7 Simulated and observed monthly total (a) water and (b) calcium fluxes for the Bear River, Utah, watershed. Fluxes result from both natural and irrigation return flows.

observed and simulated water (Figure 7a) and salt flux (Figure 7b) quantities.

Although agreement is rather good, a significant share of the calcium flux originated with natural (diffuse) sources, and was derived from direct observation using regression methods. For this reason, the model would probably not be a good predictor of water quality if basin conditions were significantly altered. However, the model does have great utility for investigating the impact of a variety of irrigation schemes on the salt content of total flow. To make such a model useful for predicting the effects of basin modification on salts originating from diffuse sources, a more completely mechanistic treatment of all aspects of hydrologic transport is required.

MECHANISTIC HYDROLOGIC TRANSPORT MODELS

An example of a mechanistic hydrologic transport model is the Hydrologic Transport Model (HTM) formulated by Huff (1968). It is based upon a combination of the Stanford Watershed Model (Crawford and Linsley, 1966) and chemistry of solutes transported by water. In addition to the disaggregation of water fluxes which ultimately produce streamflow, explicit consideration is given to the following processes:

a. Interception of wet and dryfall deposition of salts, and subsequent washoff via throughfall.
b. Ion exchange processes affecting transport of materials carried by surface runoff.
c. Erosion and transport of materials sorbed on soil particles.
d. Percolation of soil water and associated leaching of materials into the soil column.

The initial application of the HTM dealt with simulating the hydrologic transport of radionuclides (^{137}Cs and ^{90}Sr) originating from atmospheric nuclear weapons testing (Huff, 1968). Figure 8 illustrates some of the comparisons between simulated and observed radionuclide fluxes in streamflow in California. Although the results were confined to a rather special case, the implication of the study was that in principle, a mechanistic type of hydrologic transport model is feasible.

Thus, the United States IBP Eastern Deciduous Forest Biome program adopted the HTM for modeling transport of nutrients. Using the same basic assumptions, but adjusting the model for the chemical properties of nitrate (NO_3^-), similar studies were conducted at Lake Wingra, Wisconsin. Figure 9 shows some preliminary results of those simulations. It should be noted that nitrate was selected for its solubility and mobility properties, and that the runoff occurred in a storm drain system, thus representing surface runoff almost exclusively. None the less, the results again demonstrate the feasibility of the approach. Perhaps the most useful aspect of the current version of the HTM is the structure and basic

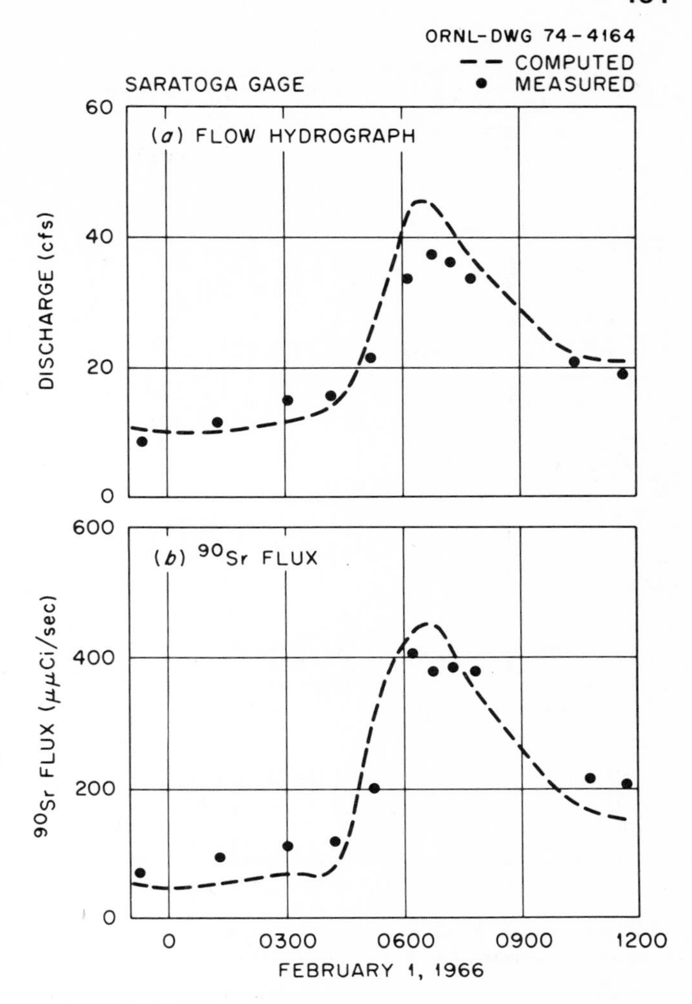

FIGURE 8 Simulated and observed hourly fluxes of (a) water and (b) ^{90}Sr for Saratoga Creek, California.

framework it provides as a foundation for more detailed future models. At present, a major revision to the HTM is underway as part of the Eastern Deciduous Forest Biome modeling work, to attempt to incorporate more chemical and biological processes than are currently included.

UTILIZATION OF HYDROLOGIC TRANSPORT MODELS

Throughout the foregoing discussion, focus has moved from parametric toward mechanistic, high-level resolution simulation models. It seems clear that the only means for developing truly predictive models lies along a similar path. Furthermore, such detailed models will be required for accurate predictive capability in situations that have not yet been observed in nature. Thus, it is appropriate that a large proportion of the modeling effort in IBP has been and will continue to be expended on developing and refining these models.

At the present state of computer technology, whole sys-

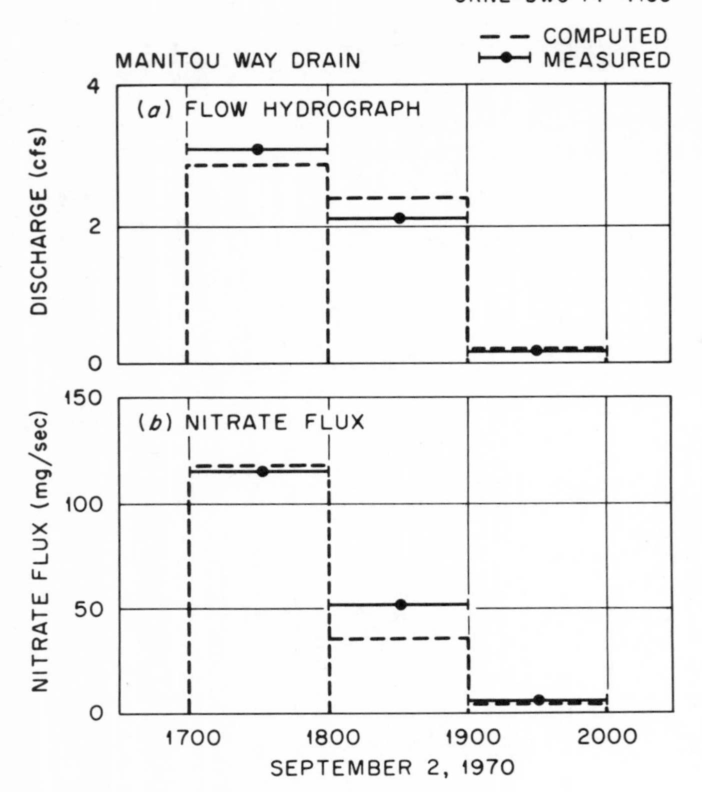

FIGURE 9 Simulated and observed fluxes of (a) water and (b) nitrate for the Manitou Way storm drain in Madison, Wisconsin. Chemical data were taken from Kluesener, 1972.

tem models of a purely mechanistic type are too expensive to operate for routine use. Furthermore, the data requirements for these models are often so great as to limit applications to very well studied situations. Thus, it seems appropriate that such detailed models be used judiciously and that the simpler stochastic or compartment-type parametric models be used for more routine applications. In this context, stochastic models in combination with deterministic models may be extremely useful for long-term studies.

CONCLUSION

There are important roles for parametric, mechanistic, and stochastic types of hydrologic transport models in ecosystem studies. However, in the final analysis, none of the modeling techniques can reach full potential unless the knowledge represented by all of the disciplines vital to ecosystem studies has been included in the comprehensive models produced. The responsibility for developing such models usually rests primarily with modelers, yet it is essential that scientists conducting basic research also make a direct contribution.

To date, the IBP studies have demonstrated that such cooperative work is possible, although establishing the communication links between disciplines and researchers and modelers has often been a slow, frustrating process. Now that many of the linkages have been formed, it is important that they be expanded and strengthened. Hydrologic transport models to date have concentrated primarily on physical processes, and have ignored biological processes to a large extent. It seems clear that fruitful advances can be made in understanding the interrelationships between hydrologic processes, mineral cycling, and productivity. However, these advances will depend upon the combined efforts of multidisciplinary teams, such as those that now exist within the IBP and the willingness of all scientists to reach beyond their disciplines.

ACKNOWLEDGMENTS

The author wishes to thank Dr. J. Paul Riley and his associates, Utah Water Research Laboratory, Logan, Utah, for their helpful suggestions and willingness to provide hydrologic simulation results for the H. J. Andrews Experimental Catchment. Thanks are also due Dr. J. Hanks, Utah State University, for discussions concerning water and salt flow in soil.

Support for the research reported here was supplied in part by the Eastern Deciduous Forest Biome US/IBP, funded by the National Science Foundation under Interagency Agreement AG-199, 40-193-69, with the Atomic Energy Commission–Oak Ridge National Laboratory.

REFERENCES

Anway, J. C., E. G. Brittain, H. W. Hunt, G. S. Innis, W. J. Parton, C. F. Rodell, and R. H. Sauer. 1972. ELM: Version 1.0. Grasslands Biome Tech. Rep. 156, Colorado State Univ. 285 p.

Crawford, N. H., and R. K. Linsley. 1966. Digital simulation in hydrology: Stanford Watershed Model IV. Stanford Univ. Dept. Civil Eng. Tech. Rep. 39, July. 210 p.

Dunne, T., and R. D. Black. 1970. Partial area contributions to storm runoff in a small New England watershed. Water Resources Res. Vol. 6, No. 5. pp. 478–490.

Dutt, G. R., M. J. Shaffer, and W. J. Moore. 1972. Computer simulation model of dynamic bio-physico-chemical processes in soils. Dept. of Soils, Water Engineering, Agricultural Experiment Station, Univ. of Arizona, Tucson. 101 p.

Endelman, F. J., M. L. Northup, D. R. Keeney, J. R. Boyle, and R. R. Hughes. 1972. A systems approach to an analysis of the terrestrial nitrogen cycle. J. Environ. Systems 2(1):3–19.

Gupta, S. C. 1972. Model for predicting simultaneous distribution of salt and water in soils. Ph.D. dissertation, Utah State Univ., Logan. 112 p.

Huff, D. D. 1968. Simulation of the hydrologic transport of radioactive aerosols. Ph.D. dissertation, Stanford Univ., January. 215 p.

Kluesener, J. W. 1972. Nutrient transport and transformations in Lake Wingra, Wisconsin. Ph.D. dissertation. U. of Wisconsin, February. p. 206.

Orlob, G. T. 1972. Mathematical modeling of estuarial systems. p. 78–128. *In* A. K. Biswas (ed.) International Symposium on Modeling Techniques in Water Resources Systems, Vol. 1. Environment Canada, Ottawa. 239 p.

Shih, G. B., R. H. Hawkins, and M. D. Chambers. 1972. Computer modeling of a coniferous forest watershed. pp. 433–452. *In*

A.S.C.E. Irrig. and Drain. Div. Spec. Conf., "Age of changing priorities for land and water." Publ. by ASCE, New York. 482 p.

Simons, T. J. 1972. Multi-layered models of currents, temperature and water quality parameters in the Great Lakes. p. 150–159. *In* A. K. Biswas (ed.) International Symposium on Modeling Techniques in Water Resources Systems, Environment Canada, Ottawa. 239 p.

Thomas, J. L., J. P. Riley, and E. K. Israelson. 1971. A computer model of the quantity and chemical quality of return flow. Publication No. PRWG 77-1, Utah Water Res. Lab., Logan, June. 94 p.

Woolhiser, D. A. 1973. Hydrologic and watershed modeling–state of the art. Trans. Amer. Soc. Agricultural Engr. 16(3):553–559.

CONTRIBUTORS

BAZILEVICH, PROF. N. I., Dokuchaev Soil Institute, Moscow, USSR.

BRYLINSKY, PROF. M., Department of Biology, Dalhousie University, Halifax, Nova Scotia.

COUPLAND, DR. R. T., Department of Plant Ecology, University of Saskatchewan, Saskatoon, Canada.

CRISP, PROF. D. J., Marine Science Laboratory, University College of North Wales, Menai Bridge, Anglesey, Wales.

CUMMINS, DR. KENNETH W., W. K. Kellogg Biological Station, Michigan State University, Hickory Corners, Michigan 49060.

DENAEYER-DE SMET, S., Chargé de Cours Associé, Université Libre de Bruxelles, Brussels, Belgium.

DINGER, DR. BLAINE E., Environmental Sciences Division, Oak Ridge National Laboratory, Building 3017, Oak Ridge, Tennessee 37830.

DUNBAR, PROF. M. J., Marine Sciences Centre, McGill University, Montreal, Quebec, Canada.

DUVIGNEAUD, PROF. PAUL, Directeur, Laboratoire de Botanique Systématique et d'Ecologie, Université Libre de Bruxelles, Brussels, Belgium.

EDWARDS, DR. NELSON T., Environmental Sciences Division, Oak Ridge National Laboratory, Building 3017, Oak Ridge, Tennessee 37830.

FEE, DR. E. J., Fisheries Research Board of Canada, Freshwater Institute, 501 University Crescent, Winnipeg 19, Manitoba, Canada.

GOLLEY, DR. FRANK B., Executive Director, Institute of Ecology, The Rockhouse, University of Georgia, Athens, Georgia 30602.

GRODZIŃSKI, DR. WLADYSLAW L., Department of Animal Ecology, Jagiellonian, University of Cracow, Cracow, Poland.

HARRIS, DR. W. FRANK, Environmental Sciences Division, Oak Ridge National Laboratory, Building 3017, Oak Ridge, Tennessee 37830.

HUFF, DR. DALE D., Department of Civil Engineering, University of Wisconsin, Madison, Wisconsin 53706.

KAUSHIK, DR. N. K., Department of Environmental Biology, University of Guelph, Guelph, Ontario, Canada.

LEAN, DR. D. R. S., Department of Zoology, University of Toronto, Toronto, Ontario, Canada.

MANN, DR. K. H., Department of Biology, Dalhousie University, Halifax, Nova Scotia.

OLSON, DR. JERRY S., Environmental Sciences Division, Oak Ridge National Laboratory, Oak Ridge, Tennessee 37830.

PARKINSON, DR. DENNIS, Department of Biology, University of Calgary, Calgary, Alberta, Canada.

PETRUSEWICZ, DR. KAZIMIERZ, Institute of Ecology, Polish Academy of Sciences, Dziekanow Lesny near Warsaw, Poland.

RODIN, DR. L. E., Komarov Botanical Institute, USSR Academy of Sciences, Leningrad, USSR.

ROZOV, DR. N. N., Dokuchaev Soil Institute, Moscow, USSR.

SCHINDLER, DR. D. W., Fisheries Institute, Freshwater Institute, 501 University Crescent, Winnipeg 19, Manitoba.

SHUGART, DR. HERMAN H., JR., Environmental Sciences Division, Oak Ridge National Laboratory, Building 3017, Oak Ridge, Tennessee 37830.

SOLLINS, DR. P., College of Forest Resources, AR-10, University of Washington, Seattle, Washington 98195.

TAMM, PROF. CARL OLOF, Department of Forest Ecology, The Royal College of Forestry, S-104 05, Stockholm 50, Sweden.

WALKER, DR. RICHARD B., Botany Department AK-10, University of Washington, Seattle, Washington 98195.

WIELGOLASKI, DR. F. E., Botanical Laboratory, University of Oslo, Oslo 3, Norway.